College Algebra

Josaphat A. Uvah   Editor

Author Team
Jalal Alemzadeh [Ch. 4 & 9]   State University of New York, Cortland
Linda Chandler [Ch. 6]   Coastal Carolina University
Ron C. Goolsby [Ch. 1 & 3]   Winthrop University
Mary Martin [Ch. 1 & 3]   Winthrop University
Tom Polaski [Ch. 8]   Winthrop University
Josaphat A. Uvah [Ch. 2 & 7]   University of West Florida
Linda Vereen [Ch. 6]   Coastal Carolina University
Deborah Vrooman [Ch. 5]   Coastal Carolina University

Eastman Science Series
a division of
weeHowf Publishing Company

ISBN 0-9652611-1-5

Book Team

George Maguire   Acquisition Editor
Deborah Vrooman   Mathematics Editor
Linda Bicker   Copy Editor
Andrew Craig   Cover Design

Eastman Science Series
a division of
weeHowf Publishing Company

CONTENTS

Preface

As the number of students requiring college algebra continues to grow, students as well as educators are faced with the task of searching for a text book to ease their burden. This Eastman Science Series edition is an attempt to produce such a book. We present the material in a way that is distinctly different from other texts, but one that still remains within the boundaries of an accepted pedagogical system. The text is built on a wealth of knowledge and understanding of the experienced educator. The myriad of tastes we provide, through the authorship of several educators, is carefully balanced against a coherent and real effort to hold the reader's attention as (s)he is lead to discover the simplicity of the subject. It is our goal to teach using simple reasoning, rather than raw mechanical manipulations. As a by-product, the reader will, hopefully, develop their critical thinking skills.

In each section of the text several examples are preceded by an easy to follow development of the topic in question. The text is punctuated with carefully selected practice exercises to provide hands-on participation in class. End of section problems provide further work to be done at home, without overwhelming the student.

We feel that the text will meet, and hopefully exceed, your needs (student and educator). We look forward to hearing your comments and suggestions, which we shall incorporate into a second edition.

Josaphat A. Uvah, Ph.D.,
Editor.
Department of Mathematics
The University of West Florida
Pensacola, Florida 32514, U.S.A.
email: juvah@uwf.edu

Ancillary materials

Solutions Manual, with multimedia CD ROM.

Chapter 1.

Fundamental Properties and Processes in Algebra

1.1. Algebraic Expressions

Algebraic expressions can contain numbers alone, a variable alone, or combinations of variables and numbers represented as sums, or differences, or products, or quotients. Examples are given in the table below.

Terms

A term can be a number, a variable or a product of a number and a variable raised to a power. **Terms** in an expression are always <u>separated</u> by **addition or subtraction**. For example:

Algebraic Expression	Terms
-9	-9
$5z$	$5z$
$3x - 7$	$3x, -7$
$2x^3 - 8x + 13$	$2x^3, -8x, 13$
$\sqrt{y}$	$\sqrt{y}$

Exercise **1.1**. *Identify the terms in the following expressions.*

(*a*) $x^3 - 2$.

Solution:

(*b*) $\frac{1}{2}y - 3x$.

Solution:

(*c*) $\sqrt{x} + 2x - 3x^5 + 19$.

Solution:

Coefficients

In the expression $5z$; 5 and z are factors. The **numerical coefficient** of a term is simply the numerical factor. In the expression $-6xy$; -6 is the numerical coefficient. The coefficient of the variable x is the product of the remaining factors, that is $-6y$.

Exercise **1.2**. *Identify the coefficients of x in the following expressions*

(a) $x + 5$.

Solution:

(b) $25z^2x$.

Solution:

(c) $2x^2$.

Solution:

1.1.i. Exponents

Definition

Base and exponent

N^m denotes N as a factor m times.

$N^m = \underbrace{N \bullet N \bullet N \bullet N \bullet \bullet \bullet N}_{m \text{ factors of } N}$

The number N is called the **base**,

and the number m is called the **exponent**.

$N^m = Base^{Exponent}$

For example, $N^4 = N \bullet N \bullet N \bullet N$. The **base is** N, and 4 is the **exponent**. Also, $(2)(2)(2) = 2^3$. Here 2 is the **base** and 3 is the **exponent**.

Definition

Negative exponents

If the exponent is negative one,

$N^{-1} = \frac{1}{N}$

In other words N^{-1} represents the reciprocal of N.

If m is a positive number, and $N \neq 0$, then

N^{-m} is the reciprocal of N^m.

$N^{-m} = \frac{1}{N^m}$

For example: $\quad 3^{-1} = \frac{1}{3} \quad \Big| \quad 10^{-4} = \frac{1}{10^4} \quad \Big| \quad x^{-5} = \frac{1}{x^5}$

Negative exponents and fractions

$$\left(\frac{a}{b}\right)^{-1} = \frac{1}{\left(\frac{a}{b}\right)} = 1 \div \frac{a}{b} = 1 \bullet \frac{b}{a} = \frac{b}{a}$$

Thus, $\left(\frac{a}{b}\right)^{-1} = \frac{b}{a}$ which is the reciprocal of $\frac{a}{b}$.

Also, $\dfrac{a^{-1}}{b^{-1}} = \dfrac{\frac{1}{a}}{\frac{1}{b}} = \dfrac{1}{a} \div \dfrac{1}{b} = \dfrac{1}{a} \bullet \dfrac{b}{1} = \dfrac{b}{a}$

Thus, $\dfrac{a^{-1}}{b^{-1}} = \dfrac{b}{a}$

In effect, a negative exponent in the numerator moves that factor into the denominator, and a negative exponent in the denominator moves that factor into the numerator.

Exercise **1.3**. *Express the following without negative exponents.*

(a) 10^{-2}.

Solution:

(b) $\frac{1}{x^{-2}}$.

Solution:

(c) $\frac{x^{-4}}{x^{-3}}$.

Solution:

The arithmetic of exponents

Exponents appear often in mathematical expressions. Therefore, it is helpful to understand the rules of arithmetic for exponents.

- | *Rule* |
  | $N^i N^j = N^{i+j}$ |

  Where N is a real number, or an algebraic expression, and i and j are integers.

Thus, to find the product, **when the bases are the same**, the exponents are added:

$$(6^3) \cdot (6^5) = (6 \cdot 6 \cdot 6)(6 \cdot 6 \cdot 6 \cdot 6 \cdot 6)$$
$$= 6^8$$

or, $(6^3) \cdot (6^5) = 6^{3+5} = 6^8$

And,

$$(2x)^4(2x)^7 = (2x)^{4+7}$$
$$= (2x)^{11}$$

- | *Rule* |
  | $N^0 = 1$ |

  Since $N^{-m} \cdot N^m = \frac{1}{N^m} \cdot N^m = \frac{N^m}{N^m} = 1$

  and, $N^{-m} \cdot N^m = N^{-m+m} = N^0$

  Thus, $N^0 = 1$, where $N \neq 0$.

- | *Rule* |
  | $\frac{N^i}{N^j} = N^{i-j}$ |

  where N is a real number, or algebraic expression, and i and j are integers.

Thus, to divide, when the bases are the same, the exponents are subtracted, (that is, the numerator exponent minus the denominator exponent).

Thus $\dfrac{N^{12}}{N^4} = \dfrac{N \bullet N \bullet N \bullet N \bullet N \bullet N \bullet N \bullet N \bullet N \bullet N \bullet N \bullet N}{N \bullet N \bullet N \bullet N}$

$$= N \bullet N \bullet N \bullet N \bullet N \bullet N \bullet N \bullet N$$

$$= N^8$$

or $\dfrac{N^{12}}{N^4} = N^{12-4}$

$$= N^8$$

Example 1.1. *Simplify* $\dfrac{27x^3}{81x^2}$.

- **Solution**:

$$\dfrac{27x^3}{81x^2} = \dfrac{3^3 x^3}{3^4 x^2}$$

$$= 3^{3-4} x^{3-2} \text{ since } \dfrac{N^i}{N^j} = N^{i-j}$$

$$= 3^{-1} x^1$$

$$= \dfrac{1}{3} x$$

$$= \dfrac{x}{3}$$

Example 1.2. *Simplify the fraction* $\dfrac{y^3 z^2}{xyz^5}$.

- **Solution**:

$$\dfrac{y^3 z^2}{xyz^5} = \dfrac{y^{3-1} \bullet z^{2-5}}{x}$$

$$= \dfrac{y^2 \bullet z^{-3}}{x}$$

$$= \dfrac{y^2}{xz^3}$$

$\boxed{\text{Note:}}$ Simplified answers are expressed with positive exponents only.

Example 1.3. *Simplify,* $\dfrac{2^3 \bullet 3^5 \bullet 5^{-1} \bullet x^{-2} y^3 z^4}{2 \bullet 3^{-1} xyz^{-1}}$.

- **Solution**:

$$\dfrac{2^3 \bullet 3^5 \bullet 5^{-1} \bullet x^{-2} y^3 z^4}{2 \bullet 3^{-1} xyz^{-1}} = 2^{3-1} \bullet 3^{5-(-1)} \bullet 5^{-1} \bullet x^{-2-1} \bullet y^{3-1} z^{4-(-1)}$$

$$= 2^2 \bullet 3^6 \bullet 5^{-1} \bullet x^{-3} \bullet y^2 \bullet z^5$$

$$= \dfrac{2^2 3^6 y^2 z^5}{5x^3}$$

$$= \dfrac{4 \bullet 729 y^2 z^5}{5x^3}$$

$$= \dfrac{2916 y^2 z^5}{5x^3}$$

Exercise 1.4. *Simplify the following:*

(a) $\dfrac{2^5 \cdot 5^{-1} \cdot x^2 y^3}{2 \cdot 3^{-1} x y^{-1}}$.

Solution:

(b) $\dfrac{2^5 \cdot 5^{-1} \cdot x^2 y^4}{2^{-2} \cdot 3 x^{-2} y^{-1}}$.

Solution:

- Rule
  $$(N^i)^j = N^{i \cdot j}$$
  where N is a real number, or an algebraic expression, and i and j are integers.

Thus, when exponents are raised to a power, multiply the exponents.

$$(4^2)^3 = (4 \cdot 4)^3$$
$$= (4 \cdot 4) \cdot (4 \cdot 4)$$
$$= 4^6$$

or $(4^2)^3 = 4^{2 \times 3}$
$$= 4^6$$

Also,

$$(3^4 x^6)^2 = 3^{4 \cdot 2} \cdot x^{6 \cdot 2}$$
$$= 3^8 x^{12}$$

- Rule
  $$(m^i n^j)^k = m^{ik} n^{jk}$$
  where N is a real number, or an algebraic expression, and i and j are integers.

For example,

$$\left(\frac{x^3}{y^5}\right)^4 = \frac{x^{3 \cdot 4}}{y^{5 \cdot 4}}$$
$$= \frac{x^{12}}{y^{20}}$$

Example 1.4. *Using the properties of exponents rewrite the expression* $\left(\frac{x^2y^3}{2y^{-1}}\right)^3$ *so that no negative exponents appear.*

- Solution:

$$\left(\frac{x^2y^3}{2y^{-1}}\right)^3 = \left(\frac{x^2y^3y}{2}\right)^3 \text{ or } \left(\frac{x^2y^{3-(-1)}}{2}\right)^3$$

$$= \left(\frac{x^2y^4}{2}\right)^3$$

$$= \frac{(x^2)^3(y^4)^3}{2^3}$$

$$= \frac{x^6y^{12}}{8}$$

$$= \frac{1}{8}x^6y^{12}$$

Exercise 1.5. *Using the properties of exponents rewrite the following expressions so that no negative exponents appear.*

(a) $\left(\frac{x^{-2}}{y^3}\right)^2$.

Solution:

(b) $\left(\frac{x^2y^3 2y^{-1}}{x^3y^2}\right)^4$.

Solution:

Example 1.5. *Using the properties of exponents rewrite the expression* $\dfrac{4w^2(2x+1)^{-3}y^4}{2^{-3}w^3(2x+1)^2y}$ *without negative exponents.*

- Solution:

$$\frac{4w^2(2x+1)^{-3}y^4}{2^{-3}w^3(2x+1)^2y} = \frac{2^2w^{2-3}(2x+1)^{-3-2}y^{4-1}}{2^{-3}}$$

$$= 2^{2-(-3)}w^{2-3}(2x+1)^{-3-2}y^{4-1}$$

$$= 2^5w^{-1}(2x+1)^{-5}y^3$$

$$= \frac{2^5y^3}{w(2x+1)^5}$$

$$= \frac{32y^3}{w(2x+1)^5}$$

Exercise 1.6. *Simplify the following expression using the properties of exponents. Express with positive exponents.*

(a) $(3^2 x^4 y^{-2})(3^{-4} x^5 y^3 z)$.

Solution:

(b) $\left(\dfrac{(x^2 + 1)^{-2} x^3 (6x)^2}{(x^2 + 1)x^{-3}} \right)$.

Solution:

(c) $\left(\dfrac{(2p)^3 p^4}{x^3 (p^2 + 1)^4} \right)^{-2}$.

Solution:

(d) $\dfrac{(2x + 1)^{-3}(x^2 + 1)^5}{x^{-4}(2x + 1)^2 (x^2 + 1)^{-2}}$.

Solution:

(e) $\dfrac{(x + 1)^2 y^{-3}(3x)^2}{(x + y)^3}$.

Solution:

(f) $\left(\dfrac{x^3 (p^2 + 1)^4}{(2p)^3 p^4} \right)^{-2}$.

Solution:

Summary 1.1.

- $N^m = \underbrace{N \bullet N \bullet N \bullet N \bullet \bullet \bullet N}_{m \text{ factors of } N.}$

- $N^{-1} = \dfrac{1}{N}$.

- $N^0 = 1,\ N \neq 0$.

- $\left(\dfrac{M}{N} \right)^{-1} = \dfrac{M^{-1}}{N^{-1}} = \dfrac{N}{M}$.

- $N^i N^j = N^{i+j}$.

- $\dfrac{N^i}{N^j} = N^{i-j}$.

- $(N^i)^j = N^{i \bullet j}$.

- $(M^i N^k)^j = M^{ij} N^{kj}$.

- $\left(\dfrac{M^i}{N^k} \right)^j = \dfrac{M^{ij}}{N^{kj}}$.

End of Section 1.1. Problems

Problem 1.1. *For the following expressions, determine the number of terms, list them individually, and identify the coefficient of each term.*

(a) $3x - 1$.

(b) $x^2y - 2xy^{-1} - x$.

(c) $2y^{-1}\sqrt{x+4} + 3x^2y$.

(d) $-3x^2 + x - 4$.

(e) $\sqrt{x}\,\sqrt{y} - 3 + x^2$.

Problem 1.2. *For the following problems, use the properties of exponents to rewrite the expressions so that only positive exponents are used and the expression is simplified.*

(a) $(3^1)(3^1)(3^1)(3^1)(3^1)(3^1)x^1x^1$.

(b) $3^2x^2x^{-1}x^3y^{-1}y^4$.

(c) $\frac{w^{-3}x^2}{wx^{-4}}$.

(d) $\frac{3^2 \cdot 4^{-3}w^3z^{-4}}{3 \cdot 4^{-5}wz^{-5}}$.

(e) $(2x^3y)^4$.

(f) $\left(\frac{x^2y^{-3}}{x^4y^2}\right)^{-2}$.

(g) $\frac{2x^5(2x)^{-4}}{xy^2}$.

Problem 1.3. *For each of the following expressions, determine whether or not the properties of exponents apply:*

(a) $\left(x^{\frac{1}{2}} + 2^1\right)^4$.

(b) $\left(x^{\frac{1}{2}} \cdot 2^1\right)^4$.

(c) $(x^2 + 3)^2$.

(d) $(3x^2)^2$.

(e) $\left(x^{\frac{1}{2}} - x\right)^2$.

(f) $\left(x^{\frac{1}{2}} - x^2\right)^2$.

(g) $\left(x^{\frac{1}{2}} \cdot x^2\right)^2$.

Problem 1.4. *For the following exercises, rewrite the expression using the properties of exponents so that no negative exponent appears in the expression.*

(a) $(2x+1)^3(2x+1)^{-1}$.

(b) $\frac{x^{-3}(2x+1)^3}{x(2x+1)^2}$.

(c) $\left(\frac{2xy^{-\frac{1}{3}}}{x^2}\right)^3$.

(d) $\frac{3x^{-1}y^{-1}}{x^{-2}y^{-3}}$.

(e) $(3x)^{-1}y^2$.

(f) $\frac{2^3(x-1)^{-3}(2x+1)^4}{2(x-1)^{-4}(2x+1)}$.

(g) $\left(\frac{3^2x^3y^4}{3xy^2}\right)^{-1}$.

1.2. Factoring

Factoring is a process of rewriting mathematical expressions into an equivalent **multiplication form using factors** that are **prime**.

| Definition |

Prime numbers

Prime numbers are numbers which are only **divisible** by **themselves** and 1.

Note: single variables are considered to be prime.

Examples of **prime numbers** are: $2, 3, 5, 7, 11, 13, 17, 19, 23...$

| **Note:** | The only even prime number is 2.

Composite numbers

Numbers that are **not prime** are called **composite numbers**.

Examples of **composite numbers** are, $4, 6, 8, 10$.

| **Note:** | The number 1 is **neither** prime or composite.

Example **1.6**. *Give the factored form of* 30.

- **Solution**: The factored form is $(2)(3)(5)$

Example **1.7**. *Is* $(8)(2)$ *the factored form of* 16.

- **Solution**: No, since 8 is not prime.

Exercise **1.7**. *Give the factored form of the following. Express your answer in exponent notation.*

(*a*) 16.

Solution:

(*b*) 9.

Solution:

(*c*) 90.

Solution:

| Definition |

The **Prime Factorization Theorem**

states that every positive integer, different from 1, is **either prime**, or can be written as a **unique product of primes**.

For example

Not prime = Product of primes

$$4 = (2)(2)$$
$$6 = (2)(3)$$
$$15 = (3)(5)$$
$$24 = (2)(2)(2)(3)$$
$$110 = (2)(5)(11)$$

Definition

Factored form

A positive integer is **factored** provided it is written as the product of prime numbers.

A negative integer is **factored** if the negative integer is written as (-1) multiplied by positive prime numbers.

Note: A prime number is already considered to be factored.

For example,

▶ $(-1)(2)(3)$ is the factored form for -6.

▶ $(-1)(2)(3)(7)$ is the factored form for -42.

▶ $(-1)(2)(2)(3)(3)(5)(13)$ is the factored form for -2340.

Note: From the above factorization it is easy to see that 13 is a divisor of 2340. That is, 2340 is equal to 13 times some other integer; the other integer would be $2 \cdot 2 \cdot 3 \cdot 3 \cdot 5$ (the factors other than 13); that is 180.

1.2.i. Determining the prime factorization of a number

To determine the **prime factorization** of a number, **divide evenly** the number by **prime** numbers until a quotient of 1 is found. Start from the smallest prime number, move to the next highest prime number when (if) the division no longer produces a whole number.

Example **1.8**. *Find the prime factorization of* 420.

• **Solution**:

$$420 \div 2 = 210$$
$$210 \div 2 = 105$$
$$105 \div 3 = 35$$
$$35 \div 5 = 7$$
$$7 \div 7 = 1$$

Thus $420 = (2)(2)(3)(5)(7)$.

Example **1.9**. *Since* $(2)(2)(3)(4)(5)$ *multiplies together to yield* 240, **is this** *the prime factorization (factored form) of* 240?

- **Solution**: No. The number 4 is not prime. The factored form for 240 would therefore be $(2)(2)(2)(2)(3)(5)$.

Exercise **1.8**. *Find the prime factorization of the following*:

    (*a*) 19.

    Solution:

    (*b*) 290.

    Solution:

1.2.ii. Use of factored forms

Using **factored forms** allows us to see the basic building blocks which make up a number. This allows us to **quickly** perform many *division problems*.

Since a number, different from zero, divided by itself is equal to 1, we can simplify the following fraction:

$$\frac{(2)(3)(5)(5)(11)(13)}{(5)(11)(13)} = \frac{(2)(3)(5)}{1}$$
$$= (2)(3)(5)$$
$$= 30$$

Example **1.10**. *Consider the following questions and calculations dealing with the number* **3,465**.

(*a*). *Does* 9 *divide* $3,465$ *as an integer with a zero remainder?*

- **Solution**: We first factor $3,465$.

$$3,465 = (5)(7)(3)(3)(11)$$
$$\Rightarrow$$
$$\frac{3,465}{9} = \frac{(5)(7)(3)(3)(11)}{(3)(3)}$$
$$= (5)(7)(11)$$
$$= 385$$

| Yes, 9 divides $3,465$. |

(*b*). *Does* 99 *divide* $3,465$ *as an integer with a zero remainder?*

- **Solution**:

$$\frac{3,465}{99} = \frac{(5)(7)(3)(3)(11)}{(3)(3)(11)}$$
$$= (5)(7)$$
$$= 35$$

Yes, 99 divides 3,465.

(c). *Does 7 divide 3,465 as an integer with a zero remainder?*

• **Solution:**

$$\frac{(5)(7)(3)(3)(11)}{7} = (5)(3)(3)(11)$$
$$= 495$$

Yes, 7 divides 3,465.

(d). *Does 25 divide 3,465 as an integer with a zero remainder?*

• **Solution:**

$$\frac{(5)(7)(3)(3)(11)}{25} = \frac{(5)(7)(3)(3)(11)}{(5)(5)}$$
$$= \frac{(7)(3)(3)(11)}{(5)}$$

Here division is **not** possible. The factors of the divisor are not all present in the factorization of the numerator. This last quotient will not yield an integer answer with an integer remainder, therefore 25 does not divide 3,465 evenly.

Exercise 1.9. *Perform the following operations, write the answer in factored form.*

(a) *Divide 5775 by 7.*

Solution:

(b) *Divide 321 by 5.*

Solution:

1.2.iii. Factoring monomials

Just as the factored form of integers aids in division, and finding the reduced form of a fraction, the factored form of a monomial aids in **dividing monomials** and **reducing fractions of monomials**.

Definition
Monomial
An expression which consists of the product of numbers and variables, is called a **monomial**. In other words, an expression that consists of only one term is a monomial.

For example $\boxed{3x}$ $\boxed{5xy}$ $\boxed{x \bullet 36 \bullet y}$ $\boxed{xyz^3}$ are monomials.

Note: $\boxed{5+x}$ $\boxed{7x^{-1} + 8y}$ $\boxed{7x^2 + 8y}$ are **not monomials**, since they have more than one term.

Example 1.11. Write the following quotient in reduced form (that is, with no common factors in the numerator or the denominator):

$$\frac{420xy^2x^2wy^2}{1386abxyx}$$

- **Solution**: Factor the numerator: $2 \bullet 2 \bullet 3 \bullet 5 \bullet x^3 \bullet y^4 \bullet w$

  The denominator factors as $2 \bullet 3 \bullet 3 \bullet 7 \bullet 11 \bullet x^2 \bullet y \bullet a \bullet b$

$$\text{Thus} \quad \frac{420xy^2x^2wy^2}{1386abxyx} = \frac{2 \bullet 2 \bullet 3 \bullet 5 \bullet 7 \bullet x^3 \bullet y^4 \bullet w}{2 \bullet 3 \bullet 3 \bullet 7 \bullet 11 \bullet x^2 \bullet y \bullet a \bullet b}$$

$$= \frac{2 \bullet 5 \bullet x \bullet y^3 \bullet w}{3 \bullet 11 \bullet a \bullet b}$$

$$= \frac{10xy^3w}{33ab}$$

▶ This is a reduced form of the fraction, but not a factored form of the fraction.

The following example illustrates these **properties** in a more complicated situation.

Example 1.12. Simplify, $\frac{180a^{-1}b^{-2}x^3y^{10}}{160cx^2b^{-4}y^2}$.

- **Solution**:

$$\frac{180a^{-1}b^{-2}x^3y^{10}}{160cx^2b^{-4}y^2} = \frac{2^2 3^2 5 x^3 y^{10} b^4}{2^5 5 cx^2 y^2 ab^2}$$

$$= \frac{3^2 xy^8 b^2}{2^3 ca}$$

$$= \frac{9b^2 xy^8}{8ac}$$

Exercise 1.10. Simplify the following quotients of monomials. What is the value of the quotient if $a = 1, b = 2, c = 5, x = 4$? Give the answer in reduced form.

(a) $\frac{a^{-3}b^2cx^{-1}}{ac^{-2}}$.

Solution:

(b) $\frac{a^2b^{-2}cx^{-3}}{c^2ab^{-1}x^{-1}}$.

Solution:

Exercise 1.11. *Complete the following:*

(a) *Factor* 5775 *as a product of primes.*

Solution:

(b) *Simplify the expression,* $\frac{x^2y^{-2}z^3}{xy}$ *, so that only positive exponents are used:*

Solution:

(c) *In (b) above let* $x = 5, y = 2,$ *and* $z = 4.$

Solution:

(d) *Divide* 5775 *by* $7 \cdot 5$ *and write the answer in factored form.*

Solution:

(e) *Multiply* 5775 *by* 32 *and write the product in factored form.*

Solution:

(f) *Does* 9 *divide* 1680 *with a remainder of zero? (i.e., does it divide evenly?)*

Solution:

Exercise 1.12. *Complete the following:*

(a) *Multiply* $(11 \cdot 3 \cdot 5^2 \cdot 7) \cdot 35$ *using the properties of exponents. Give the answer in factored form.*

Solution:

(b) *Simplify,* $\frac{a^2b^{-2}cx^{-3}}{c^2ab^{-1}x^{-1}}$. *What is the value of the quotient if* $a = 1, b = 2, c = 5, x = 4$? *Give the answer in reduced form.*

Solution:

Summary 1.2.

- **Prime numbers** are numbers which are only **divisible** by **themselves** and 1.
  - ▶ Single variables are considered to be prime.
- **Factoring** aids in division and the reduction of fractions (numeric and monomial) to their simplest terms.
- A **monomial** is a product of numbers and variables, and is considered to be factored if it is written as a product of primes.
- **Exponential notation** allows one to represent multiple products in a short form.
  - ▶ And also to indicate that a number is to be divided (negative exponents).

End of Section 1.2. Problems

Problem 1.5. Factor the following:

(a) 144.   (b) 129.   (c) −302.   (d) 78.

(e) 924.   (f) 1287.   (g) 3567.

Problem 1.6. Divide 288 by 6, and leave the answer in prime factored form.

Problem 1.7. Divide 306 by 17, and leave the answer in prime factored form.

Problem 1.8. Divide −525 by 6, and leave the answer in prime factored form.

Problem 1.9. Is 7 a prime factor of 105?

Problem 1.10. Does 9 divide 486 with remainder zero?

Problem 1.11. Using division and the properties of exponents, simplify the following expressions.
Leave all exponents positive.

(a) $\dfrac{26x^2y^3rs^3}{13x^3y^4rs}$.   (b) $\dfrac{(2^{\frac{1}{2}}x^{\frac{1}{2}})^6 r^2}{2xr^{-1}}$.   (c) $\left(\dfrac{12x^{-2}y^{-2}}{r^{-1}s^{-3}}\right)^{-1}$.

Problem 1.12. Evaluate the expression $\dfrac{26x^2y^3rs^3}{13x^3y^4rs}$ when $x = 1$, $y = 3$, $r = 2$, and $s = 1$.

1.3. Properties of Polynomials and Order of Operations

A polynomial is a mathematical expression, formed by using one or more operations of addition, subtraction and multiplication on a variable (x) with real constants. Examples of polynomials are given in the table below.

Polynomials	Number of term
$3x + 1$	2
$-\frac{1}{2}x^2 + 6x - 1$	3
$2x$	1

Note: There are **no restrictions** on the numbers that may be substituted for the **value** of x (the value of the variable) in a polynomial. We can use this property to identify polynomials.

Example **1.13.** *Is the following expression,* $\dfrac{(x-1)(x-2)}{(x-2)}$*, a polynomial?*

- **Solution:** No. If $x = 2$, then

$$\frac{(x-1)(x-2)}{(x-2)} = \frac{(x-1)(2-2)}{(2-2)}$$
$$= \frac{(x-1)(0)}{(0)}$$

Since division by zero is not possible, x cannot be equal to 2. Since we now have a **restriction** on the value of the **variable** this expression **cannot** be a **polynomial**.

Example **1.14.** *Is the expression,* $\dfrac{(x-1)(x^2+1)}{x^2+1}$*, a polynomial?*

- **Solution:** This expression reduces to $x - 1$, which is a <u>polynomial</u>. The denominator, $x^2 + 1$ is never zero; since for any value of x, x^2 is always positive or zero.

Exercise **1.13.** *Are the following expressions polynomials?*

(a) $\dfrac{(x-1)(x^2+1)}{x^3+1}$.

Solution:

(b) $\dfrac{1}{x^2+1}(x^2-1)(x^2+1)$.

Solution:

An extension of the vocabulary dealing with polynomials

Definition

A polnomial in standard form

$a_N x^N + a_{N-1} x^{N-1} + ... + a_2 x^2 + a_1 x + + a_0$

where $a_0, a_1, a_3, a_4, ..., a_N$ are real numbers, and x is the variable.

$a_0, a_1, a_3, a_4, ..., a_N$ are the **coefficients** and may be
either positive, negative or zero.

Note In a polynomial, negative, or fractional exponents are **not allowed**.

Thus, when $N = 2$, we have the standard form: $a_2 x^2 + a_1 x^1 + a_0$. A polynomial of this type would be, for example, $2x^2 + 10x + 5$.

1.3.i. A description of a polynomial as a product of primes

Definition

Polynomials:

Polynomials that are factored over the reals already **factored** are
called **prime polynomials**.

For example:

- A polynomial with one term that is a **constant** is considered **prime**. This polynomial is called a **constant polynomial**
  - ▶ Any number is an example of a constant polynomial.

- Expressions of the form $\boxed{mx + b}$ where m and b are real numbers with no common factors, are **prime polynomials**.
  - ▶ $2x - 3$ **is** a prime polynomial.
  - ▶ $2x + 2$ is **not** a prime polynomial, since $2x + 2 = 2(x + 1)$.
    - ◀ That is, the terms have common factors.

Exercise **1.14.** *Which of the following polynomials are prime.*

(*a*) $2x$.

Solution:

(*b*) $3x + 6$.

Solution:

(*c*) $3x + 7$.

Solution:

The **degree of a polynomial**, in **one variable**, is the **highest power** of the variable in the polynomial. The following table illustrates this point.

Polynomial	Degree
$P(\mathbf{x}) = 4 = 4x^0$	0
$P(x) = 3x + 1 - 2x + 1$	1
$P(x) = 2x + 3x^2 - 1$	2
$P(\mathbf{x}) = -x^3 + 1 + x^4$	4

Exercise **1.15**. *Give the degree of the following polynomials.*

(*a*) $2x^2 + x$.

Solution:

(*b*) $3x + 6$.

Solution:

(*c*) $3x^4 + 7$.

Solution:

1.3.ii. Arithmetic properties of polynomials

Polynomials have rules of arithmetic just as there are rules of arithmetic for numbers. The distributive property applies equally to real numbers and polynomials.

Definition

The Distributive Property

$x \bullet (y + z) = x \bullet y + x \bullet z$

Or,

$(y + z) \bullet x = y \bullet x + z \bullet x$

for any real numbers x, y and z.

For example,

$$2(3 + 4) = 2 \bullet 3 + 2 \bullet 4$$
$$2(7) = 6 + 8$$
$$14 = 14$$

Note: We can calculate the quantity $2(3 + 4)$ either by operating inside the parentheses first, or by using the right side of the distributive property.

For example, $3 \bullet 7 + 8 \bullet 7$, can be simplified as follows:

$$3 \cdot 7 + 8 \cdot 7 = (3 + 8)7$$
$$= 11 \cdot 7$$
$$= 77$$

• Here the common factor, 7, is factored out, and thus simplifies the arithmetic.

1.3.ii.a. Addition of Polynomials and the Distributive Property

Example **1.15.** *Add* $2x$ *and* $3x$.

• **Solution**: Apply the distributive property.
$$2x + 3x = 2 \cdot x + 3 \cdot x$$
$$= (2 + 3) \cdot x$$
$$= 5 \cdot x$$
$$= 5x.$$

Example **1.16.** *Add* $-x^2$ *and* $5x^2$.

• **Solution**:
$$-x^2 + 5x^2 = (-1 + 5)x^2$$
$$= 4 \cdot x^2$$
$$= 4x^2$$

Notice that in the previous examples, we have combined two monomials by recognizing the common values in each term and using the distributive property. This **only works** if the **powers of the variable are the same**. Consequently, $3x^2 + 4x^2$ **can** be simplified to $7x^2$. However, $3x^3 + 4x^2$ **cannot** be combined into a single monomial, because x^3 and x^2 are **not like terms**; the powers are different.

Exercise **1.16.** *Add the following monomials*:

(a) $-2x^2$ *and* $3x^2$.

Solution:

(b) $2x^3$ *and* $-4x^3$.

Solution:

Example **1.17.** *Add* $(3x^2 - 4x + 1)$ *and* $(x^2 + x - 3)$.

• **Solution**: This requires regrouping to collect terms of like degree and then adding the coefficients using the distributive property:
$$(3x^2 - 4x + 1) + (x^2 + x - 3) = (3x^2 + x^2) + (-4x + x) + (1 + (-3))$$
$$= (3 + 1)x^2 + (-4 + 1)x + (-2)$$
$$= 4x^2 - 3x - 2$$

1.3.ii.b. Subtraction of polynomials and the Distributive Property

The distributive property can be written with subtraction as the operation instead of addition:

$$x \bullet (y - z) = x \bullet y - x \bullet z$$

Example 1.18. *Solve the following subtraction*: $(3x^2 - 4) - (2x - 4)$.

- **Solution**:

$$(3x^2 - 4) - (2x - 4) = (3x^2 - 4) + (-1)(2x - 4)$$
$$= 3x^2 - 4 + (-1)2x - (-1)4$$

This operation is often missed

$$= 3x^2 - 4 - 2x + 4$$
$$= 3x^2 - 2x$$

Exercise 1.17. *Solve the following subtraction*:

(*a*) $(x^2 - 3x) - (6x + 4)$.

Solution:

(*b*) $(x^3 - 2x^2) - (6x^3 + x)$.

Solution:

1.3.iii. Multiplication of polynomials and the Distributive Property

The procedure for multiplying polynomials is the same as that used to multiply numbers.

Example 1.19. *Use the distributive property to solve the multiplication*: 23×48.

- **Solution**: The first step is to apply the distributive property. Remember: $x \bullet (y + z) = x \bullet y + x \bullet z$.

Thus, $x \times (y + z) = (x \times y) + (x \times z)$
$$(23) \times (40 + 8) = (23 \times 40) + (23 \times 8)$$
$$= ((20 + 3) \times 40) + ((20 + 3) \times 8)$$
$$= [(20 \times 40) + (3 \times 40)] + [(20 \times 8) + (3 \times 8)]$$
$$= (800 + 120) + (160 + 24)$$
$$= 920 + 184 = 1004$$

Exercise **1.18.** *Use the distributive property to solve the multiplications:*

(a) $2x(8x + 1)$.

Solution:

(b) $-5x(10x - 2)$.

Solution:

1.3.iii.a. Multiplying two binomials (the FOIL Method)

- | Definition |
  Binomial:
  A polynomial with two terms.
  For example $\boxed{6x + 3}$ $\boxed{3x^2 - 5}$

Example **1.20.** *Solve the multiplication of* $2u + 3$ *and* $4u + 8$.

- **Solution**:

$$(2u + 3) \times (4u + 8) = ((2u + 3) \times 4u) + ((2u + 3) \times 8)$$
$$= [(2u \times 4u) + (3 \times 4u)] + [(2u \times 8) + (3 \times 8)]$$
$$= (8u^2 + 12u) + (16u + 24)$$
$$= 8u^2 + 28u + 24$$

- There is an easy reminder of the order in which to perform these multiplications. Again, consider $(2u + 3)(4u + 8)$
  - ▶ First terms $(2u \times 4u)$
  - ▶ Outside terms $(2u \times 8)$
  - ▶ Inside terms $(3 \times 4u)$
  - ▶ Last terms (3×8)
    - ◀ The first initials of these phrases form the acronym FOIL.

The distributive property and the process of multiplication apply equally well to multiplying polynomials with three terms, four terms, or any number of terms.

Example **1.21**. *Perform the multiplication*: $(6x - 3)(2x - 1)$.

- **Solution:**

$$
\begin{array}{cccc}
\text{First} & \text{Outer} & \text{Inner} & \text{Last}
\end{array}
$$
$$(6x - 3) \times (2x - 1) = (6x \times 2x) + (6x \times -1) + (-3 \times 2x) + (-3 \times -1)$$
$$= (12x^2 - 6x) + (-6x + 3)$$
$$= 12x^2 - 12x + 3$$

Another method using the distributive property is illustrated below:
$$(6x - 3)(2x - 1) = (6x - 3)(2x) + (6x - 3)(-1)$$
$$= (6x)(2x) - 3(2x) + 6x(-1) - 3(-1)$$
$$= 12x^2 - 6x - 6x + 3$$
$$= 12x^2 - 12x + 3$$

Exercise **1.19**. *Perform the multiplication*:

(a) $(-6x - 3)(4x + 9)$.

Solution:

(b) $(2x + 2)(x - 1)$.

Solution:

1.3.iii.b. Multiplying two trinomials

In **multiplying** a polynomial of **degree** n **times** one of **degree** m, the **total** number of **multiplications** will be mn.

Example **1.22**. *Multiply* $(4x^2 + 3x - 6)$ *and* $(-2x^2 - 5x + 1)$.

- **Solution:**

$$(4x^2 + 3x - 6)(-2x^2 - 5x + 1) = (4x^2 + 3x - 6)(-2x^2) + (4x^2 + 3x - 6)(-5x)$$
$$+ (4x^2 + 3x - 6)(1)$$
$$= (4x^2)(-2x^2) + (3x)(-2x^2) + (-6)(-2x^2)$$
$$+ (4x^2)(-5x) + (3x)(-5x) + (-6)(-5x)$$
$$+ (4x^2)(1) + (3x)(1) + (-6)(1)$$
$$= -8x^4 - 6x^3 + 12x^2 - 20x^3 - 15x^2 + 30x + 4x^2 + 3x - 6$$
$$= -8x^4 - 26x^3 + x^2 + 33x - 6$$

Exercise **1.20.** *Perform the following arithmetic operations:*

(*a*) $(x^2 + 3x + 1) - (-2x^2 + 6x + 4)$.

Solution:

(*b*) $(6x - 3)(2x - 1)$.

Solution:

(*c*) $(x + 3)(x + 3)(x + 3)$.

Solution:

(*d*) $(x^5 + x^2 - 4x + 3)(x^3 + 2x^2 - 3)$.

Solution:

(*e*) $(-3x + 4)(-x + 1)$.

Solution:

(*f*) $(x - 1)(x^2 + x + 1)$.

Solution:

(*g*) $(x^5 - 2x + 1)(x^4 + 2x^3 + 4x + 5)$.

Solution:

(*h*) $(2x + 1) - (4x - 1)$.

Solution:

(*i*) $x^2(9x)$.

Solution:

Summary 1.3.

- Addition of polynomials and multiplication of polynomials are based on the distributive property.
- **The Distributive Property**:
  - ▶ For real numbers x, y and z
  $$x \bullet (y + z) = x \bullet y + x \bullet z$$
- In **adding polynomials**, the only **terms** that can be **combined** are those with the **same degree** (same power on the variable). That is, $x^2 + 2x^2$, but not $x^2 + 3x^4$.
- In **multiplying binomials**, there are four multiplications. These can be remembered using the acronym "**FOIL**".
- In **multiplying** a polynomial of **degree** n **times** one of **degree** m, the **total** number of **multiplications** will be mn.

End of Section 1.3. Problems

Problem 1.13. *Identify which of the following mathematical expressions are polynomials:*

(a) $3x + 1$.   (b) $2x - \frac{1}{2}$.   (c) $-\frac{1}{2}x + \pi x + 1$.   (d) $3x^2 - 6x + 4$.

(e) $x^3 - x - 4$.   (f) $x^6 + 0x^5 + x^4 - 10$.   (g) $\frac{x-1}{x}$.   (h) $\frac{x^2}{x-1}$.

(i) $\sqrt{x} - 1$.   (j) $\sqrt[3]{x-1}$.

Problem 1.14. *Why is $x^2 + x + 3$ a polynomial?*

Problem 1.15. *Simplify the following expressions by combining like terms:*

(a) $-2x^2 - 3x + x^2$.   (b) $-\frac{1}{2}x^3 + 1 + x^3$.   (c) $2x^2 - 3x + 4x - x^2$.

(d) $\frac{1}{3}x^3 + x - 3 + 1$.   (e) $x^2 - (3x^2 - 1)$.

Problem 1.16. *Simplify the following expressions by combining like terms:*

(a) $x^2 - 3x + 2(2x + 1)$.   (b) $(x^3 - x^2) - 2(-x^3 + 1)$.   (c) $x - (3 - 2x)$.   (d) $x^2 - 3\left(\frac{1}{3}x^2 + 1\right)$.

Problem 1.17. *Simplify the following expressions by combining like terms:*

(a) $(6x^2 - 4x + 1) - (2x^2 - x)$.   (b) $\frac{1}{2}(2x^2 - 4x) + x^2$.   (c) $\left(3x^3 - \frac{1}{3}x + 1\right) + \left(x^3 - \frac{2}{3}x - 1\right)$.

(d) $4(2x - 1) - 8x$.   (e) $6x^2 - x^2 - (2x - 4)$.

Problem 1.18. *Why is $\sqrt{x} + x$ not a polynomial?*

Problem 1.19. *For the following polynomials, determine the degree of the polynomial:*

(a) $f(x) = 3x + 1$.   (b) $f(x) = -1x^3 - 2$.   (c) $f(x) = 0x + 4$.

(d) $f(x) = 3$.   (e) $f(x) = 1 - 3x^5$.

Problem 1.20. *Perform the following operations:*

(a) *Add* $-3x$, 1, *and* x^2.   (b) *Multiply* $3x$ *and* $x + 1$.   (c) *Subtract* $3x$ *from* $x^4 - 1$.

Problem 1.21. *A student performs the following incorrect procedure:*

$$3x - (2x + 1) = 3x - 2x + 1$$

Explain the mistake the student has made.

Problem 1.22. *In the following exercises, use the distributive property:*

(a) $3(4 + 6)$.   (b) $x^2(-x + 1)$.   (c) $-3x(x^2 - 4x + 1)$.

(d) $(x^2 - 4x + 1)(-3x)$.   (e) $2x\left(\frac{1}{4}x^3 + 2x - 1\right)$.   (f) $(x^2 + x + 1)(-x)$.

Problem 1.23. *Perform the following multiplications using FOIL:*

(a) $(x + 1)(x - 3)$.   (b) $(a - b)(a + b)$.   (c) $(3x + 1)(3x + 1)$.   (d) $(-x + 2)(x + 5)$.

(e) $(2x + 1)(-x + 5)$.   (f) $\left(\sqrt{2}x - 1\right)\left(\sqrt{2}x + 1\right)$.   (g) $(1 - u)(u + 1)$.   (h) $(3x - 4)(x - 1)$.

(i) $(1 - x)(1 - x)$.   (j) $(u + w)(r + s)$.   (k) $(z - 1)(3 - 2z)$.   (l) $(-2x + 1)(4 + x)$.

Problem 1.24. *Using the distributive property, perform the following mutiplications:*

(a) $(x^2 + x + 1)(x - 1)$.   (b) $(2x^3 + x - 3)(-x + 2)$.   (c) $(-x^2 + 3x - 1)(x^2 + 2x + 1)$.

(d) $(3x + 1)(9x^2 - 3x + 1)$.   (e) $\left(x^2 + \sqrt{2}x + 1\right)\left(x^2 - \sqrt{2}x + 1\right)$.   (f) $(x^2 + 1)(x - 1)$.

(g) $(1 - x)(1 + x + x^2 + x^3)$.

1.4. The Distributive Property and Factoring

In the previous section the Distributive Property was discussed. For example,

$$x(a + b) = x(a) + x(b)$$
$$= ax + bx$$

If the process is reversed, **factoring** is the result.

$$ax + bx = (a)x + (b)x$$
$$= (a + b)x$$

Example **1.23.** *Using the distributive property factor* $x^2y + x^2z$.

- **Solution**: Identify the largest monomial common to both terms and factor it out. In this case, x^2, thus

$$x^2y + x^2z = x^2(y + z)$$

Example **1.24.** *Factor the expression* $x^2(x + 1) - 2(x + 1)$.

- **Solution**: To apply the distributive property, it must first be made clear that this is the sum of two terms, so we rewrite the expression as

$$x^2(x + 1) + (-2)(x + 1) = (x + 1)x^2 + (x + 1)(-2).$$

Now apply the distributive property.

$$x^2(x + 1) - 2(x + 1) = (x + 1)x^2 + (x + 1)(-2)$$
$$= (x + 1)(x^2 + (-2))$$
$$= (x + 1)(x^2 - 2)$$

Exercise **1.21.** *Using the distributive property factor .*

(a) $x^3y + x^2z$.

Solution:

(b) $2x - 2$.

Solution:

Example **1.25.** *Factor* $5x^3 + 5x^2$.

- **Solution**: Factor out the largest common factor to both terms in the sum $(5x^2)$.

$$5x^3 + 5x^2 = 5x^2(x) + 5x^2(1)$$
$$= 5x^2(x + 1)$$

Example **1.26.** *"Factor", or "force",* x^2 *as a factor of* $2x^2 + x - 3$.

- **Solution**: We force the factorization in the following manner:

$$2x^2 + x - 3 = x^2 \left(\frac{2x^2}{x^2} + \frac{x}{x^2} - \frac{3}{x^2} \right)$$
$$= x^2 \left(2 + \frac{1}{x} - \frac{3}{x^2} \right)$$

Note: The process of choosing the term to factor becomes the critical step as the expression becomes more complex. This example allows flexibility in using the distributive property.

Exercise **1.22.** *"Factor", or "force",* x^2 *as a factor of*:

(*a*) $x^2 + 11x + 18$.

Solution:

(*b*) $6x^2 + 13x - 5$.

Solution:

Next is a method for using the process with negative exponents.

Example **1.27.** *Factor the expression* $-2x^{-\frac{1}{2}} + x^{\frac{3}{2}}$.

- **Solution**: The common base for the exponents in each term is x, with exponents $-\frac{1}{2}$ and $\frac{3}{2}$. **Note:** $-\frac{1}{2}$ is "smaller" than $\frac{3}{2}$. Thus, the factor $x^{-\frac{1}{2}}$ is used:

$$-2x^{-\frac{1}{2}} + x^{\frac{3}{2}} = x^{-\frac{1}{2}} \left(\frac{2x^{-\frac{1}{2}}}{x^{-\frac{1}{2}}} + \frac{x^{\frac{3}{2}}}{x^{-\frac{1}{2}}} \right)$$
$$= x^{-\frac{1}{2}} \left(2 + x^{\frac{3}{2}} x^{-\frac{1}{2}} \right)$$
$$= x^{-\frac{1}{2}} \left(2 + x^{\frac{4}{2}} \right)$$
$$= x^{-\frac{1}{2}} (2 + x^2)$$

Exercise **1.23.** *Factor the following equations.*

(*a*) $3x + 3y$.

Solution:

(*b*) $2zx^3 - 6x^3$.

Solution:

(*c*) $(x+1)^2(3x) + (x+1)^2 5$

Solution:

(*d*) $x^{-\frac{1}{2}}(x+3) + x^{\frac{1}{2}}$.

Solution:

(*e*) $2x - 2^2$.

Solution:

(*f*) $x^{-\frac{1}{3}}(3x-4) + x^{\frac{2}{3}}(-3x)$

Solution:

(*g*) $\dfrac{1}{\sqrt{x^2+1}}(x+1) - \dfrac{1}{\sqrt{x^2+1}}$.

Solution:

(*h*) $(x+1)^3 x^2 - (x+1)^2 x^3$.

Solution:

(*i*) $\sqrt{x+3}\,x - (x+3)^{-\frac{1}{2}} x^2$.

Solution:

(*j*) $4x^3 + 4x^2$.

Solution:

(*k*) $\dfrac{x}{4} - \dfrac{7}{4}$.

Solution:

(*l*) $-3x^2 + 8x - 4$

Solution:

Summary 1.4.

- The **Distributive Property** is a useful tool that describes **multiplication of polynomials** when read from left to right.
  - ▶ When read from **right to left**, it gives a method of **factoring**.
  - ▶ If the factor includes a base with an exponent, use the "smallest" exponent.

End of Section 1.4. Problems

Problem 1.25. Factor the term 2 from the expression $6x^2 + 2$.

Problem 1.26. Factor the term xy from the expression $xy^2 + xy$.

Problem 1.27. Factor the term -1 from the expression $-x + 3$.

Problem 1.28. Factor the term x^2 from the expression $3x^3 + x^2$.

Problem 1.29. Factor the term $x(x+2)$ from the expression $(x+2)x^3 + (x+2)x$.

Problem 1.30. Factor the term xy from the expression $-2xy + x^2y$.

Problem 1.31. Factor x^2 from $-x^2 + 1$.

Problem 1.32. Factor $2x^2$ from $2x^2 + x - 3$.

Problem 1.33. Factor x^3 from $-x^3 - 3x^2 + 1$.

Problem 1.34. For the following expressions, determine a base common to each term and indicate the least exponent for that base.

(*a*) $3x^2 + 4x^3$.   (*b*) $-3(x+1)^{-1} + 4(x+1)^2$.   (*c*) $-8x^{-\frac{1}{2}} + x^4$.   (*d*) $-x^3 + 2x^2 + x^4$.

1.5. Factoring Second Degree Polynomials

> **Definition**
>
> A polynomial $P(x)$ can be **factored** provided that it can be written as the multiplication of the polynomials $g(x)$ and $h(x)$, where both $g(x)$ and $h(x)$ have degree less than the degree of $P(x)$.

The factorization of a polynomial is **complete** provided it is a **product of prime polynomials**. (Note Zero degree polynomials and first degree polynomials are considered **prime polynomials**.) Determining the factorization of the **non-prime, second degree** polynomials is an important tool in algebra.

This theme is developed below.

1.5.i. Use of the Quadratic Formula in identifying prime second degree polynomials.

> **Definition**
>
> **The Quadratic Formula**
>
> Let a, b, c be real numbers and
> $$P(x) = ax^2 + bx + c$$
> be a second degree polynomial in the variable x.
> The only two values for x which will cause $P(x) = 0$ are given by:
> $$x = \frac{-b + \sqrt{b^2 - 4ac}}{2a} \text{ and } x = \frac{-b - \sqrt{b^2 - 4ac}}{2a}$$
> or, $\dfrac{-b \pm \sqrt{b^2 - 4ac}}{2a}$.
> These two equations are known together as the **quadratic formula**.

Thus, if $x = \dfrac{-b + \sqrt{b^2 - 4ac}}{2a}$ or $x = \dfrac{-b - \sqrt{b^2 - 4ac}}{2a}$, then $ax^2 + bx + c = 0$.

Example **1.28.** *Determine the values for* x *that make* $P(x) = 3x^2 - 2x - 1 = 0.$

- **Solution**: From $ax^2 + bx + c;$ $a = 3,$ $b = -2,$ and $c = -1.$ Substituting these three values into the formulas above, gives

$$x = \frac{-(1)(-2) + \sqrt{(-2)^2 - 4 \cdot 3 \cdot (-1)}}{2 \cdot 3} = \frac{2 + \sqrt{16}}{6} = \frac{6}{6} = 1$$

and

$$x = \frac{-(-2) - \sqrt{(-2)^2 - 4 \cdot 3 \cdot (-1)}}{2 \cdot 3} = \frac{2 - \sqrt{16}}{6} = \frac{-2}{6} = -\frac{1}{3}$$

Check

$$P(x) = 3x^2 - 2x - 1 = 0$$
$$P(1) = 3(1)^2 - 2(1) - 1$$
$$= 0$$

and

$$P(1) = 3\left(\frac{1}{3}\right)^2 - 2\left(\frac{1}{3}\right) - 1$$
$$= 3\left(\frac{1}{9}\right) - \left(\frac{2}{3}\right) - 1$$
$$= \frac{1}{3} + \frac{2}{3} + 1$$
$$= 0$$

Exercise **1.24.** *Determine the values for* x *that make*:

(a) $P(x) = 2x^2 + 6x - 20 = 0.$

Solution:

(b) $P(x) = -6x^2 + 4x + 2 = 0.$

Solution:

- | Definition |

The **discriminant determines primeness**.

The **discriminant of** $P(x) = ax^2 + bx + c$ is the quantity $\boxed{b^2 - 4ac.}$

One of the following two cases must occur:

1. If $\boxed{b^2-4ac}$ is strictly **negative**, then $P(x)$ is **prime**,

and **is in factored form**.

(Note: This case corresponds to the case when the Quadratic Formula fails to yield answers which are real numbers; there is a negative value under the square root.)

2. If $\boxed{b^2-4ac}$ is **zero or positive**, then $P(x)$ is **not prime**.

When $x = r,$ and $x = s,$ where r and s are the two values of x derived from the Quadratic Formula, then the **factorization** of $P(x)$ is $\boxed{a(x-r)(x-s)}$.

In the next examples, this theorem will be used to either determine that a second degree polynomial $P(x)$ is **prime**, or **provide** the **prime factorization** of it.

Example **1.29.** *Determine if* $P(x) = 3x^2 + 2x + 1$ *is prime or factorable?*

- **Solution**: In this case, $a = 3$, $b = 2$, and $c = 1$.

  Thus $b^2 - 4ac = 2^2 - 4(3)(1) = -8$. Since the discriminant is negative, the polynomial $P(x) = 3x^2 + 2x + 1$ is **prime** and already **factored**.

Example **1.30.** *Determine if* $P(x) = x^2 - 1$ *is prime or factorable?*

- **Solution**: In this case, $a = 1$, $b = 0$, and $c = -1$, so the discriminant $b^2 - 4ac = 0^2 - 4(1)(-1) = 4$. Thus $P(x) = x^2 - 1$ is <u>not prime</u>. Consequently, $P(x)$ can be factored. Using the equations of the quadratic formula $\dfrac{-b \pm \sqrt{b^2 - 4ac}}{2a}$ gives values

$$x = \frac{-0 + \sqrt{4}}{2 \cdot 1} = 1 = \text{r} \text{ and } x = \frac{-0 - \sqrt{4}}{2 \cdot 1} = -1 = \text{s}.$$

This means that $P(x) = x^2 - 1$ in the form $P(x) = \text{a}(\text{x} - \text{r})(\text{x} - \text{s})$

$$= (1)(x - 1)(x - (-1))$$
$$= (x - 1)(x + 1)$$

Exercise **1.25.** *Determine if the following are prime or factorable?*

(**a**) $P(x) = 3x^2 + 2x - 1$.

Solution:

(**b**) $P(x) = x^2 + x - 20$.

Solution:

Example **1.31.** *Determine if* $P(x) = 9x^2 - 6x + 1$ *is prime or factorable?*

- **Solution**: $a = 9$, $b = -6$, and $c = 1$.

  The discriminant Determine the values for x that make $\boxed{b^2 - 4ac}$ $= (-6)^2 - 4(9)(1) = 36 - 36 = 0$. Consequently, $P(x)$ can be factored. The values for which $P(x)$ is zero are

$$x = \frac{-b \pm \sqrt{b^2 - 4ac}}{2a} = \frac{-6 + \sqrt{0}}{2 \cdot 9} = -\frac{1}{3} = \text{r} \text{ and } x = \frac{-6 - \sqrt{0}}{2 \cdot 9} = \frac{-1}{3} = \text{s}.$$

- These values for x are not different, but they still yield a factorization, from $\boxed{\text{a}(\text{x} - \text{r})(\text{x} - \text{s})}$ (where r and s are x values derived from the **quadratic formula**), of $P(x) = 9\left(x - (-\frac{1}{3})\right)\left(x - (-\frac{1}{3})\right)$. Checking is accomplished by multiplying the factors together and verifying that the result is indeed the original polynomial.

1.5.ii. Another method of factoring second degree polynomials

The **quadratic formula** <u>**always**</u> works to determine whether or not a quadratic polynomial (a second degree polynomial) will factor. However, this process requires several steps which can be somewhat lengthy, even for simple problems.

Another method, which is less laborious, may be applied if certain conditions are met. This method is based upon the acronym used for the multiplication of two first degree polynomials ("**FOIL**"). This method is a trial and error process and works almost **exclusively** on those second degree equations which have a **discriminant which is a perfect square**.

- Recall that "**FOIL**" stands for First, Outside, Inside, and Last.

  For example, $(2x - 1)(3x + 4)$, gives

$$\overset{F}{(2x \cdot 3x)} + \overset{O}{(2x \cdot 4)} + \overset{I}{(-1 \cdot 3x)} + \overset{L}{(-1 \cdot 4)} = 6x^2 + 8x - 3x - 4$$
$$= 6x^2 + 5x - 4$$

  ▶ Thus, if you multiply **two, first degree** polynomials together, say $m_1x + b_1$ and $m_2x + b_2$, then the **result is a second degree** polynomial which may be written in the form $\boxed{ax^2 + bx + c}$

$$\text{That is} \quad \boxed{(m_1x + b_1)(m_2x + b_2)} = m_1m_2x^2 + b_1m_2x + b_2m_1x + b_1b_2$$
$$= m_1m_2x^2 + (b_1m_2 + b_2m_1)x + b_1b_2$$
$$= \boxed{ax^2 \qquad\qquad +bx \qquad\qquad +c}$$

- The goal of this method is to use the values of a, b, and c to guess appropriate values of m_1, m_2, b_1, and b_2. (The **FOIL method of factoring**.)

  ▶ If you compare the second and third lines of the above equations, you can see that
  $$\boxed{a = m_1m_2,} \quad \boxed{b = b_1m_2 + b_2m_1,} \quad \text{and} \quad c = b_1b_2.$$

The above equations provide a starting point for the process.

Example **1.32.** *Factor the polynomial* $P(x) = -6x^2 + 4x + 2$.

- **Solution:** The discriminant is 64, $\boxed{b^2 - 4ac} = 4^2 + 4 \cdot 6 \cdot 2$, so the polynomial can be factored. (Since the discriminant is not negative, and therefore not prime.) Since $\boxed{a = m_1m_2}$ and $\boxed{c = b_1b_2}$, choose $a = -6$ and $c = 2$, we need to select numbers m_1, m_2, b_1, and b_2 so that $-6 = m_1m_2$ and $2 = b_1b_2$.

  Now make an "educated" guess for m_1, m_2, b_1, and b_2 and then check that the middle terms work out. Suppose we first guess $m_1 = -3$, $m_2 = 2$, $b_1 = 1$, and $b_2 = 2$. We make this selection because $-3 \cdot 2 = 6$ and $2 \cdot 1 = 2$. Check, from $\boxed{(m_1x + b_1)(m_2x + b_2)}$, thus $(-3x + 1)(2x + 2)$. Using "FOIL" we get the polynomial $-6x^2 - 4x + 2$.

  This is **not the polynomial we started with**, so our initial guesses were <u>wrong</u>. However, the first and last terms are correct. Noting that the sign on the middle term is wrong, we can try a new guess.

  Suppose $m_1 = 3$, $m_2 = -2$, $b_1 = 1$, and $b_2 = 2$. From $\boxed{(m_1x + b_1)(m_2x + b_2)}$ we get $(3x + 1)(-2x + 2)$.

$$\text{Multiplying gives} \quad (3x + 1)(-2x + 2) = -6x^2 + 6x - 2x + 2$$
$$= -6x^2 + 4x + 2$$
$$= \text{original polynomial}$$

  Thus, the original polynomial factors as $(3x + 1)(-2x + 2)$.

  $\boxed{\text{Note:}}$ **This method is useful only when the discriminant is a perfect square.**

Exercise **1.26.** *Factor the polynomial* $P(x) = -4x^2 + 6x - 2$.

Solution:

Example **1.33.** *Factor the polynomial* $P(x) = x^2 + 2x - 24$.

- **Solution**: Observe that the discriminant $\boxed{b^2 - 4ac}$ is $(2)^2 - 4 \cdot 1 \cdot (-24) = 100$, which is a perfect square, therefore we will try the **FOIL method**.

  Since $a = 1$, and $\boxed{a = m_1 m_2}$, choose $m_1 = 1$, and $m_2 = 1$. Since $c = -24$, and $\boxed{c = b_1 b_2}$, choices for b_1 and b_2 include (2 and 12), (1 and 24), (4 and 6), or (3 and 8) with one of the two numbers required to be <u>negative.</u>

  The middle term should be small, therefore try $m_1 = 1$, $m_2 = 1$, $b_1 = 6$, and $b_2 = -4$

  Checking our guess, from $\boxed{(m_1 x + b_1)(m_2 x + b_2)}$, we see that

  $$(x + 6)(x - 4) = x^2 - 4x + 6x - 24$$
  $$= x^2 + 2x - 24$$

  Thus <u>our</u> factorization is correct.

  $\boxed{\text{Note:}}$ The quadratic formula <u>would</u> also work for the example above.

Example **1.34.** *Factor the polynomial* $P(x) = 3x^2 + x + 1$.

- **Solution**: First check the discriminant $\boxed{b^2 - 4ac}$. For this polynomial, the discriminant is -11, therefore the original polynomial is already prime and cannot be factored further .

Exercise **1.27.** *Factor the following expression* $P(w) = 3w^2 + 5w + 1$.

- $\boxed{\text{Note:}}$ The discriminant $\boxed{b^2 - 4ac}$ is $5^2 - 4 \cdot 3 \cdot 1 = 13$. Since 13 is not a perfect square, the method discussed is not a viable method. Since the discriminant is positive, the polynomial is not prime; therefore use the Quadratic Formula to find the factorization.

Solution:

Exercise **1.28.** *Factor the following expressions* .

(*a*) $P(w) = 5w^2 + 3w - 1$.

Solution:

(*b*) $P(w) = 17w^2 - 3w + 1$.

Solution:

1.5.iii. A comparison between the foil method and the quadratic formula method

To see the connection between the **FOIL** method we have just demonstrated and the **quadratic formula** let us consider the following polynomial:

$$P(x) = 16x^2 - 8x - 3.$$

■   The discriminant of $P(x)$ $\boxed{b^2 - 4ac}$ is $(-8)^2 - 4(16)(-3) = 25$.

▶   Since this is positive, the polynomial will factor. It is a perfect square as well, therefore we can use the **FOIL method**.

▶   Choose $m_1 = 4$, $m_2 = 4$, $b_1 = 1$, and $b_2 = -3$, to give the correct values for the first term and the last term. From $\boxed{(m_1 x + b_1)(m_2 x + b_2)}$ gives, $(4x + 1)(4x - 3)$. Multiplying gives:

$$(4x + 1)(4x - 3) = 16x^2 - 12x + 4x - 3$$
$$= 16x^2 - 8x - 3$$

▶   Confirm that the middle term is correct.

■   To compare this result with the factorization we get from the Quadratic Formula, we need to adjust our factors from the **FOIL method** to have a coefficient of 1 in front of the x. This is necessary because it is in this format $\boxed{a(x - r)(x - s)}$ (where r and s are values of x derived from the Quadratic Formula) in which answers are stated using the Quadratic Formula.

▶   Using the techniques demonstrated in the previous section, we see that

$$(4x + 1) = 4\left(x + \tfrac{1}{4}\right) = 4\left(x - \left(-\tfrac{1}{4}\right)\right).$$
$$\text{and} \quad (4x - 3) = 4\left(x - \tfrac{3}{4}\right)$$

▶   In the $\boxed{a(x - r)(x - s)}$ format

$$16x^2 - 8x - 3 = (4x + 1)(4x - 3)$$
$$= 4\left(x - \left(-\tfrac{1}{4}\right)\right)4\left(x - \tfrac{3}{4}\right)$$
$$= 16\left(x - \left(-\tfrac{1}{4}\right)\right)\left(x - \tfrac{3}{4}\right)$$
$$= a(x - r)(x - s)$$

▶   Using the Quadratic Formula directly, we obtain

$$x = \frac{-(1)(-8) - \sqrt{(-8)^2 - 4 \cdot 16 \cdot (-3)}}{2 \cdot 316} = \frac{8 - 16}{326} = -\frac{1}{4} = r$$

$$\text{and } x = \frac{-(1)(-8) + \sqrt{(-8)^2 - 4 \cdot 16 \cdot (-3)}}{2 \cdot 16} = \frac{8 + 16}{32} = \frac{3}{4} = s$$

▶   From $a(x - r)(x - s)$ this yields a factorization of $16\left(x - \left(-\tfrac{1}{4}\right)\right)\left(x - \tfrac{3}{4}\right)$.

Thus, either process yields the same answer.

Exercise 1.29. *Determine if the following equations are prime or factorable.*

(a) $P(x) = 3x^2 - 2x + 1.$

Solution:

(b) $P(x) = \frac{1}{2}x^2 - 3x + 1.$

Solution:

(c) $P(x) = x^2 + 5.$

Solution:

(d) $P(x) = -x^2 + 3x + 4.$

Solution:

Exercise 1.30. *Determine if the following equations are prime or factorable. If they are factorable, determine the factorization.*

(a) $P(x) = x^2 - 2x + 1.$

Solution:

(b) $P(x) = -4x^2 + 1$ *using the foil method.*

Solution:

(c) $P(x) = 4x^2 + 26x - 14.$ *Identify the values which are yielded by the quadratic formula (labelled r and s).*

Solution:

(d) $P(x) = 12x^2 + 11x - 4.$ *Identify the values which are yielded by the quadratic formula (labelled r and s).*

Solution:

(e) $P(x) = 9x^2 + 6x + 1$. *Using the FOIL method.*

(f) $P(w) = 6x^2 + 11x + 3$.

Solution:

Solution:

Summary 1.5.

* There are two methods for factoring a second degree polynomial:
  - A method based on the **quadratic formula**.
  - A method based upon "**FOIL**" **multiplication**.
    - ▶ The discriminant will tell us if the polynomial is **prime** and, if it is **not prime**, and if it is practical to use the **FOIL** method.
    - ▶ The Quadratic Formula method always works.
    - ▶ If it is possible to use both methods, each will yield the same answer. The FOIL method is slightly quicker if it is available.

End of Section 1.5. Problems

Problem 1.35. *For the following quadratics, find the discriminant, determine that it is a perfect square, and use this to factor the quadratic by the FOIL method.*

(a) $4x^2 - 16$.   (b) $x^2 - 2x + 1$.   (c) $-x^2 - 6x + 7$.   (d) $2x^2 - 13x + 21$.

(e) $-9x^2 + 6x - 1$.   (f) $5x^2 - 11x - 12$.   (g) $6x^2 - x - 2$.   (h) $6x^2 - 19x + 3$.

(i) $-2x^2 + 15x - 7$.   (j) $4x^2 - 12x + 9$.

Problem 1.36. *For the following second degree polynomials, determine the values of a, b, and c and use these values to determine the value $b^2 - 4ac$. Use the value of $b^2 - 4ac$ to decide if the polynomial is prime or factorable over the reals.*

(a) $2x^2 - 3x + 1$.   (b) $-x^2 + 1$.   (c) $x^2 - 3$.

(d) $6x^2 + x - 1$.   (e) $2x^2 + 1$.   (f) $x^2 + 3x + 1$.

(g) $-x^2 + 2x + 3$.   (h) $2x^2 + x$.   (i) $x^2 + x + 1$.

(j) $x^2 - 6x + 9$.

Problem 1.37. *Use the quadratic formula to determine the zeros of of the following expressions. In this process, identify r and s for $P(x)$ and rewrite $P(x)$ in the form $a(x - r)(x - s)$.*

(a) $P(x) = x^2 - 5x + 6$.   (b) $P(x) = -2x^2 + 7x - 3$.   (c) $P(x) = x^2 - 9$.

(d) $P(x) = 10x^2 + 3x - 1$.   (e) $P(x) = x^2 - 5$.

End of <u>Chapter</u> 1. Problems

Problem 1.38. *Identify the terms and coefficients for the expression*:

(a) $-x^3 + 3x^2 - x + 1$. (b) $-3\sqrt{x}\,y^2 - 4x + y$. (c) $2x^{\frac{1}{2}}z^{-\frac{1}{2}} + x + 13$.

Problem 1.39. *Simplify the following expressions, so that only positive exponents are used.*

(a) $3^{-4}x^2$. (b) $\left(5^{\frac{1}{2}}x^2y^{\frac{3}{2}}\right)^4$. (c) $(5x)^{\frac{1}{2}}$. (d) $5x^{-\frac{1}{2}}$.

(e) $3^3 3^2 5^{-1} 5^2 x^3 x^{-7}$. (f) $\frac{x^2 z^{-1} w^{-2}}{2x^3 z^2 w^{-4}}$. (g) $\left(\frac{w^{-3}p^6 2^6}{w^6 p^{-3} 2^3}\right)^{\frac{1}{3}}$.

Problem 1.40. *Factor 98 into its prime factorization.*

Problem 1.41. *Which of the following numbers divide* 2340 *with a zero remainder?*

(a) 4 (b) 65 (c) 9. (d) 13. (e) 8. (f) 36.

Problem 1.42. *Simplify the following expressions, so that no negative exponents appear.*

(a) $\frac{x^{-1}y^2}{x^2 y^{-3}}$. (b) $\frac{(2xy)^3}{4xy}$. (c) $\left(3^{\frac{1}{2}}x^{\frac{1}{3}}y\right)^{-6}$. (d) $\left(\frac{xy}{zw}\right)^{-1}$.

(e) $\left(\frac{x^{-3}z^{-1}w}{s^{-2}}\right)^{-1}$. (f) $\left(\frac{1}{3} \cdot \frac{xy}{x^{-2}}\right)^{-1}$. (g) $\left(3x^{\frac{1}{2}}yz^3\right)^4$.

Problem 1.43. *Let* $x = 2$, $y = 3$, *and* $z = 7$. *Rewrite* $\frac{2^3 \cdot 3^{-1} \cdot 7^2}{2 \cdot 3^2 \cdot 7}$ *in terms of* x, y, *and* z *so that no negative exponents appear.*

Problem 1.44. *Complete the following chart, indicating whether or not the expression is a polynomial. If is a polynomial, give the degree.*

	Polynomial	Degree
$3x + 1$	Yes	
$\sqrt{3}\,x + 5$		
$-x^2 + x + 1$		
$\sqrt{x} + 12$		
$\frac{1}{2}x^6 + 5$		
$\frac{1}{x}$		
$-5x^{\frac{2}{3}} + 1$		

Problem 1.45. *Add* $-\frac{1}{2}x^2 + 1$ *and* $\frac{3}{2}x^2 - x + 1$.

Problem 1.46. *Add* $-x^2 - 3x - 1$ *and* $-x^2 + 4x - 3$.

Problem 1.47. *Subtract* $x^2 - 4$ *from* $-x + 2$.

Problem 1.48. *Perform the following operations*:

(a) $2x(1 - x^2)$. (b) $xy(a + b)$. (c) $-2x(4x - x^3)$. (d) $(1 + x + 3x^2)(-3x)$.

Problem 1.49. Multiply the following using FOIL:

(a) $(1+3x)(1+3x)$. (b) $(2-7x)(4x+1)$. (c) $\left(\frac{1}{2}x-\frac{1}{3}\right)\left(\frac{1}{2}x+\frac{1}{3}\right)$. (d) $(4x-7)(2x+3)$.

(e) $(-2x+6)(x+1)$. (f) $(x+1)^2$. (g) $(2x-1)^2$.

Problem 1.50. A student recognizes that $(23)(45) = (2(10)+3)(4(10)+5)$. Letting $x = 10$, the student multiplies $(2x+3)(4x+5)$. What expression does the student obtain when multiplying these two polynomials? What is the value of the expression when $x = 10$?

Problem 1.51. Consider $(14)(19) = (10+4)(20-1)$. If you let $x = 10$, rewrite the right hand multiplication in terms of x.

Problem 1.52. Consider $(11)(17) = (14-3)(14+3)$. Let $x = 14$; rewrite the multiplication as an expression of x.

Problem 1.53. Multiply the following:

(a) $(2x-1+x^2)(x^2-1)$. (b) $(x+1)^3$.

Problem 1.54. Factor the term $-x^2$ from the expression $-x^2 + x^3 + 3x^4$.

Problem 1.55. Factor the term -1 from the expression $-2 + x$.

Problem 1.56. Factor the term $\sqrt{x+1}$ from the expression $\sqrt{x+1}\,(x^2) - 3\sqrt{x+1}$.

Problem 1.57. Factor the term $\frac{1}{2}x$ from the expression $\frac{1}{2}x + x^2$.

Problem 1.58. Factor the term pq from the expression $pq^3 + pq^3$.

Problem 1.59. Factor the term $-2x^3$ from the expression $-2x^3 + x - 1$.

Problem 1.60. Factor the term $x^{-\frac{1}{3}}$ from the expression $2x^{\frac{2}{3}} - x^{-\frac{1}{3}}$.

Problem 1.61. Factor the term $(2x-1)^{-1}$ from the expression $3(2x-1)^{-1} + (2x-1)^2$.

Problem 1.62. Factor the term $x^{-\frac{1}{2}}$ from the expression $\sqrt{x}\,(x+1) + x^{-\frac{1}{2}}$.

Problem 1.63. Identify the base common to each of the terms in the given sum or difference; also identify the least power to which that base is raised.

(a) $y^2x + y^3 - y^{-1}$. (b) $(x+1)^3y + (x+1)y^2 + (x+1)^3y^4$. (c) $\sqrt{x}\,y + x^3$.

(d) $x^{\frac{1}{2}}y + x^{-\frac{1}{2}}(4)$. (e) $x^2y^3z^{-1} - 2xy^4z^{-2}x^{-1}y^3z$.

Problem 1.64. For the following second degree polynomials, determine the values of a, b, and c and use these values to determine the value $b^2 - 4ac$. Use the value of $b^2 - 4ac$ to decide if the polynomial is prime or factorable over the reals.

(a) $9x^2 - 12x + 4$. (b) $x^2 - x + 1$. (c) $-3x^2 - 2x + 1$.

(d) $2x^2 + 4x - 1$. (e) $x^2 - 5$.

Problem 1.65. Find the zeros of the following quadratic expressions using the quadratic formula.

(a) $P(x) = x^2 - 2x + 1$. (b) $P(x) = x^2 - 3$. (c) $P(x) = 6x^2 - 5x + 1$.

(d) $P(x) = -2x^2 + 5x - 3$. (e) $P(x) = -2x^2 + 7x - 3$.

Problem 1.66. Find the factorization of the following using the zeros found in question **1.65**. above.

(a) $P(x) = x^2 - 2x + 1$.

(b) $P(x) = x^2 - 3$.

(c) $P(x) = 6x^2 - 5x + 1$.

(d) $P(x) = -2x^2 + 5x - 3$.

(e) $P(x) = -2x^2 + 7x - 3$.

Problem 1.67. Determine whether or not the following are prime or factorable. If it is factorable, use FOIL to factor it.

(a) $P(x) = -5x^2 + 11x - 2$.

(b) $P(x) = 9x^2 - 1$.

(c) $P(x) = 6x^2 + 11x - 2$.

(d) $P(x) = 2x^2 + 3x + 2$.

(e) $P(x) = 8x^2 - 8x + 2$.

Chapter 2.

Equations and Inequalities in One Variable

2.1. Introduction to Equations

Any mathematical statement involving equality ($=$) is referred to as an equation. Thus, an equation is an expression with **two sides** - a **left hand side** (**LHS**) and a **right hand side** (**RHS**) - connected by equality. Consider the following statements involving the variable (unknown) x:

1. $2(3x - 1) = 6x - 2$. (The equation is **true** for **any value** of x.)
2. $x - \frac{1}{x} = \frac{x^2 - 1}{x}$. (The equation is **true** for **any value** of x **except zero**.)
3. $x + 2 = 9$. (The equation is **true** if x is 7 **only**.)
4. $(x + 1)(x - 2) = 0$. (The equation is **true** if x is **either** -1 **or** 2.)
5. $2(x + 1) = 2(x - 2)$. (The equation is **not true** for **any value** of x (i.e. it is **always false**).)

The observations above motivate the following classification of equations:

> Definition
>
> **Identity**:
>
> For a given equation in the variable x, the equation is called an **identity** if it is true for all allowable values of x
>
> **Conditional equation**
>
> The equation is called a **conditional equation** if it is true for some, but not all, allowable values of x.
>
> **Contradiction**
>
> The equation is a **contradiction** if it is always false.

2.1.i Solution Set

Example 2.1. *Solve the following equation,* $2(x - 1) = 7$.

Solution: If $x = 3$, then

$$
\begin{aligned}
LHS &= 2(x - 1) \\
&= 2(3 - 1) \\
&= 2 \cdot 2 \\
&= 4 \neq 7 \ (RHS)
\end{aligned}
$$

3 *does not satisfy* the equation. In contrast, if $x = \frac{9}{2}$, we substitute into the equation to see that $\frac{9}{2}$ *satisfies* the equation:

$$
\begin{aligned}
LHS &= 2(x - 1) \\
&= 2(\frac{9}{2} - 1) \\
&= 2 \cdot \frac{7}{2} = 7
\end{aligned}
$$

We say that $\frac{9}{2}$ is a **solution**, or a **root** of the equation, while 3 is not a solution.

In general, for a given equation, a solution is a value of the variable that satisfies the equation. The **set** S of all solutions of the equation is called the **solution set** for the equation. Thus, to solve an equation is synonymous with finding its solution set.

2.1.ii Equivalent Equations

Several **conditional equations** may have the **same solution set**. Such equations are said to be

equivalent.

For instance consider the equations:

- ◀ $2x = 10$
- ◀ $\frac{1}{5}x = 1$
- ◀ $x - 2 = 3$
- ◀ $3(x - 1) = x + 7$

Check to see that $x = 5$ satisfies each equation.

In fact, $S = \{5\}$ is the solution set for each of the equations. Thus, they are all **equivalent** equations. However, the equation $x(x - 5) = 0$ has the solution set $\boxed{S = \{0, 5\},}$ and so it is not equivalent to any of the preceding equations even though 5 is a solution for each of the equations. (Why?)

2.1.iii Solving Equations

Remember that we have classified equations in terms of whether **all, some, or none** of the allowable values of the unknown are **elements** of the **solution set** for the equations. $\boxed{\text{Note}}$ that those classifications themselves do not tell us *how* to solve a given equation. In fact, let it be clear that we may not be able to conclude as to which class a given equation belongs until we have found its solution set!

For the purpose of mastering strategies and studying methods of solution, we identify equations by *type*.

Although there are many more types of equations, in this chapter we consider the following types of equations in one variable x:

$\boxed{\textit{Definition}}$ *TYPES OF EQUATIONS*

Linear equations - these can be rearranged in the form $ax = b$.

Quadratic equations - these have the standard form $ax^2 + bx + c = 0$, where $a \neq 0$.

Rational equations that lead to linear or quadratic equations

- these have numerators and denominators that are polynomials.

Radical equations that lead to linear or quadratic equations

- these have radicals (rational exponents).

Absolute value equations that lead to linear or quadratic equations.

Methods for solving equations differ from type to type. Thus, in order to solve an equation, it is important to first identify its type, and then simplify the equation to a desirable form.

The following are properties of equations:

$\boxed{\textit{Property}}$

Addition/Subtraction :

If C is well-defined and $A = B$ then

$A + C = B + C$ (equivalent equations).

Multiplication/Division

If α is a nonzero quantity and $A = B$ then

$\alpha A = \alpha B$ (equivalent equations).

The addition/subtraction property permits us to add or subtract *any* well-defined quantity to *both* sides of an equation. On the other hand, the multiplication/division property permits us to multiply or divide both sides of an equation by a *non-zero* quantity to obtain an equivalent equation. In essence, the two properties guide all four operations of arithmetic: addition, subtraction, multiplication, and division. Consequently, we need these properties when solving equations.

Summary

- ■ **Key Words**
  - ► **Identity** - an equation that is true for all permissible values of the unknown.
  - ► **Conditional Equation** - one that holds for only some of the permissible values.
  - ► **Contradiction** - an equation that is false for all values of the unknown.
  - ► **Solution Set** - the set of all values that satisfy the equation.
  - ► **Equivalent Equations** - have the same solution set.

2.2. Linear Equations

> **Definition**
>
> Linear Equations:
>
> An equation in the variable x is said to be **linear** if the **degree** of each term is **one** or **zero**.
> This means simply that the **exponent** of x in each term is essentially **one** or **zero**.

For instance:

- $3x - 5 = x + 7$ is a **linear** equation.
  $\sqrt{x} - 3 = 2x - 1$ is **not linear** because of the radical term.
  $\dfrac{x}{x - 1} = 2$ is **not linear** because of the rational term.
  $x^2 - 5x + 6 = 0$ is **not linear** because of the quadratic term.

2.2.i. Solution Set

In order to solve a **linear** equation, we simplify the equation to obtain the standard form: $\boxed{ax = b.}$
If a is nonzero, we then use the **division property** of equations to obtain the solution:

$$\frac{ax}{a} = \frac{b}{a}$$
$$x = \frac{b}{a}$$

As the last statement indicates, we expect exactly one solution most of the time. However, this is not always the case.

In general, we are able to decide from the standard form if the equation is an **identity**, a **contradiction**, or a **conditional equation** using the following criteria:

1. If $a \neq 0$, the equation is **conditional** and there is one solution $\frac{b}{a}$; hence, $\boxed{S = \{\frac{b}{a}\}.}$

2. If $a = 0$ and $b = 0$, the equation is an **identity** and the solution is all allowable values of x $\boxed{S = \{x/x \in R\}.}$

3. If $a = 0$ but $b \neq 0$, the equation is a **contradiction**; there is no solution; $\boxed{S = \phi.}$

The following examples further clarify the preceding points:

Example 2.2. *Solve for* x: $\frac{7}{6}(x+1) = \frac{4}{5}(x+3) - \frac{1}{2}$.

- **Solution**: Note that the equation is linear. In order to get to the standard form we first "clean-up" the fractions by eliminating the denominators. The **least common multiple** (**LCM**) of $\{6, 5, 2\}$ is 30. We multiply both sides by 30:

$$30 \cdot \frac{7}{6}(x+1) = 30 \cdot \frac{4}{5}(x+3) - 30 \cdot \frac{1}{2}$$

then expand

$$35(x+1) = 24(x+3) - 15$$
$$35x + 35 = 24x + 72 - 15$$

and use the addition/subtraction property

$$35x + 35 - 24x - 35 = 24x + 57 - 24x - 35$$

to get the standard form

$$11x = 22$$
$$ax = b$$

Since $a \neq 0$, we have exactly one solution $x = 2$. Observe that in the standard form, $ax = b$, $a = 11$ is nonzero. Thus, equation is **conditional** and has exactly one solution. That is: $\boxed{S = \{2\}.}$

Example 2.3. *Solve for* x, $2(5x - 3) + 3(1 - 2x) = 2(2x + 3) - 9$.

- **Solution**: Here again, the equation is linear and we simplify to get to the standard form:

$$10x - 6 + 3 - 6x = 4x + 6 - 9$$
$$4x - 3 = 4x - 3$$

Notice that the two sides are identically the same regardless of the value of x. This is enough to conclude that the equation is an **identity**. If we attempt further to put it in standard form, we obtain

$$4x = 4x$$

or the trivial truth:

$$-3 = -3 \text{ or,}$$
$$0 = 0$$
$$ax = b$$

Equivalently, the standard form is

$$0x = 0$$

Notice that $a = 0 = b$. Since the left hand side is equal to the right hand side regardless of the value of x, every real number is a solution and the equation is an identity. The solution set is $\boxed{S = \{x/x \in R\}.}$

Exercise 2.1. *Solve each of the following equations. In each case classify the equation as a conditional equation, an identity, or a contradiction. Clearly state the solution set.*

(a) $2(3x - 5) + 3(x + 2) = 7x + 6$.

Solution:

(b) $\frac{2}{3}(2x + 1) - \frac{1}{2}(x - 3) = -5$.

Solution:

Example **2.4.** *Solve for* x, $3(x-1) - 2(x+1) = x - 7$.

- **Solution**: Again, we look for the standard form:

$$3x - 3 - 2x - 2 = x - 7$$
$$x - 5 = x - 7$$

Unlike the preceding example, the two sides are not identically the same. Further simplification yields

$$0x = -2$$
$$ax = b \text{ or,}$$
$$0 = -2$$

which is a **contradiction** since $0 \neq -2$. In the standard form, note that $a = 0$ but $b = -2 \neq 0$. Thus, there is no solution and the solution set is $\boxed{S = \phi.}$

Exercise **2.2.** *Solve each of the following equations. In each case classify the equation as a conditional equation, an identity, or a contradiction. Clearly state the solution set.*

(*a*) $0.5x - 0.2(3x - 1) = 0.1(7 - x)$.   (*b*) $5(2x + 1) + 9(3 - x) = x + 32$.

Solution: **Solution:**

2.2.ii Equations Leading to Linear Equations

Some equations that are not linear may lead to linear equations when we simplify them. We shall consider two such types of equations here. In both cases, we stress that we are not required or expected to know what the equation leads to by mere inspection. However, we must emphasize that there is a definite strategy to be followed. Given an equation, we must attempt to arrive at an identifiable type of equation by simplifying the given equation.

Higher Degree Equations

Some higher degree **polynomial** equations do **reduce** to **linear equations**. As an illustration, let us consider the equation:

$$(2x - 3)(3x + 1) = (6x - 5)(x - 1)$$

Observe that the products on each side of the equation involve x^2 so that the equation is not linear. In order to identify a type, we multiply the brackets

$$6x^2 - 7x - 3 = 6x^2 - 11x + 5$$

The equation is a quadratic equation. However, by adding $-6x^2$ to both sides of the equation, we obtain the linear equation

$$-7x - 3 = -11x + 5$$

Collecting like terms in standard form, we have

$$4x = 8$$

And a solution is $x = 2$. Just to be certain, we check the solution in the original equation (this is desirable but not necessary for this type of equation because there were no restrictions in the original equation): if $x = 2$,

$$LHS = (2x - 3)(3x + 1)$$
$$= (4 - 3)(6 + 1)$$
$$= 1 \cdot 7$$
$$= 7$$

On the other hand
$$RHS = (6x - 5)(x - 1)$$
$$= (12 - 5)(2 - 1)$$
$$= 7 \cdot 1$$
$$= 7$$

Thus, the two sides are equal when $x = 2$, and this is a solution. Could there be any other solutions?

- **Note:** If a higher degree polynomial equation is reduced to a lower degree equation by multiplying given factors and/or collecting like terms only, then the two equations are equivalent.

This tells us that if we reduce a higher order equation, as in the above example, then both equations have the same solution set.

Exercise 2.3 Solve the following. Specify the solution set.

(a) $(x + 3)(x - 2) = (x + 1)(x - 7)$.

Solution:

(b) $(x + 2)^2 - (x - 2)^2 = 8x + 3$.

Solution:

Rational Equations

Consider the equation $\frac{x}{x - 1} - 3 = \frac{5}{x - 1}$.

The **difference** between this and any of the preceding equations lies in the fact that the unknown x appears in the denominator. Since division by zero is meaningless, in this example, this says that $x - 1$ cannot be zero. In other words, x cannot be 1. Observe that this is a **restriction** on the solution set.

Example 2.5. For $\frac{x}{x - 1} - 3 = \frac{5}{x - 1}$ solve for x.

- **Solution:** We simplify by multiplying both sides by the LCM of the denominators:
$$(x - 1)\left[\frac{x}{x - 1} - 3\right] = \left[\frac{5}{x - 1}\right](x - 1)$$

That is:
$$x - 3(x - 1) = 5.$$

Can you see that this resulting equation is linear? We now solve it.
$$x - 3x + 3 = 5$$
$$-2x = 2$$
$$x = -1$$

Here, we *must check* that $x = -1$ is a solution for the *original equation*. If $x = -1$,

$$LHS = \frac{x}{x-1} - 3$$
$$= \frac{-1}{-1-1} - 3$$
$$= \frac{1}{2} - \frac{6}{2}$$
$$= -\frac{5}{2}$$

On the other hand,

$$RHS = \frac{5}{x-1}$$
$$= \frac{5}{-1-1}$$
$$= -\frac{5}{2}$$

Since $LHS = RHS$ if $x = -1$, it is a solution. In fact, $S = \{-1\}$. It is important that the solution is not a restricted value for the equation.(Why?) The following example is revealing!

Example 2.6. Solve for x, $\frac{2x}{x-3} - 1 = \frac{6}{x-3}$.

- **Solution**. As in the preceding example, we note that x cannot be 3 (a restriction) and we eliminate the denominators by multiplying by the LCM $x - 3$:

$$(x-3)\left[\frac{2x}{x-3} - 1\right] = \left[\frac{6}{x-3}\right](x-3)$$

which is a linear equation

$$2x - (x-3) = 6$$

We solve

$$2x - x + 3 = 6$$
$$x = 3$$

As before, we must check the original equation. When $x = 3$,

$$LHS = \frac{2x}{x-3} - 1$$
$$= \frac{6}{3-3} - 1$$
$$= \frac{6}{0} - 1$$

We must stop here because $\frac{6}{0}$ is meaningless (undefined). Observe that indeed $x = 3$ is the solution for the resulting linear equation but it is not valid in the original equation. Remember that 3 is a **restricted value** for the original equation. As a result, the equation has no solution and $S = \phi$. The value 3 is called an **extraneous solution**, a solution for the resulting equation which is not a solution for the original equation.

> **Note:** The solution set for a **rational equation** is a subset of the solution set of the resulting polynomial equation obtained by multiplying the original equation by the LCM of the denominators. Thus, a **solution** of the **resulting equation** is **not necessarily** a solution of the **original equation**.

This is precisely why we need to check the potential solutions in the original equation. In fact, any potential solution that is also a **restricted value** is an **extraneous solution**.

Exercise 2.4 Solve each of the following. Specify the solution set.

(a) $\frac{x+3}{x+2} - 5 = \frac{x}{x+2}$.

Solution:

(b) $\frac{3x}{x-1} + x = \frac{x^2+2}{x-1}$.

Solution:

2.2.iii   Literal Equations

In applications, we often come across formulas that involve several parameters. For instance, in mechanics if a body is projected vertically upwards with **initial velocity** v_0, under ideal conditions, its velocity after time t is given by the linear equation

$$v = -gt + v_0$$

where g is the acceleration due to gravity. In the linear equation, the subject, v, is expressed in terms of three other quantities $-g, t$, and v_0. We may wish to make t the subject:

$$v = -gt + v_0$$
$$v - v_0 = -gt$$
$$gt = v_0 - v$$
$$t = \frac{v_0 - v}{g}$$

We have solved for the variable t. Note that the statement "solve for t" simply demands that we make t the subject. In other words, write t in terms of the other quantities. The above is an example of a **literal equation**, an equation that **connects two or more variables** (or parameters).

Example **2.7**. *Solve for* r *in the equation* $A = P(1 + rt)$.

- **Solution**. We may expand the bracket, then solve:

$$P(1 + rt) = A$$
$$P + rtP = A$$
$$rtP = A - P$$
$$r = \frac{A - P}{tP}$$

Notice that we have made r the subject by expressing r in terms of A, P, and t.

Exercise **2.5**. *Solve for* x.

(a) $\frac{x}{x+2} = \frac{a+b}{c}$.

Solution:

(b) $s = \frac{1}{2}[2a + (x-1)d]$.

Solution:

Summary

- **Key Words**
  - ▶ **Linear Equation** - a first degree polynomial equation.
  - ▶ **Rational Equation** - a fractional equation with polynomial numerators and denominators.
  - ▶ **Extraneous Solution** - an invalid solution that is not a solution for the original equation.
- **Strategy**
- In order to solve a given linear equation we may take the following steps:
  - ▶ Write down any restrictions
  - ▶ if the equation involves fractions, eliminate denominators (use the LCM)
  - ▶ simplify to the standard form $ax = b$

- ▶ if a is nonzero, solve
- ▶ If the unknown is in the denominator of the original equation, you must check the solution using the original equation (making sure that the solution is not a restricted value).
- ▶ Specify the solution set.

End of Section 2.2 Problems

Problem 2.1 *In each case, determine if the statement is **true** or **false**:*

(a) $x = -3$ *is a solution of* $4x + 5 = -7$.

(b) $x = \frac{3}{2}$ *is a solution of* $2x - 5 = 3 - 8x$.

(c) $x = \frac{3}{4 - p}$, $p \neq 4$ *is equivalent to* $4x = 3 + px$.

(d) $x = \frac{2}{p + 5}$, $p \neq -5$ *is equivalent to* $5x = 2 + px$.

(e) $x = 6$ *is a solution of* $\frac{3}{x} - 1 = \frac{1}{2} - \frac{6}{x}$.

(f) $\frac{8x + 1}{x - 2} + 4 = \frac{7x + 3}{x - 2}$ *has no solution.*

Problem 2.2 *Solve each equation.*

(a) $2x + 7x - 5 = 3x + 13$.　(b) $3x - 5x + 4 = x - 9$.　(c) $0.4x - 0.7x = 2.2$.

(d) $\frac{1}{2}x + \frac{2}{3}x = 5$.　　　(e) $\frac{3}{4}x - \frac{2}{3}x = 3 - \frac{5}{6}x$.　(f) $1.25x + 3 = 6 - 0.15x$.

Problem 2.3 *Solve each equation:*

(a) $3(2 - x) = 4(x - 1)$.　　　　(b) $3x + 2 - 2(x - 1) = 3(2x + 3)$.

(c) $2(x + 3) + x = 3(x + 2)$.　　　(d) $5(2x - 1) - 3x = 5 - 7x$.

(e) $6(3 - x) + 2(4x - 5) = x - (3 - x)$.　(f) $4(3x - 1) - 5(3x - 2) = 2(x + 3) - 5x$.

Problem 2.4 *Solve each equation*

(a) $(x - 3)(x + 2) = x(x + 3)$.　　(b) $(x - 4)(x - 2) = (x + 1)(x - 5)$.

(c) $(2x + 1)(x - 3) = 2x(x - 3) + 1$.　(d) $(3x - 1)(x + 1) - (3x + 2)(x - 3) = -4$.

(e) $(3x + 2)(x + 1) - x(3x + 1) = 0$.　(f) $(x - 2)^2 - (x + 3)^2 = -3$.

Problem 2.5 *Solve for x;*

(a) $\frac{1}{2}(3x - 1) + \frac{1}{4}(x - 3) = \frac{1}{2}$.　(b) $\frac{1}{3}(2x + 1) + \frac{1}{4}(x + 1) = -\frac{1}{6}$.

(c) $\frac{1}{x - 2} + \frac{1}{2} = \frac{3}{x - 2}$.　　　(d) $\frac{x}{x - 1} - 1 = \frac{3}{x - 1}$.

(e) $\frac{2}{x - 2} + 1 = \frac{x + 2}{x - 2}$.　　(f) $\frac{4}{x - 4} - 2 = \frac{1}{x - 4}$.

Problem 2.6 *Solve for the indicated variable:*

(a) *Solve for* F: $C = \frac{5}{9}(F - 32)$.　(b) *Solve for* R_2: $\frac{1}{R} = \frac{1}{R_1} + \frac{1}{R_2}$.

(c) *Solve for* x: $\frac{3x}{x - a} = \frac{5}{a}$.　　(d) *Solve for* a: $s = \frac{1}{2}[2a + (x - 1)d]$.

(e) *Solve for* r: $A = P(1 + rt)$.　　(f) *Solve for* L: $S = \frac{a - rL}{L - r}$.

2.3. Complex Numbers

Definition

Complex numbers:

A **complex number** z is a number of the form $\boxed{z = a + bi}$

where a and b are real numbers and $i = \sqrt{-1}$.

The **number** i is often referred to as the **imaginary square root** of -1 since it has no position on the real number line.

Real numbers:

Every **real number** x is also a **complex number** since it can be expressed in standard form $\boxed{x = x + 0i.}$

Thus, the **set of real numbers** is a **subset of the set of complex numbers**. If $z = a + bi$, the real number a is called the *real part* of z, and b is called the *imaginary part* of z.

Definition

Conjugate of a complex number (z):

If $z = a + bi$, we define the **conjugate** of z by $\bar{z} = a - bi$.

Clearly, the conjugate of z is also a complex number.

For example:

▶ $z = 3 - 2i \Rightarrow \bar{z} = 3 + 2i.$

▶ $z = 3 + 2i \Rightarrow \bar{z} = 3 - 2i.$

▶ $z = -3 - 2i \Rightarrow \bar{z} = -3 + 2.$

Note: A **real number** is its own conjugate.

2.3.i. Powers of i

By the above definition, $i = \sqrt{-1}$. Thus, we use the laws of exponents to obtain:

$$i^0 = 1$$
$$i^1 = i$$
$$i^2 = \left(\sqrt{-1}\right)^2 = -1$$
$$i^3 = i^2 i = -1i = -i$$
$$i^4 = i^2 i^2 = (-1)^2 = 1$$

The cycle of integer powers of i is now complete since

$$i^4 = 1 = i^0$$

We may now use this table to find **any integer power** of i.

Example **2.8**. *Evaluate* i^{55}.

- **Solution**:

$$i^{55} = i^{52} \cdot i^3$$
$$= (i^4)^{13} \cdot i^3$$
$$= i^3 \text{ since } i^4 = 1$$
$$= -i \text{ from above}$$
$$= \sqrt{1}$$

$\boxed{\text{Note}}$ 52 is the largest multiple of 4 that is less than 55. The relation $\boxed{i^{55} = i^3}$ is noteworthy since 3 is the remainder when 55 is divided by 4.

> The **algorithm for reducing positive integer powers of** i,
>
> that is, a mechanism used to simplify i^n:
>
> 1. Divide n by 4,
> 2. if the remainder is r, then $i^n = i^r$. (note that r is $0, 1, 2,$ or 3),
> 3. complete the simplification of i^r (the only possibilities are ± 1 or $\pm i$).

Example **2.9**. *Using the algorithm above, simplify* i^{96}.

- **Solution**: $i^{96} = i^0 = 1$, since dividing 96 by 4 leaves no remainder.

Example **2.10**. *Using the algorithm above, simplify* i^{1006}.

- **Solution**: $i^{1006} = i^2 = -1$, because dividing 1006 by 4 leaves 2 as the remainder.

Exercise **2.6** *Simplify each of the following.*

(a) $2i^{15} - 5i^3 + 4i^{60}$.

Solution:

(b) $2i^{26} + 3i^{54} + 4i^{75} + i^{13}$.

Solution:

2.3.ii. Operations

Our aim here is to introduce the **arithmetic operations** on the set of complex numbers.

2.3.ii.a. Addition

- If $z = a + bi$ and $w = c + di$, the **sum** of the two complex numbers is defined by
$$z + w = (a + c) + (b + d)i.$$
Thus we simply add real and imaginary parts, respectively.

Example **2.11.** *If* $z = 3 - 7i$ *and* $w = -1 - i$, *solve* $z + w$.

- **Solution:**

$$z + w = (3 - 7i) + (-1 - i)$$
$$= 2 - 8i$$

Exercise **2.7.** *Simplify the following (leave your answer in standard form* $a + bi$*). Take* $z = -5 + 4i; \; w = 3 - 7i$:

(*a*) $3z - 2w + \bar{w}$.

Solution:

(*b*) $4\bar{w} - 3w + z$.

Solution:

2.3.ii.b. Multiplication

- Multiplication of complex numbers is defined in a natural way. If $z = a + bi$ and $w = c + di$ the product

$$zw = wz$$
$$= (a + bi)(c + di)$$
$$= ac + adi + bci + bdi^2 \text{ (remember } i^2 = -1.)$$
$$= (ac - bd) + (ad + bc)i$$

Example **2.12.** *Solve* $-2(3 - i)$.

- **Solution:** $-2(3 - i) = -6 + 2i$

Example **2.13.** *Solve* $(2 + 3i)(5 - 2i)$.

- **Solution:**

$$(2 + 3i)(5 - 2i) = 10 - 4i + 15i - 6i^2 \text{ (remember } i^2 = -1.)$$
$$= 10 + 11i + 6$$
$$= 16 + 11i$$

- | *Identity* |

  Let $z = a + bi$, then,
  $$z\bar{z} = (a + bi)(a - bi)$$
  $$= (a)^2 - (bi)^2$$
  $$= a^2 + b^2$$

Conjugate-Product Property

- The **product of a complex number** and its **conjugate** is a **real number**.

  Specifically, the **product** is the **sum of squares** of the **real** and **imaginary** parts of the complex number.

This means that if $z = 3 - 4i$, then $\boxed{z\bar{z} = 3^2 + (-4)^2 = 25}$

This helps us to define division by a complex number in the next section.

Exercise **2.8** *Simplify each of the following (leave your answer in standard form $a + bi$).*
Take $z = -5 + 4i$; $w = 3 - 7i$:

(a) $z^2 + zw$.

Solution:

(b) $z\bar{z} - 2\bar{w}$.

Solution:

2.3.ii.c. Division

- Let $z = a + bi$ and $w = c + di$. Then the quotient $w \div z$ is defined relative to division in the set of real numbers:

$$\frac{w}{z} = \frac{c + di}{a + bi}$$

To rewrite in the standard form we multiply the numerator and the denominator by $\bar{z}$

$$\begin{aligned}
\frac{w}{z} &= \frac{c + di}{a + bi} \\
&= \frac{c + di}{a + bi} \cdot \frac{a - bi}{a - bi} \\
&= \frac{(c + di)(a - bi)}{a^2 + b^2} \\
&= \frac{(ca + bd) + (ad - bc)i}{a^2 + b^2}
\end{aligned}$$

We can now put the last expression in standard form:

$$\frac{c + di}{a + bi} = \frac{(ca + bd)}{a^2 + b^2} + \frac{(ad - bc)}{a^2 + b^2}i$$

Example **2.14.** *Write* $\frac{2+7i}{-3+4i}$ *in the form* $a+bi$.

- **Solution**

$$\frac{2+7i}{-3+4i} = \frac{(2+7i)}{(-3+4i)} \cdot \frac{(-3-4i)}{(-3-4i)}$$
$$= \frac{-6-8i-21i+28}{9+16}$$
$$= \frac{22-29i}{25}$$
$$= \frac{22}{25} - \frac{29}{25}i$$

It is important to multiply both numerator and denominator of the given fraction by the conjugate of the denominator. The new numerator is a product of two complex numbers, and as such it is also a complex number. In contrast, the above identity guarantees that the new denominator will be a nonzero real number. Since division by a nonzero real number is well defined, the process always works.

Exercise **2.9.** *Simplify the following (leave your answer in standard form* $a+bi$*). Take* $z = -5 + 4i$; $w = 3 - 7i$:

(a) $\frac{w}{z}$.

Solution:

(b) $\frac{\bar{z}}{w}$.

Solution:

Summary

- **Key Words**
  - ▶ **Complex number** - has the standard form $z = a + bi$.
  - ▶ **Conjugate** - $\bar{z} = a - bi$.
  - ▶ **Identity** - $z\bar{z} = a^2 + b^2$.
  - ▶ **Powers of i**- $i^n = i^r$ if $n \div 4$ has remainder r.

End of Section 2.3. *Problems*

Problem 2.7 *Simplify each of the following. Leave your answer in standard form:* $a + bi$

(a) $i^2 - 3i^3 + 2i^4$. *(b)* $3 - \sqrt{-9} + \sqrt{-4}$.

(c) $4 - \sqrt{-32} + \sqrt{-50}$. *(d)* $5 - 2i^{25} + 3i^{39}$.

(e) $5i^{97} + 3i^{44} - i^{17}$. *(f)* $\sqrt{25} + \sqrt{-98} - \sqrt{-50}$.

Problem 2.8 *Compute and write the answer in the form* $a + bi$

(a) $(3+4i)(-2+5i)$. *(b)* $(2-7i)^2 - (3-2i)(1+i)$.

(c) $\overline{(3-2i)}(-1-i)$. *(d)* $i(4i-5) - (2-2i)^2$.

(e) $(7-i)i^3 + 2i(3+5i)$. *(f)* $(i-1)(i+1) - \overline{(-3+2i)}$.

Problem 2.9 *Let* $z = -4 + 5i$ *and* $w = 3 - i$. *Compute:*

(a) $\frac{z}{w}$. *(b)* $z^2 + zw$.

(c) $z\bar{w} + \bar{z}w$. *(d)* $\frac{w}{z} - 3w$.

(e) $\frac{z\bar{w}}{zw}$. *(f)* $\frac{z+w}{z-2w}$.

2.4. Quadratic Equations

> **Definition**
>
> **Quadratic equation**:
>
> A **quadratic equation** in the variable x is a **second degree** polynomial equation of the form $\boxed{ax^2 + bx + c = 0}$ where, $a \neq 0$.

There are **three methods** for solving a quadratic equation:

1. **Factoring**.
2. **Completing the square.**
3. The **quadratic formula**.

We shall discuss all three methods.

2.4.i. Factoring

The method of factoring works because of the **product rule**:

$$\text{If } A \cdot B = 0, \text{ then either } A = 0 \text{ or } B = 0.$$

An illustration follows:

Example 2.15. *Solve for* x, $x^2 - x - 6 = 0$.

- **Solution**. Notice that the equation is quadratic and in standard form. We factor the left hand side

$$(x + 2)(x - 3) = 0$$

We now use the product rule to set each factor to zero

$$x + 2 = 0 \text{ and } x - 3 = 0$$

Solving the linear equations, we get

$$x = -2 \text{ and } x = 3$$

Let us check the solutions: when $x = -2$

$$x^2 - x - 6 = (-2)^2 - (-2) - 6$$
$$= 4 + 2 - 6$$
$$= 0$$

Similarly, when $x = 3$

$$x^2 - x - 6 = 3^2 - 3 - 6$$
$$= 9 - 3 - 6$$
$$= 0$$

The solution set is $\boxed{S = \{-2, 3\}.}$

It is important to observe that the following equations are equivalent, and that they also have the same standard form:

$$x^2 - x - 6 = 0$$
$$x^2 - x = 6$$
$$x^2 - 6 = x$$
$$x + 6 = x^2$$

Exercise **2.10.** *Solve by factoring:*

(a) $x^2 + x - 12 = 0$.

Solution:

(b) $20x^2 + 11x - 3 = 0$.

Solution:

Two Terms Only

Some quadratic equations may have only two terms.

Example **2.16.** *Solve for* x, $3x^2 + 5x = 0$.

- **Solution.** Notice that the constant term is missing. We can (and should) factor the common factor x on the left hand side

$$x(3x + 5) = 0$$

As before, set each factor to zero

$$x = 0 \text{ and } 3x + 5 = 0$$

Now solve the linear equations

$$x = 0 \text{ and } x = -\frac{5}{3}$$

Here, the solution set is $\boxed{S = \{0, -\frac{5}{3}\}.}$

Terms containing $x^1 (=x)$ are missing.

Example **2.17.** *Solve:* $4x^2 - 9 = 0$.

- **Solution.** We factor the left hand side as the difference of two squares

$$(2x + 3)(2x - 3) = 0$$

Again, we use the product rule, then we solve

$$2x + 3 = 0 \text{ and } 2x - 3 = 0$$
$$x = -\frac{3}{2} \text{ and } x = \frac{3}{2}.$$

The solution set is $\boxed{S = \{\pm\frac{3}{2}\} = \{-\frac{3}{2}, \frac{3}{2}\}.}$

The following procedure is often used for such equations, but the justification rests with the method of factoring above. Reconsider the given equation above ($4x^2 - 9 = 0$), but arrange it in the form

$$4x^2 = 9$$

eliminate 4 from the left side by division

$$x^2 = \frac{9}{4}$$

now *extract* square roots of both sides, *remembering to insert* $\pm$ on the right side

$$\sqrt{x^2} = \pm\sqrt{\frac{9}{4}}$$

now, simplify both sides to get

$$x = \pm\frac{3}{2}$$

We must recognize that there are two solutions in the solution set $\boxed{S = \{\pm\frac{3}{2}\} = \{-\frac{3}{2}, \frac{3}{2}\}.}$

- We also **caution** that the *only* quadratic equations to be solved in this fashion are those of the **form** $\boxed{ax^2 = c}$

Exercise **2.11**. *Solve*:

(**a**) $9x^2 = -7x.$

Solution:

(**b**) $3x^2 + 5 = 0.$

Solution:

2.4.ii. Completing the Square

Recall that the square of a binomial (two terms) expands as follows
$$(x + k)^2 = x^2 + 2kx + k^2$$
$$(x - k)^2 = x^2 - 2kx + k^2$$
If we examine the two sides of both equations (identities) closely, we observe that:

a. The two identities differ by a sign in exactly one place. If the binomial is a sum $(x + k)$, then all the signs on the right are positive. If, however, the binomial is $(x - k)$, then the middle term on the right side carries a negative sign.

b. The coefficients of powers of x on the right side are $\{1, \pm 2k\ k^2\}$. Note their relation to the binomial $\{x \pm k\}$ on the left. The coefficients $\{2k, k^2\}$ are the significant ones.

Given an expression of the form $x^2 + bx + c$, the observations above imply that the given **expression** is a **perfect square** only if $\boxed{b = 2k \text{ and } c = k^2 \text{ for a number } k.}$

Example **2.18**. *Is* $x^2 + 6x + 9$ *is a perfect square?*

- **Solution**: We first locate
$$b = 6 \text{ and } c = 9$$
then we set
$$2k = b = 6$$
$$\Rightarrow k = 3$$

We now check $k^2 = 3^2$ to see if this gives c (in this case, the value $k = 3$ does square to give $c = 9$). We conclude that
$$x^2 + 6x + 9 = (x + 3)^2$$

Which is a **perfect square**.

Example **2.19**. *Make* $x^2 + 8x + 10$ *into a perfect square.*

- **Solution**: As before,
$$k = \frac{b}{2} = 4 \text{ but } k^2 = 4^2 = 16$$

Since $k^2 = 16$ is different from $c = 10$, we know that $x^2 + 8x + 10$ is not a perfect square. However, notice that we can add and subtract what we need in order to get a perfect square:

$$x^2 + 8x + 10 = \underbrace{x^2 + 2(4)x + (4)^2} - 16 + 10$$

$$= (x + 4)^2 - 16 + 10$$

$$= (x + 4)^2 - 6$$

We say that we have **completed the square** for the given expression.

Example **2.20**. *Complete the square for the following expression* $x^2 - 12x - 1$.

- **Solution:**

$$x^2 - 12x - 1 = \underbrace{x^2 - 2(6)x + (6)^2} - 36 - 1$$

$$= (x - 6)^2 - 37.$$

Example **2.21**. *Complete the square for the following expression* $x^2 - 5x + 1$.

- **Solution:**

$$x^2 - 5x + 1 = x^2 - 2\left(\frac{5}{2}\right)x + \left(\frac{5}{2}\right)^2 - \frac{25}{4} + 1$$

$$= \left(x - \frac{5}{2}\right)^2 - \frac{25}{4} + \frac{4}{4}$$

$$= \left(x - \frac{5}{2}\right)^2 - \frac{21}{4}$$

- **In general**, in order to complete the square on $\boxed{x^2 + \beta x + \lambda,}$ we add and subtract $\boxed{\left(\frac{\beta}{2}\right)^2}$ to the expression (we call this quantity the adjustment term). Next, we combine the first three terms into a perfect square, then we simplify the remaining two terms.

$$x^2 + \beta x + \lambda = x^2 + \beta x + \underbrace{\left(\frac{\beta}{2}\right)^2 - \left(\frac{\beta}{2}\right)^2} + \lambda$$

$$= \underbrace{x^2 + \beta x + \left(\frac{\beta}{2}\right)^2} - \frac{\beta^2}{4} + \lambda$$

$$= \left(x + \frac{\beta}{2}\right)^2 \underbrace{- \frac{\beta^2}{4} + \lambda}$$

$$= \left(x + \frac{\beta}{2}\right)^2 \underbrace{- \frac{\beta^2}{4} + \frac{4\lambda}{4}}$$

$$= \left(x + \frac{\beta}{2}\right)^2 + \frac{4\lambda - \beta^2}{4}$$

This completes the process.

Exercise **2.12**. *Complete the square for each of the following*:

(a) $x^2 + 8x - 3$.

Solution:

(b) $x^2 + 3x + 7$.

Solution:

(c) $x^2 - \frac{4}{3}x + 1$.

Solution:

(d) $x^2 - 5x - 2$.

Solution:

2.4.ii.a. The Method of Completing the Square

In order to **solve** a **quadratic equation** by completing the square, we recall that

$$z^2 = a$$
$$z = \pm\sqrt{a}$$

For $(x + 1)^2 = 9$, the solution can be obtained by simplifying thus

$$(x + 1) = \pm\sqrt{9} = \pm 3$$

this produces two linear equations

$$x + 1 = 3 \text{ and } x + 1 = -3$$

Hence, the solutions are $x = 2$ and $x = -4$.

Example **2.22.** *Solve for* x, $x^2 - 6x + 2 = 0$.

- **Solution.** First we remove the constant term from the left side

$$x^2 - 6x = -2$$

Next, we add the adjustment $(\frac{6}{2})^2$ to both sides of the equation because we need $(\frac{6}{2})^2$ to make the left side a perfect square:

$$x^2 - 6x + 3^2 = -2 + 9$$

The left hand side is now a perfect square

$$(x - 3)^2 = 7$$

We now solve

$$x - 3 = \pm\sqrt{7}$$
$$x = 3 + \sqrt{7} \text{ and } x = 3 - \sqrt{7}$$

The solution set is $\boxed{S = \{3 + \sqrt{7},\, 3 - \sqrt{7}\}.}$ It is acceptable to write $S = \{3 \pm \sqrt{7}\}$ provided we understand that there are two solutions here, one for each sign.

Exercise **2.13.** *Solve each of the following equations by completing the square*:

(a) $x^2 - 2x - 4 = 0$.

Solution:

(b) $2x^2 + 4x - 1 = 0$.

Solution:

Example 2.23. *Solve by completing the square,* $2x^2 + 3x + 4 = 0$.

- **Solution.** We proceed as in the preceding example but only after dividing through by 2

$$x^2 + \frac{3}{2}x = -2$$

we add the adjustment term

$$x^2 + \frac{3}{2}x + \left(\frac{3}{4}\right)^2 = \frac{3}{16} - 2$$

compound the left and simplify the right

$$\left(x + \frac{3}{4}\right)^2 = -\frac{29}{16}$$

now, extract square roots

$$x + \frac{3}{4} = \pm\sqrt{-\frac{29}{16}}$$

$$= \pm\frac{\sqrt{29}}{4}i$$

solve for x

$$x = -\frac{3}{4} \pm \frac{\sqrt{29}}{4}i$$

The solutions can be expressed in the equivalent form: $x = \dfrac{-3 \pm \sqrt{29}\,i}{4}$

Exercise 2.14. *Solve each of the following equations by completing the square:*

(a) $x^2 + 2x + 5 = 0$.

Solution:

(b) $3x^2 - x - 2 = 0$.

Solution:

2.4.iii. The Quadratic Formula

We now use the method of completing the square to derive a formula for solving quadratic equations. To solve the equation

$$ax^2 + bx + c = 0$$

we note that $a \neq 0$ so we divide throughout by a

$$x^2 + \frac{b}{a}x + \frac{c}{a} = 0$$

then we eliminate the constant term from the left side

$$x^2 + \frac{b}{a}x = -\frac{c}{a}$$

now add the adjustment term to both sides

$$x^2 + \frac{b}{a}x + \left(\frac{b}{2a}\right)^2 = \frac{b^2}{4a^2} - \frac{c}{a}$$

form the square on the **left** and simplify the right

$$\left(x + \frac{b}{2a}\right)^2 = \frac{b^2}{4a^2} - \frac{4ac}{4a^2}$$

$$= \frac{b^2 - 4ac}{4a^2}$$

This has the form $z^2 = \alpha$ so, taking square roots of both sides, we obtain

$$x + \frac{b}{2a} = \pm\sqrt{\frac{b^2 - 4ac}{4a^2}}$$

$$= \pm\frac{\sqrt{b^2 - 4ac}}{2a}$$

now solve for x and simplify

$$x = -\frac{b}{2a} \pm \frac{\sqrt{b^2 - 4ac}}{2a}$$

- The denominators on the right are the same, hence $\boxed{x = \frac{-b \pm \sqrt{b^2 - 4ac}}{2a}}$

- This is the **quadratic formula**.

Example **2.24.** *Solve* $5x^2 - 2x = 3$ *using the quadratic formula.*

- **Solution**: In order to use the formula, we must correctly identify the coefficients $\{a, b, c\}$. Thus it is crucial to rearrange the equation in standard form:

$$5x^2 - 2x - 3 = 0$$

Comparing with the standard form $ax^2 + bx + c = 0$, we see that $a = 5$, $b = -2$, and $c = -3$. We then substitute these values into the formula

$$x = \frac{-b \pm \sqrt{b^2 - 4ac}}{2a}$$

$$= \frac{-(-2) \pm \sqrt{(-2)^2 - 4 \cdot 5(-3)}}{2 \cdot 5}$$

$$= \frac{2 \pm \sqrt{4 + 60}}{10}$$

now simplify

$$x = \frac{2 \pm \sqrt{64}}{10}$$

$$= \frac{2 \pm 8}{10}$$

We must realize here that there are two solutions

$$x = \frac{2 + 8}{10} = \frac{10}{10} = 1 \text{ and}$$

$$x = \frac{2 - 8}{10} = -\frac{6}{10} = -\frac{3}{5}$$

The solution set is $\boxed{S = \{1, -\frac{3}{5}\}}$. Let us check our solutions:

if $x = 1$, then

$$5x^2 - 2x - 3 = 5(1)^2 - 2(1) - 3$$

$$= 0$$

if $x = -\frac{3}{5}$, then

$$5x^2 - 2x - 3 = 5\left(-\frac{3}{5}\right)^2 - 2\left(-\frac{3}{5}\right) - 3$$

$$= 5 \cdot \frac{9}{25} + \frac{6}{5} - \frac{15}{5}$$

$$= \frac{9}{5} + \frac{6}{5} - \frac{15}{5}$$

$$= 0$$

As you can see, both values are solutions.

Exercise 2.15. *Solve each of the following equations using the quadratic formula:*

(a) $2x^2 - 4x - 1 = 0.$

Solution:

(b) $x^2 - 2x + 6 = 0.$

Solution:

(c) $3x^2 + x - 2 = 0.$

Solution:

(d) $5 - 8x - 4x^2 = 0.$

Solution:

2.4.iii.a. Nature of Roots; the Discriminant

The solutions of the quadratic equation $ax^2 + bx + c = 0$, where a, b, c are real numbers, may be classified into **three types**:

1. **Two real** and **distinct** roots
2. **One real**, **repeated (double)** root
3. **Two complex** roots.

The following examples illustrate the point:

Example 2.25. *Solve* $5x^2 - 2x - 3 = 0.$

- **Solution**. We have seen that

$$x = \frac{-b \pm \sqrt{b^2 - 4ac}}{2a} = \frac{2 \pm \sqrt{64}}{10}$$

$$S = \{1, -\tfrac{3}{5}\}$$

The equation has **two real** and **distinct** roots.

Example 2.26. *Solve* $x^2 + 6x + 9 = 0.$

- **Solution**. Note that we can **factor** the left hand side:

$$(x + 3)(x + 3) = 0,$$

so that $x = -3$ **twice**. Thus we have **one real, repeated (double) root**.
Alternatively, we use the quadratic formula:

$$x = \frac{-6 \pm \sqrt{6^2 - 4 \cdot 9}}{2}$$

$$= \frac{-6 \pm \sqrt{36 - 36}}{2}$$

$$= -\frac{6 \pm 0}{2} = -3$$

Thus, $\boxed{S = \{-3\}.}$

Example **2.27.** Solve $x^2 - 2x + 5 = 0$.

• **Solution.** In this case,

$$x = \frac{2 \pm \sqrt{4 - 20}}{2}$$

$$= \frac{2 \pm \sqrt{-16}}{2} = \frac{2 \pm 4i}{2}$$

$$= 1 \pm 2i$$

As you can see, the solution set $S = \{1 + 2i, 1 - 2i\}$ consists of **two complex conjugates**. This is not an accident, **complex solutions come in conjugate pairs**.

■ The above three examples show that the nature of the solutions, of $ax^2 + bx + c = 0$, is determined by the sign of the **radicand** $\boxed{b^2 - 4ac}$. Let us re-examine the solutions:

▶ *For* $5x^2 - 2x - 3 = 0$, $\boxed{b^2 - 4ac} = 64 > 0$

and the equation has *two distinct real roots.*

▶ *For* $x^2 + 6x + 9 = 0$, $\boxed{b^2 - 4ac} = 0$

and the equation has *one repeated real root.*

▶ *For* $x^2 - 2x + 5 = 0$, $\boxed{b^2 - 4ac} = -16 < 0$

and the equation has *two complex roots.*

$\boxed{\text{Definition}}$

The Discriminant and the Type of Roots:

$\boxed{b^2 - 4ac}$ is called the **discriminant**.

In general, if the coefficients $\{a, b, c\}$ are real,

then the quadratic equation $\boxed{ax^2 + bx + c = 0}$ has:

A. **Two distinct real** roots if $\boxed{b^2 - 4ac} > 0$

B. **One repeated real** root if $\boxed{b^2 - 4ac} = 0$

C. **Two complex conjugate** roots if $\boxed{b^2 - 4ac} < 0$

Exercise **2.16.** *Determine the nature of roots for each equation (Do not solve).*

(*a*) $3x^2 + 2x + 1 = 0$. (*b*) $2x^2 = 11x - 5$. (*c*) $9x^2 + 4 = 12x$. (*d*) $x^2 - 5x + 6 = 0$.

Solution: **Solution:**   **Solution:**   **Solution:**

Summary

■ A **quadratic equation** has standard form $\boxed{ax^2 + bx + c = 0.}$ We may **solve a quadratic equation** by **factoring**, by **completing the square**, or by using the **quadratic formula**:

■ In practice, it is advisable to *first* try solving by **factoring** as it is easy to check our factorization by multiplication.

▶ If the polynomial does not factor, then we apply the quadratic formula. The method of completing the square is often used when we do not remember the formula precisely and the polynomial does not factor.

- In order to successfully solve a quadratic equation by the **Quadratic formula**, we must remember the formula precisely $x = \dfrac{-b \pm \sqrt{b^2 - 4ac}}{2a}$ we must identify the coefficients $\{a, b, c\}$ correctly; and we must substitute and simplify the expression.

- Solutions of quadratic equations are: **real and distinct**, **real repeated**, or **complex conjugates**, depending on whether the **discriminant** $b^2 - 4ac$ is positive, zero, or negative, respectively. Thus, if the aim is not to solve but to simply determine the nature of the solutions, we use the discriminant.

- In order to **complete the square** for the expression $x^2 + \beta x + \lambda$,
  - ▶ the adjustment $\left(\frac{\beta}{2}\right)^2$ is needed to form a perfect square
  - ▶ the adjustment is added and subtracted to nullify the effect of addition
  - ▶ three terms form the square $\left(x + \frac{\beta}{2}\right)^2$
  - ▶ the remaining terms are simplified into $\dfrac{4\lambda - \beta^2}{4}$

End of Section 2.4. *Problems*

Problem 2.10 *Solve each equation*:

(a) $3x^2 + 2x = 0.$ (b) $3x^2 = 5x.$
(c) $7x^2 - 3 = 0.$ (d) $3x^2 + 1 = 0.$
(e) $4x^2 = -3x.$ (f) $9x^2 = 4.$

Problem 2.11. *Solve by factoring*:

(a) $x^2 - 3x - 10 = 0.$ (b) $x^2 - 6x + 8 = 0.$
(c) $20x^2 - 19x + 3 = 0.$ (d) $2x^2 + 5x = -3.$
(e) $2x^2 - 5x - 3 = 0.$ (f) $3x^2 + 8x = 3.$

Problem 2.12 *Solve by completing the square*:

(a) $x^2 - 6x + 15 = 0.$ (b) $x^2 + 5x = 9.$
(c) $2x^2 - 8x - 3 = 0.$ (d) $3x^2 + x - 8 = 0.$
(e) $5x^2 - 2x - 1 = 0.$ (f) $2x^2 - 3x - 1 = 0.$

Problem 2.13 *Solve using the quadratic formula*:

(a) $3x^2 + 2x = 0.$ (b) $-2x^2 + 3x + 5 = 0.$
(c) $2x^2 + 5x + 2 = 0.$ (d) $9x^2 - 18x + 8 = 0.$
(e) $2x^2 + 4x - 3 = 0.$ (f) $6x^2 + 17x = 3.$

Problem 2.14 *Solve each equation by any* **two** *methods*

(a) $x^2 + 5x + 6 = 0.$ (b) $x^4 - 13x^2 + 36 = 0.$
(c) $4x^4 - 5x^2 - 9 = 0.$ (d) $2x^2 + 5x + 4 = 0.$
(e) $3x^2 - 4x + 1 = 0.$ (f) $4x^3 + 2x^2 + 3x = 0.$

Problem 2.15 *Without solving, determine the nature of the roots of each equation*:

(a) $2x^2 + 5x + 2 = 0.$ (b) $x^2 + 3x + 6 = 0.$
(c) $6x^2 + 17x = 3.$ (d) $4x^2 - 12x + 9 = 0.$
(e) $9x^2 - 18x + 8 = 0.$ (f) $25x^2 + 10x + 1 = 0.$

2.5. Inequalities

An inequality in the variable x is a mathematical expression involving x and one of the following order operations:

$$\{<, \leq, >, \geq\}$$

We shall show that inequalities differ from equations in fundamental ways. For instance, the addition/subtraction property of equations also holds for inequalities, but the multiplication/division property of equations must be modified for inequalities.

2.5.i. Linear Inequalities

There are **two properties** of inequalities that differentiate them from equations. We examine these through illustrative examples.

First property

Consider the inequality

$$2x > 6$$

and the corresponding equation

$$2x = 6$$

The solution set for the equation is the singleton set $S = \{3\}$. On the other hand, any real number greater than 3 satisfies the inequality. Consequently, the solution set is the set of real numbers each of which is greater than 3. Examples of solutions for the inequality are $3.001, 3.5, 10,$ and 10^{15}. We cannot list members of this set. (Why not?) Thus we must represent the set by other well-defined means.

- Two notations are appropriate here:
  - $S = \{x \in R / x > 3\}$.
    - ▶ This is the **set builder notation** and it reads: "**the set of all real numbers** x **such that** x **is greater than** 3".
  - The same set can also be represented using the **interval notation**
    - ▶ $S = (3, \infty)$.
- Aside from presentation, it also makes sense to **graph** the solution set on the **number line**:

$$\hat{0} \quad \hat{3}$$

> | Property |
> | --- |
>
> The **linear inequality** has **infinitely** many solutions that cannot be enumerated,
> unlike the corresponding **linear equation** which has exactly **one** solution.

Second property

Consider the valid statement

$$2 > -3$$

If we multiply both sides by 3, the resulting inequality is

$$6 > -9$$

Notice that the inequality has maintained its original direction. This is *always* the case if we multiply (or divide) an inequality by a positive quantity. However, if we multiply (or divide) the original inequality by (-1), we obtain

$$-2 < 3$$

Observe that we needed to **reverse the direction** of the **original inequality** in order to get a valid statement.

> **Rule**
>
> To multiply (or divide) an inequality by a negative quantity the **direction** of the **inequality** must be **reversed**.

The implication of this property of inequalities is that *we cannot multiply (or divide) an inequality by an unknown quantity such as a variable*. Thus, for instance, the inequalities

$$\frac{x}{-2} > 1 \text{ and } x < -2$$

are equivalent. We multiplied the first inequality by -2 and changed the direction of the inequality, so as to obtain the second.

However, the inequalities

$$\frac{x-2}{x} < 0 \text{ and } x - 2 < 0$$

are **not equivalent** even though multiplying through the first inequality by x results in the second. **We cannot multiply the inequality by** x **since we do not know (before hand) the sign of** x. $\boxed{\text{Note}}$ that $x = 0$ is a solution for the second one, but it is a **restricted value** for the first. Now, recall that the solution set for an inequality requires special representation.

The following example shows the two forms and graphs:

Expression	Set-builder Notation	Interval Notation
$2 < x < 5$	$\{x \in R / 2 < x < 5\}$	$(2, 5)$
$2 \leq x < 5$	$\{x \in R / 2 \leq x < 5\}$	$[2, 5)$
$x > -1$	$\{x \in R / x > -1\}$	$(-1, \infty)$
$x \leq 2$	$\{x \in R / x \leq 2\}$	$(-\infty, 2]$

Exercise **2.17**. *Solve each inequality. Use interval notation to represent the solution set and graph the set.*

(a) $2(x+1) + 5 > 11$.

Solution:

(b) $3(2x-1) - 2(x+5) \geq 2x + 7$.

Solution:

(c) $\frac{2}{3}(6-x) - \frac{3}{4}(5-2x) \leq \frac{1}{6}(3-x)$.

Solution:

(d) $2(3x-4) - 3(4x+1) < 7$.

Solution:

2.5.ii. Combinations of linear inequalities

It is sometimes necessary to combine two or more inequalities. For instance, the compound inequality $-2 < x \leq 3$ is a combination of two inequalities $-2 < x$ **AND** $x \leq 3$.

We should recognize the combined inequality as the interval $I = (-2, 3]$ which is the set of real numbers between -2 and 3, excluding -2 but including 3. Equivalently, the two intervals combined to give $I = (-2, 3]$, are $(-2, \infty)$ **AND** $(-\infty, 3]$.

The graph of both intervals on the same number line shows that I is the **intersection** of the two intervals (inequalities).

Example **2.28**. *Given the sets* $I_1 = [-2, 3)$, $I_2 = [-5, 1]$, *determine the sets* $A = I_1 \cap I_2$ *and* $B = I_1 \cup I_2$. *Show graphs for both solutions.*

- **Solution**. Since $A = I_1 \cap I_2$, it is the set of points that belong to both I_1 **AND** I_2. Thus

$$A = [-2, 3) \cap [-5, 1]$$
$$= [-2, 1]$$

On the graph, A is the portion of the number line where both sets are defined.

- On the other hand, $B = I_1 \cup I_2$ is the set of points that are members of I_1 **OR** I_2 or both. Hence,

$$B = [-2, 3) \cup [-5, 1]$$
$$= [-5, 3).$$

Geometrically, B comprises of portions of the number line where any of the two sets is defined.

Example **2.29**. *Solve the inequality* $-5 < 3 - 2x \le 7$. *Use interval notation for the solution set and graph the solution set.*

- **Solution**. Since the combined has three sides (left, center, and right), we must aim for a solution of the form $\alpha < x < \beta$, remembering to include equality on one side. Observe that this form requires us to eliminate 3 and -2 from the middle part of the given inequality. Although we may do this in any order, in this example it is easier to eliminate 3 and -2, in this order, as follows:

$$-5 < 3 - 2x \le 7$$

Add -3 to all three sides and simplify

$$-8 < -2x \le 4$$

Divide through by -2 and change the direction of the inequalities

$$4 > x \ge -2$$

This is the same as

$$-2 \le x < 4$$

The solution set is $\boxed{S = [-2, 4).}$

Exercise 2.18. *Do each of the following and graph the solution set:*

(a) If $A = [-4,4)$ and $B = (0,7)$,
determine the sets $A \cap B$ and $A \cup B$.

Solution:

(b) If $A = (-\infty,-2]\cup(2,7)$; $B = [-5,5]$,
determine the sets: A **AND** B; A **OR** B.

Solution:

(c) *Solve* $-7 < 3x - 1 \le 8$.

Solution:

(d) *Solve* $-1 \le 3 - 2x \le 9$.

Solution:

End of Section 2.5. Problems

Problem 2.16 *Express the given inequality in interval notation.*

(a) $x > 3$.   (b) $x \le 2$.

(c) $-3 \le x < 4$.   (d) $-1 < x \le 5$.

(e) $x \le 1$ OR $x > 3$. (f) $-2 \le x \le 7$.

Problem 2.17 *Solve each inequality. Express the result in interval notation and graph the solution set.*

(a) $3x - 8 + 2x \le 4x - 10$.   (b) $2x\_12 + 6x \ge 7x + 6$.

(c) $2(3x - 4) - 3(x + 1) > 11$.   (d) $4(3 - x) + 3(3x - 2) \le 2(x + 7)$.

(e) $2(5x - 2) + 7(x - 1) \ge x + 15$. (f) $3(2x + 5) - 4(2 - x) < 5 + 2(x - 3)$.

Problem 2.18 *Solve each inequality and graph the solution set.*

(a) $-3 < 2x - 5 < 7$.   (b) $-1 \le 2x + 3 < 1$.

(c) $-3 < 1 - 2x \le 7$. (d) $-3 \le 2x - 7 \le 5$.

(e) $-5 \le 4 - 3x < -2$. (f) $1 \le 1 - 4x < 9$.

Problem 2.19 *Find the values of x that satisfy the given inequality. Graph the set.*

(a) $\frac{1}{2}x - \frac{2}{3}(x - 1) \ge 1$.   (b) $\frac{3}{4}(y + 2) + \frac{1}{3}(x = 1) \le 3$.

(c) $\frac{1}{3}(x - 1) + \frac{1}{5} < \frac{1}{5}(x + 2) - \frac{1}{3}$. (d) $\frac{5}{2x - 7} \ge 0$.

(e) $\frac{3}{2x + 5} < 0$.   (f) $\frac{3}{2(x - 1)} \ge \frac{5}{x - 1}$.

2.6. Nonlinear Inequalities

An inequality that is not linear is referred to as **nonlinear**. Although nonlinear inequalities are of several types, we shall be concerned with two types only: **higher degree polynomial inequalities** and **rational inequalities**. The reasoning behind the method of solution is the same for the two types.

2.6.i. Analytic Method

Consider the inequality

$$x^2 - x > 2$$

At first glance, it may be difficult to see how to proceed without guessing. However, since the inequality is quadratic, it seems natural to rearrange it in the form

$$x^2 - x - 2 > 0$$

We may be able to see more clearly if we interpret the inequality as "the left hand side is *greater than zero*". This becomes more obvious since we can **factor** the left side:

$$(x + 1)(x - 2) > 0$$

The statement has now taken the form "$A \cdot B$ is positive". We know that the product of two factors is positive only if the two factors have the same sign. That is: they are either **both** positive (positive **AND** positive) or both negative (negative **AND** negative). Thus, we shall look for x such that either:

$$\{x + 1 > 0\} \textbf{ AND } \{x - 2 > 0\}$$

OR

$$\{x + 1 < 0\} \textbf{ AND } \{x - 2 < 0\}$$

We simplify the four inequalities:

$$\{x > -1\} \textbf{ AND } \{x > 2\}$$

OR

$$\{x < -1\} \textbf{ AND } \{x < 2\}.$$

Recall that the conjunction **AND** is the intersection of sets. Thus, we have

$$\{x > -1\} \bigcap \{x > 2\} = \{x > 2\}$$

OR

$$\{x < -1\} \bigcap \{x < 2\} = \{x < -1\}$$

These may be seen more clearly from the graphs:

So far, we have determined the two intersections. We now use the connective - **OR**:

$$\{x > 2\} \bigcup \{x < -1\}$$

Since the two sets are disjoint (have no common elements), there can be no further simplification and this is our **solution set**; in interval notation, $\boxed{S = (-\infty, -1) \bigcup (2, \infty)}$

The graph of S is as follows:

We summarize the preceding development and make further observations.

In summary: For the inequality $x^2 - x > 2$:

1. Put it in the *standard form*: $(x + 1)(x - 2) > 0$.

2. Since the "left side is positive" in this form, our pursuit of a solution set is based on *signs* of **factors** rather than on values of the variable x.

3. The numbers -1 and 2 play a critical role as boundaries of the sets that combine toform the solution set. We call these *critical points*.

4. **Critical points can be obtained from the standard form by setting each factor to zero. That is:** $x + 1 = 0$ and $x - 2 = 0$.

Example 2.30. *Solve the inequality* $(x + 1)(x - 2) \geq 0$.

- **Solution**: Comparing this with the preceding example, we have two separate problems:

$$(x + 1)(x - 2) > 0$$

and

$$(x + 1)(x - 2) = 0$$

The solution set of the inequality is (from above) $\boxed{S_1 = (-\infty, -1) \bigcup (2, \infty).}$

The *solution set* of the *equation* is the *set of critical points* $\boxed{S_2 = \{-1, 2\}.}$

Thus, x is a solution if x is a member of either S_1 **OR** S_2. Thus the **solution set** for $(x + 1)(x - 2) \geq 0$ is $\boxed{S = S_1 \bigcup S_2 = (-\infty, -1] \bigcup [2, \infty).}$

$\boxed{\text{Note}}$ that we included the critical points in the solution set because of **equality** in the given inequality.

$$-1 \quad 0 \quad 2$$

For **rational inequalities**, for example

$$\frac{x + 1}{x - 2} > 0 \text{ and } \frac{x - 2}{x + 1} > 0$$

we note that they have the same factors as the standard form $(x + 1)(x - 2) > 0$. Consequently, they all have the same critical points and the same solution set.

However, neither of the inequalities

$$\frac{x + 1}{x - 2} \geq 0 \text{ or } \frac{x - 2}{x + 1} \geq 0$$

has the same solution set with $(x + 1)(x - 2) \geq 0$. This is because *critical points* obtained from *factors* in the *denominator cannot* be part of the *solution set* as they make the **denominator zero**. Hence for

$$\frac{x + 1}{x - 2} \geq 0$$

the solution set is $\boxed{S = (-\infty, -1] \cup (2, \infty).}$

Note that the solution set includes -1 but excludes 2.

$$-1 \quad 0 \quad 2$$

Similarly, the inequality

$$\frac{x-2}{x+1} \geq 0$$

has the solution set $\boxed{S = (-\infty, -1) \bigcup [2, \infty).}$

$$-1 \quad 0 \quad 2$$

2.6.ii. The Test-Point Method

This method is based on the preceding discussion.

> **Definition**
>
> **Inequality in standard form**:
>
> An inequality is said to be in standard form if
>
> -it has **zero** on **one side**, and
>
> -the **side** with the **variable** is completely **factored**.

> **Definition**
>
> A real number α is a **critical point** for the rational quantity
>
> $\boxed{\dfrac{P(x)}{Q(x)}}$ if either $P(\alpha) = 0$ or $Q(\alpha) = 0$.

The numbers $\pm\infty$ are (trivial) for any rational quantity. Furthermore, two critical points $\{\alpha < \beta\}$ for a rational expression are said to be **adjacent** if the expression has **no critical point** in the **interval** (α, β).

For example consider the rational inequality

$$\frac{(x+1)(x-2)}{(x-4)(x-7)} \leq 0$$

By the above definition, the inequality is in standard form. For nontrivial critical points, we set the numerator and denominator to zero:

$$(x+1)(x-2) = 0$$
$$(x-4)(x-7) = 0$$

and solve. We list the critical points (in ascending order):

$$\{-1, 2, 4, 7\}$$

Observe that -1 and 2 are **adjacent critical points** because there is no critical point inside the interval $(-1, 2)$. However, 2 and 7 are not adjacent critical points since the critical point 4 is in the interval $(2, 7)$.

> **Theorem**
>
> For the inequality $\dfrac{P}{Q} > 0$, where P and Q are polynomials.
>
> Let α be a real number such that $c_1 < \alpha < c_2$, where c_1 and c_2 are adjacent critical points of the rational quantity $\dfrac{P}{Q}$.
>
> **1**. If α **satisfies** the inequality, then **every number** in the interval $c_1 < x < c_2$ satisfies the inequality.
>
> **2**. If α **does not satisfy** the inequality, then **no number** in the interval $c_1 < x < c_2$ satisfies the inequality.

Example 2.31. *Solve the inequality* $\frac{(x+1)(x-3)}{(x+5)(x-1)} \leq 0$.

- **Solution.** The given inequality has zero on the right hand side and the left hand side is a quotient of factors. Hence it is already in standard form. We proceed to find the critical points by setting the numerator and denominator to zero:

$$(x+1)(x-3) = 0$$
$$(x+5)(x-1) = 0$$

The critical points are: $\{-5, -1, 1, 3\}$.

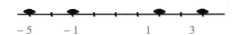

$$-5 \qquad -1 \qquad\qquad 1 \qquad 3$$

- These divide the number line into five disjoint intervals:

$$(-\infty, -5), (-5, -1), (-1, 1), (1, 3), (3, \infty)$$

Observe that the intervals are **bounded** by **adjacent critical points**. According to the above theorem, if a number that lies between two adjacent critical points (i.e. within an interval) satisfies the inequality, then all numbers in that interval are solutions. Thus, to find the solution set, we must test each interval against the inequality.

In order to test an interval, we choose *any* point inside the interval and substitute for x in the inequality. For example, in the interval $(-5, -1)$, we may choose -3 (or any other number between -5 and -1). This chosen number is our **test point**. Substituting -3 into the left hand side of the inequality, we obtain

$$\begin{aligned} LHS &= \frac{(x+1)(x-3)}{(x+5)(x-1)} \\ &= \frac{(-3+1)(-3-3)}{(-3+5)(-3-1)} \\ &= \frac{(-2)(-6)}{(2)(-4)} \end{aligned}$$

What needs to happen for -3 to satisfy the inequality? Looking back at the inequality in standard form, we see

$$LHS \leq 0$$

this reads "the left hand side is negative or zero". Thus we are interested principally in the sign but not the value of the left side when $x = -3$. Since there are three negative quantities in the left hand quotient when $x = -3$, we conclude that

$$LHS = \frac{(-2)(-6)}{(2)(-4)} < 0$$

Thus -3 satisfies the inequality and so does any number in the interval $(-5, -1)$. Thus the interval $(-5, -1)$ is **part** of the **solution set**.

In order to decide whether an interval is in the solution set, we must choose a test point and test the inequality. We may show our work in an organized manner using a table:

Test Point	$x-3$	$x+1$	$x-1$	$x+5$	**Quotient** $\left[\dfrac{(x+1)(x-3)}{(x+5)(x-1)}\right]$
-7	$-$	$-$	$-$	$-$	$+$
-3	$-$	$-$	$-$	$+$	$-$
0	$-$	$+$	$-$	$+$	$+$
2	$-$	$+$	$+$	$+$	$-$
10	$+$	$+$	$+$	$+$	$+$

- $\boxed{\text{It cannot be over-emphasized}}$ that **critical points** <u>**cannot**</u> **be test points** because a critical point makes one of the factors in the quotient zero. The **essence** of a **test** is to determine the **sign** (negative or positive) of the **quotient** at the test point. In the table above, note, for example that the test point 0 is picked from the interval $(-1, 1)$ defined by the adjacent critical points -1 and 1. From the table the sign of each factor when $x = 0$ is recorded in the cell and the sign of the quotient is $+$ since the quotient contains two negatives. Since the inequality requires "left side is negative", the test-point $x = 0$ does not satisfy the inequality. Accordingly, the interval from which the test point was chosen cannot be part of the solution set. The table shows that the test points -3 and 2 satisfy the inequality, hence the intervals $(-5, -1)$ and $(1, 3)$ are in the solution set.

Since the original inequality is "$\leq$", we now address the question of equality.

$$\frac{(x+1)(x-3)}{(x+5)(x-1)} = 0 \Rightarrow (x+1)(x-3) = 0$$

i.e. $x = -1$ or $x = 3$

Observe that these are precisely the critical points of the numerator of the quotient in standard form. The solution set is $\boxed{S = (-5, -1] \cup (1, 3].}$

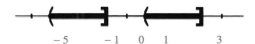

$$-5 \qquad -1 \quad 0 \quad 1 \qquad 3$$

Exercise 2.19. *Solve each inequality using the Test Point Method. Graph your solution set.*

(a) $x^2 + x - 12 \leq 0.$

Solution:

(b) $3x^2 + 5x - 2 > 0.$

Solution:

Example 2.32. *Solve the inequality* $\frac{x}{x-1} \geq \frac{2}{x+3}.$

- Solution. Observe that the given inequality is not in standard form. First, we eliminate terms to get zero on the right hand side:

$$\frac{x}{x-1} - \frac{2}{x+3} \geq 0$$

next, we compound the left side into one fraction and factor if possible:

$$\frac{x(x+3) - 2(x-1)}{(x-1)(x+3)} \geq 0$$

$$\frac{x^2 + x + 2}{(x-1)(x+3)} \geq 0$$

this is now in standard form since the numerator cannot be factored. Now, we set the numerator and denominator to zero to get the critical points:

$$x^2 + x + 2 = 0$$

$$(x-1)(x+3) = 0$$

In the quadratic equation $x^2 + x + 2 = 0$, the **discriminant** $\boxed{b^2 - 4ac} = 1 - 8$ is negative. This **implies** that there are **two complex roots**. Hence, the numerator has no critical points. This means that the numerator has the same sign for all real values of x. Thus, the critical points for the rational expression are 1 and -3.

$$-3 \qquad \qquad 0 \quad 1$$

- The intervals under consideration are $(-\infty, -3), (-3, 1), (1, \infty)$.

  We choose a test point from within each interval and test the inequality:

Test point	$x - 1$	$x + 3$	$x^2 + x + 2$	Quotient $\dfrac{x^2 + x + 2}{(x - 1)(x + 3)}$
-5	$-$	$-$	$+$	$+$
0	$-$	$+$	$+$	$-$
5	$+$	$+$	$+$	$+$

- Observe that there is no reason to include the quadratic factor in the test because it is positive for all real numbers. However, if a factor is negative for all real numbers, we need to use its sign to correctly determine the sign of the quotient for each test point. The table above shows that the test points ± 5 satisfy the inequality (the quotient is positive) and, accordingly, the intervals $(-\infty, -3)$ and $(1, \infty)$ are in the solution set. Finally, the numerator of the quotient $(x^2 + x + 2)$ has no critical points which means that there are no real numbers for which the inequality equals zero. Therefore, the solution set is $S = (-\infty, -3) \bigcup (1, \infty)$.

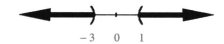

$$-3 \quad 0 \quad 1$$

Summary

- In order to use the Test-Point Method to solve a nonlinear inequality, we
  - ▶ first put it in standard form with zero on one side of the inequality (e.g. *LHS* is positive);
  - ▶ get critical points by setting numerator/denominator to zero and solving;
  - ▶ n critical points divide that number line into $n + 1$ intervals;
  - ▶ choose test points, one from within each interval;
  - ▶ test the inequality in standard form.
  - ▶ Intervals from which test points satisfy the inequality are in the solution set.
  - ▶ If the inequality involves equality, the critical points of the numerator are also solutions.

End of Section 2.6. *Problems*

Problem 2.20 *Solve each inequality. Graph the solution set.*

(a) $x^3 \geq 4x$.

(b) $x^2 - x - 12 > 0$.

(c) $2x^2 + 3x \leq 5$.

(d) $x^3 + 2x^2 - 8x \geq 0$.

(e) $3x^2 - 4x > 4$.

(f) $2x^3 + 5x^2 + 3x \leq 0$.

Problem 2.21 *Solve and graph the solution set.*

(a) $(x - 1)(x + 2)(x - 5) > 0$.

(b) $(2x - 1)(x - 4)(x + 3) < 0$.

(c) $(x + 3)(3x - 1)(x + 7) \leq 0$.

(d) $(x^2 - 4)(x + 5) \geq 0$.

(e) $\frac{x - 1}{x + 3} < 0$.

(f) $\frac{2x + 3}{x - 2} \geq 0$.

Problem 2.22 *Solve and graph the solution set.*

(a) $\frac{x^2 - x - 6}{x + 5} < 0$.

(b) $\frac{x^2 + x - 12}{x^2 - 4} \geq 0$.

(c) $\frac{(x - 1)(x + 2)}{(x - 4)(x + 5)} \leq 0$.

(d) $\frac{2x}{x + 3} \leq 1$.

(e) $\frac{1}{x - 2} - x \geq 2$.

(f) $\frac{2x}{x + 1} - \frac{1}{x - 1} \leq \frac{4}{x^2 - 1}$.

2.7. Equations and Inequalities Involving Absolute Value

We shall explore a geometric interpretation of the absolute value of a real number. Using such interpretation, we will develop simple but practical models for solving equations and inequalities. First observe that

$$\text{-if } x = 3, \text{ then } |x| = 3 = x$$
$$\text{-if } x = 0, \text{ then } |x| = 0 = x$$
$$\text{-if } x = -3, \text{ then } |x| = 3 = -x$$

In addition, there is no real number x for which $|x| = -3$. These observations lead us to a definition and the meaning of absolute value:

| Definition |

The **absolute value** of a **real number** z, denoted by $|z|$ is defined by the equation:

$$|z| = \begin{cases} z \text{ if } z \geq 0 \\ -z \text{ if } z < 0 \end{cases}$$

Compare each of the three cases above and see that they fit the definition. On the number line, note that the real numbers 3 and -3 are 3 units from 0. In general, *the absolute value of z is the distance from z to 0 on the number line.*

$$-3 \qquad 0 \qquad 3$$

We can now develop **three models** using the fact that the absolute value is distance from zero. In each model we assume that z is a variable.

2.7.i. Equality Model

Suppose that we are looking for real numbers z for which $|z| = 5$. This means that the distance from z to zero is 5. It is clear that z is either 5 or -5. This leads us to the generalization (model)

$$|z| = c \Rightarrow \begin{cases} z = c \text{ or} \\ z = -c \end{cases}$$

We use the model to see how to proceed in an equation involving absolute value.

Example **2.33.** *Solve the equation* $|2x - 3| = 5$.

Solution. By the model for equality, we have two equations (here, our $z = 2x - 3$ and $c = 5$)

$$\text{Either } 2x - 3 = 5$$
$$\text{or } 2x - 3 = -5$$

Solving these equations, we obtain

$$2x = 8, \text{ i.e. } x = 4$$
$$2x = -2, \text{ i.e. } x = -1$$

The solution set is $S = \{-1, 4\}$. We may check the solutions against the original equation

If $x = -1$, then

$$|2x - 3| = |-2 - 3| = |-5| = 5$$

If $x = 4$, then

$$|2x - 3| = |8 - 3| = |5| = 5$$

It is crucial to note that the solution set is neither ± 4 nor ± 1. Although the model is

$$|z| = 5 \Rightarrow z = \pm 5$$

if we substitute $z = 2x - 3$ into the model and we only solve one equation, say $2x - 3 = 5$, we obtain the one correct solution $x = 4$. *This does not imply that $x = -4$ is also a solution.* In fact if $x = -4$, then

$$|2x - 3| = |-8 - 3| = |-11| = 11$$

Thus, in the use of the model we are restricted to getting rid of the absolute value only, then we proceed to solve the two resulting equations independently.

Example 2.34. *Solve the equation* $|x^2 - 5x| = 6$.

Solution. As in the preceding example, we use the model to obtain two independent equations

$$x^2 - 5x = 6$$

and

$$x^2 - 5x = -6$$

The fact that the resulting equations are quadratic does not alter how we use the model. We now solve the equations

$$x^2 - 5x = 6$$
$$x^2 - 5x - 6 = 0$$
$$(x + 1)(x - 6) = 0$$
$$x = -1 \text{ or } 6$$

Similarly,

$$x^2 - 5x = -6$$
$$x^2 - 5x + 6 = 0$$
$$(x - 2)(x - 3) = 0$$
$$x = 2 \text{ or } 3$$

Thus, the solution set is $\boxed{S = \{-1, 2, 3, 6\}}$.

Check to see that each of these is a solution for the original absolute value equation.

> Remark
>
> In the preceding example, the **absolute value equation**
>
> is of the **form** $\boxed{|ax^2 + bx| = c}$
>
> This will naturally lead to **two quadratic equations**.
>
> Some of the solutions for the resulting quadratic equations may be complex numbers.
>
> Although, these are also valid solutions, we shall **restrict** ourselves to **real solutions** for which we have an interpretation.

Exercise **2.20.** *Solve the following and graph the solution set where applicable*:

(a) $|3 - 2x| = 9$.

Solution:

(b) $|x^2 + 2x| = 8$.

Solution:

2.7.ii. Inequality Model

Next, we examine a model of the type $|z| < 5$.

This says that we are looking for real numbers whose distance from zero is less than 5. A look at the number line shows that all numbers between -5 and 5 fit this description. The line segment in question is the interval $(-5, 5)$, which we can also write as

$$-5 < z < 5$$

Thus, the model for this type of inequality is

$$|z| < c$$
$$\Rightarrow -c < z < c$$

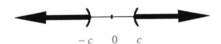

Example **2.35.** *Solve*: $|5 - 2x| < 7$.

- **Solution.** Using the model with $z = 5 - 2x$ and $c = 7$, we write

$$-7 < 5 - 2x < 7$$

As in the first case, our use of the model must stop here. We solve the resulting compound inequality in the usual way:

$$-12 < -2x < 2$$
$$6 > x > -1$$

The solution set is $S = (-1, 6)$.

In the preceding example we should note that

$$|5 - 2x| = |2x - 5|$$

solving

$$-7 < 2x - 5 < 7$$

yields the same result as above.
In addition, if we wish to solve

$$|5 - 2x| \leq 7$$

the effect of equality must be shared in the corresponding compound inequality:

$$-7 \leq 5 - 2x \leq 7$$

and the solution set is $\boxed{S = [-1, 6]}$.

Observe that the difference between the solution set for the **strict inequality** and that of the **non-strict inequality** is that the end points of the interval are also elements of the solution set (only) in the non-strict case.

Exercise 2.21. *Solve the following and graph the solution set where applicable*:

(*a*) $|3x - 2| \leq 7$.

Solution:

(*b*) $|2x + 5| < 3$.

Solution:

2.7.iii. Another Inequality Model

For a final model involving absolute value, consider $|z| > 3$. The distance from z to zero is greater than 3, hence z must be a member of either $(-\infty, -3)$ **OR** $(3, \infty)$. We may also write

$$\{z < -3\} \bigcup \{z > 3\}$$

Here, some people may prefer $\{-z > 3\}$ to $\{z < -3\}$ but these are equivalent statements and should not cause any confusion. In general, if $c > 0$, then

$$|z| > c \Rightarrow \{z > c\} \bigcup \{z < -c\}$$

Example 2.36. *Solve*: $|3x - 2| > 7$.

- **Solution.** We use the model to transform the absolute value inequality into a pair of inequalities:
$$\{3x - 2 > 7\} \bigcup \{3x - 2 < -7\}$$
that is,
$$\{x > 3\} \bigcup \left\{x < -\tfrac{5}{3}\right\}$$
Thus the solution set is $\boxed{S = (-\infty, -\tfrac{5}{3}) \bigcup (3, \infty).}$

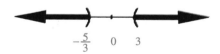

- As before, the solution set for the non-strict inequality
$$|3x - 2| \geq 7$$
includes the finite end points of the intervals in the solution set so that $\boxed{S = (-\infty, -\tfrac{5}{3}] \bigcup [3, \infty).}$

Exercise **2.22**. *Solve the following equation and graph the solution set where applicable*:

(*a*) $|3x + 2| > 11$.

Solution:

(*b*) $|2x - 3| \geq 5$.

Solution:

Summary

- **Key Words**
  - ▶ **Absolute Value** - distance from zero on the number line.
- **Three models**:
  1. Equality method, $|x| = c \Rightarrow x = c$ or $x = -c$.
  2. Inequality i method, $|x| < c \Rightarrow -c < x < c$.
  3. Inequality ii method, $|x| > c \Rightarrow x > c$ or $x < -c$.
     - ▶ We use the models to interpret the absolute value but we solve the resulting problems independently.
     - ▶ For non-strict inequality, we attach all finite end points to the solution set.

End of *Section* 2.7. Problems

Problem **2.23** *Solve and check.*

(*a*) $|x - 3| = 5$. (*b*) $|2x + 1| = 3$.

(*c*) $|3 - 2x| = 7$. (*d*) $\left|\frac{2}{3} - 5x\right| = 4$.

(*e*) $|x^2 - 2x| = 8$. (*f*) $|2x - 3| = |3x - 2|$.

Problem **2.24** *Solve and graph the solution set.*

(*a*) $|x - 1| < 3$. (*b*) $|2x + 3| \leq 7$.

(*c*) $|3 - 2x| < 5$. (*d*) $|3x + 2| \leq 4$.

(*e*) $\left|\frac{2}{3}x - 1\right| \leq \frac{1}{2}$. (*f*) $\left|3x - \frac{2}{5}\right| < 1$.

Problem **2.25** *Solve and graph the solution set*:

(*a*) $|x + 3| > 2$. (*b*) $|2x + 1| \geq 5$.

(*c*) $|3x - 4| \geq 2$. (*d*) $|5x + 3| > 18$.

(*e*) $\left|\frac{3}{4}x - 1\right| \geq 2$. (*f*) $|2 - 5x| > 0$.

Problem **2.26** *Solve and graph where applicable.*

(*a*) $|5x + 1| = x + 6$. (*b*) $|x - 2| = x - 2$.

(*c*) $\left|\frac{4}{x - 4}\right| \leq 1$. (*d*) $\frac{|x - 4|}{3} < 2$.

(*e*) $\left|\frac{3}{x - 1}\right| > 1$. (*f*) $\frac{|2x + 1|}{3} \geq -2$.

2.8 Applications/Word Problems

Although there are numerous real-life situations that can be expressed by linear equations, we shall discuss only a few in this section. The aim is to show some of the connections between equations and everyday life and to develop a strategy for solving word problems.

■ One such strategy is to:

▶ read the whole problem;

▶ determine what is to be found and name this as the variable;

▶ clearly label what the variable represents in the problem;

▶ write down any formulas (if any) connecting the variable and any other parameters.

▶ read the problem again; this time, write down an equation based on given information;

▶ solve the equation that you set up.

2.8.i. Number Problems

Example **2.37**. *The sum of two numbers is* 30. *Three times the smaller number is twice the larger number. Find the two numbers.*

• **Solution**: Let

$$x = \text{the smaller number}$$

Then, the larger number is $30 - x$.

Notice that we have chosen our variable to be one of the quantities that we are looking for. In this way when we solve for x, we are almost done. Now interpreting the relation between the numbers, we have

$$3x = 2(30 - x).$$

We now solve the linear equation:

$$3x = 60 - 2x$$
$$5x = 60$$
$$x = 12.$$

This is the smaller number as we had clearly stated. The larger number is $30 - x = 30 - 12 = 18$.

Thus, the numbers are 12 and 18.

Example **2.38**. *The product of two consecutive integers is* 12. *Find the integers.*

• Solution. Let

$$x = \text{one integer}$$

Then, the second inter is $x + 1$.

Note that we could have named the second integer as $x - 1$. The product is

$$x(x + 1) = 12$$

The emerging equation is quadratic. We now solve it.

$$x^2 + x = 12$$
$$x^2 + x - 12 = 0$$
$$(x + 4)(x - 3) = 0$$
$$x = -4 \text{ or } x = 3.$$

It is important to observe that the integersrs are not -4 and 3 as these are not consecutive. However,

$$x = -4 \Rightarrow x + 1 = -3$$
$$x = 3 \Rightarrow x + 1 = 4$$

Thus, there are two solutions here: the consecutive integers are:

-4 and -3 or 3 and 4.

Example 2.39. *The units digit of a two-digit number is 5 less than the tens digit. If the number is 6 more than 7 times the sum of the digits, find the number.*

- Solution. Let the units digit be

$$x = \text{the tens digit.}$$

Then the units digit is $x - 5$;

The number is $10x + (x - 5)$

We are given that

$$10x + (x - 5) = 6 + 7(x - 5 + x)$$

Solving the equation,

$$10x + x - 5 = 6 + 14x - 35$$
$$-3x = -24$$
$$x = 8$$
$$x - 5 = 3$$

Thus, the number is 83.

Exercise 2.23. *The sum of three numbers is 36. The second number is 4 more than the first while the third number is twice the first number. Find the numbers.*

Solution:

Exercise 2.24. *The units digit of a two-digit number is 4 less than the tens digit. If the product of the digits is 12, find the number.*

Solution:

2.8.ii. Motion Problems

Example 2.40. *A car travels 20 miles per hour faster than a truck. If the two vehicles are traveling in the same direction and the car travels 240 miles in the same time that the truck travels 180 miles, find the speeds of the two vehicles.*

- **Solution:** Here, we must remember that constant rate motion is governed by the formula

$$d = rt$$

where d, r, t, are distance, rate (speed), time, respectively. Let

·_89_·

$$x = \text{rate of the car in mph.}$$
$$x - 20 = \text{rate of the truck in mph.}$$

Note that the time taken for the car to travel 240 miles is
$$\frac{d}{r} = \frac{240}{x}$$
Similarly, the time taken for the truck to travel 200 miles is
$$\frac{d}{r} = \frac{180}{x - 20}$$
We are given that the two"times" are the same, hence the second equation.
$$\frac{240}{x} = \frac{180}{x - 20}$$
We simplify the equation : $240(x - 20) = 180x$
We solve : $240x - 240(20) = 180x$
i.e. $60x = 4800$

Thus, we have
$$x = 80 \text{ and } x - 20 = 60.$$

The speed of the car is 80 mph, and the speed of the truck is 60 mph.

Exercise **2.25**. *Seza started a* 320-*mile journey in bad weather and drove at a certain speed for three hours before meeting good weather. She then increased her speed by twenty miles per hour and made the rest of the journey in two hours. Find the two speeds at which Seza drove.*

Solution:

2.8.iii. Mixtures and Percentages

Example **2.41**. *Kevin borrowed a part of* $7,000 *at* 6% *simple interest and the remainder at* 8%. *After* 3 *years, Kevin pays a total of* $1380 *interest on the loan. How much did Kevin borrow at each rate?*

- **Solution**: First, we note that the important relation here is
$$I = PRT,$$
where I, P, R, T, are the interest, principal, rate, and time, respectively. Let
$$x = \text{amount @ 8\%}$$
$$700 - x = \text{amount @ 6\%}.$$
Thus, the interests in three years are given by
$$.08(3)x \text{ and } .06(3)(7000 - x)$$
and the total interest is
$$.24x + .18(7000 - x) = 1380$$
we now solve the linear equation:

$$24x + 18(7,000 - x) = 138,000$$
$$24x + 126,000 - 18x = 138,000$$
$$6x = 12,000$$
$$x = 2,000$$
$$7,000 - x = 5,000$$

$$\boxed{\$5,000 \text{ was borrowed at } 6\%, \$2,000 \text{ was borrowed at } 8\%.}$$

Example **2.42**. *A nurse has two alcohol solutions: one at 6% concentration, the other at 20% concentration. How much of each solution must she mix to produce 49 ml. of a 12% alcohol solution?*

• **Solution**: Let

$$x = \text{amount of the 6\% solution in the mixture (in ml.)}$$
$$49 - x = \text{amount of the 20\% solution in the mixture (in ml.).}$$

Then,

quantity of pure alcohol in x ml. @ 6% is $.06x$;

quantity of pure alcohol in $49 - x$ ml. @ 20% is $.2(49 - x)$;

quantity of pure alcohol in 49 ml. @ 12% is $.12(49)$.

In mixture problems, we equate like quantities. Thus the equation is
$$.06x + .2(49 - x) = .12(49)$$
We may multiply the second equation by 100 to clear the fractions:
$$6x + 20(49 - x) = 12(49)$$
$$6x + 980 - 20x = 588$$
$$14x = 392$$
$$x = 28$$
$$49 - x = 21.$$

That is: the nurse must mix:

$\boxed{28 \text{ml. of the } 6\% \text{ solution with } 21 \text{ml. of the } 20\% \text{ solution to get } 49 \text{ml. of solution @ } 12\% \text{ concentration.}}$

Exercise **2.26**. *A certain amount of a 70% alcohol solution is to be added to another quantity of a 15% alcohol solution to produce a 40 ml. alcohol solution at 37% concentration. How many ml. of each solution must be used to produce the mixture?*

Solution:

2.8.iv. Miscellaneous Problems

Example **2.43.** Work Two cranes working together can unload a ship in 4 hours. If the faster crane working alone requires 6 hours less than the slower crane to do the same job, how long does it take each crane to do the job.

- **Solution**. In all problems involving work, we seek to see the amount of work done in one unit of time. Let

$$x = \text{time it takes the faster crane to do the job}$$

Then, it takes the slower crane $x + 6$ hours to do the job.

Thus, in one hour,

the faster crane does $\frac{1}{x}$ of the job;

the slower crane does $\frac{1}{x+6}$ of the job.

These are the respective work-rates of the cranes. In four hours, the cranes do

$$\frac{4}{x} \text{ and } \frac{4}{x+6} \text{ of the job.}$$

We may create a table similar to the one below so as to be organized:

Cranes	Rate (r)	Time (t)	Work $= rt$
slower	$\frac{1}{x+6}$	4	$\frac{4}{x+6}$
faster	$\frac{1}{x}$	4	$\frac{4}{x}$

The sum of fractions of the job that they do independently in 4 hours is the whole job. That is:

$$\frac{4}{x} + \frac{4}{x+6} = 1.$$

We now solve the resulting equation

$$4(x+6) + 4x = x(x+6)$$
$$4x + 24 + 4x = x^2 + 6x$$
$$x^2 - 2x - 24 = 0$$
$$(x-6)(x+4) = 0$$
$$x = -4 \text{ or } x = 6.$$

Since x is number of hours, it cannot be negative. Thus,

It takes the faster crane 6 hours to do the job;

It takes the slower one 12 hours to do the job.

Example **2.44.** Value Nena's piggy bank contains $3.20 in quarters and nickels. If she has 2 more quarters than nickels, how many coins of each type are there?

- **Solution**. Let

$$x = \text{number of quarters.}$$

Then, $x - 2 = \text{number of nickels.}$

We formulate a table for convenience:

Coins	Number (n)	Value per coin (v)	Total value $= nv$
quarters	x	25	$25x$
nickels	$x - 2$	5	$5(x-2)$

Here, it is important that all the values are in the same unit of measure (cents in this case). We now set up the equation:

$$25x + 5(x-2) = 320.$$

That is:

$$25x + 5x - 10 = 320$$
$$30x = 330$$
$$x = 11$$
$$\text{and } x - 2 = 9$$

Thus, there are 11 quarters and 9 nickels.

Example **2.45.** Geometry *The length of a rectangle is 6 meters more than its width. If the area of the rectangle is 216 square meters, find the dimensions of the rectangle.*

● **Solution.** Let

$$x = \text{length of the rectangle.}$$
$$\text{Then } x - 6 = \text{width of the rectangle.}$$

Since the area is the product of the length and the width, we have

$$x(x - 6) = 216$$
$$\text{i.e. } x^2 - 6x = 216$$

We solve the quadratic equation:

$$x^2 - 6x - 216 = 0$$
$$(x - 18)(x + 12) = 0$$
$$x = 18 \text{ or } - 12.$$

Since length is nonnegative, we discard the negative value.

The length is 18 meters. | The width is 12 meters.

Exercise **2.27.** *Starting at 12 noon, Jossy starts to mow a field that would take him 9 hours to complete. One hour later, his friend Tony joins him with a tractor and they take 2 hours to complete the job. How many hours would it take Tony alone to do the job with the tractor?*

Solution:

Exercise **2.28.** *Minnie has* 44 *coins totaling* $4.60 *in quarters, dimes and nickels. If she has* 3 *more dimes than quarters, how many coins of each type are there?*

Solution:

Exercise **2.29.** *George wishes to construct an open cardboard box with a square base,* 6 *inches tall, and a capacity of* 384 *cubic inches. Determine the size (square) of cardboard sheet that George should buy.*

Solution:

Summary

- In order to apply systems of linear equations to solve problems,
    - ▶ read the problem completely;
    - ▶ choose your variables to be the quantities you wish to find;
    - ▶ use any formulas connecting the variables;
    - ▶ write down equations based on given information;
    - ▶ solve the system.
    - ▶ For number problems involving digits, note that a two digit number has value
        $$10(\text{tens digit}) + 1(\text{units digit}).$$
        - ◀ For instance, $37 = 10(3) + 1(7)$.
- For constant rate motion problems, the guiding formula is
    $$\text{distance} = \text{rate} \times \text{time}.$$
- The formula $I = PRT$ connects the interest I, the principal P, the rate R, and the time T.
- For mixture and value problems, remember to equate like quantities.
- Work done per unit of time is essential for work problems.

End of Section 2.8. Problems

Problem 2.27. *Find three consecutive integers whose sum is* 21.

Problem 2.28. *The sum of two numbers is* 46. *If three times the first number is* 7 *less than two times the second number, find the two numbers.*

Problem 2.29. *The product of two consecutive positive integers is* 2 *more than* 7 *times the next consecutive number. Find the two integers.*

Problem 2.30. *The sum of the digits of a three-digit number is* 17. *The units digit is* 2 *more than the hundreds digit and* 95 *times the hundreds digit is* 54 *less than the number. Find* *the number.*

Problem 2.31. *The difference of the speeds of two cars is 10 m.p.h.. The cars are 800 miles apart and are traveling towards each other. If they meet in 5 hours, find the speeds of the cars.*

Problem 2.32. *On a 36 mile journey, Kevin drove the first 18 miles at an average speed of 54 mph. During the second half of the journey, he reduced his speed by one-third of his original speed. What was Kevin's average speed for the total distance traveled?*

Problem 2.33. *. A pharmacist has two iodine solutions of 30% and 70% concentration. How much of each solution should be mixed to produce 50 ml. of a 54% iodine solution?*

Problem 2.34. *A bank teller receives a deposit of 41 bills totaling $530. If the bills are $10 and $20 bills, how many of each denomination did she receive?*

Problem 2.35. *A retailer mixes coffee beans which sell for $1.20 per pound with coffee beans selling for $1.80 per pound to produce a blend that sells for $1.60 per pound. Determine the quantity of each type of coffee needed to produce 36 pounds of the blend.*

Problem 2.36. *How much of an 17% alcohol solution must be added to 40ml. of a 30% alcohol solution to produce a 25% solution?*

Problem 2.37. *An investor put $12,000 in two accounts. The first of these pays 8% annual interest while the other has a return of 10% annually. If her total annual income is $1040, how much did she invest at each rate?*

Problem 2.38. *A painter and his assistant working together finish a job in 4 hours. The painter working alone can finish the job in 6 hours less than the assistant working alone. How long would it take each person to do the job alone?*

Problem 2.39. *It takes Aaron 39 hours longer than it takes Ben to do a job. If Aaron and Ben work together, they can finish the job in 40 hours. How long would it take each man alone to do the job?*

Problem 2.40. *A club trip cost a group $1600. If there had been 8 fewer members, the trip would have cost each member $10 more. How many club members are in the group?*

Problem 2.41. *A trader mixes two kinds of nuts, one worth $1.20 per pound and the other $1.80 per pound. If the trader has 120 pounds of the mixture which is worth $1.40 per pound, how much of each type of nut are there in the mixture?*

Problem 2.42. *The length of a rectangular plot is 25 meters more than its width. If the perimeter of the plot is 330 meters, what are the dimensions of the plot?*

Problem 2.43. *Two opposite sides of a square are each increased by 3 feet while the other two sides are each decreased by 4 feet. If the area of the rectangle is 23 square feet less than the area of the square, find the side of the square and the dimensions of the rectangle.*

End of <u>Chapter 2</u>. Problems

Problem 2.44. *Solve the following linear equations*

(a) $11x + 2 = 6x - 3$. (b) $7x + 8 = 3x - 4(x - 1)$.

(c) $5(x - 3) = 2(x - 4)$. (d) $5 - (2 - x) = 3x - 4(x + 1)$.

(e) $6(x - 4) + 2(x + 5) = 8x - 19$. (f) $3(x - 2) - 2(1 - 3x) = 15$.

Problem 2.45. *Solve the following quadratic equations*

(a) $3x^2 + 4x - 4 = 0$. (b) $4x^3 + 4x^2 - 15x = 0$.

(c) $2x^2 + 2x - 5 = 0$. (d) $2x^2 - 5x + 4 = 0$.

(e) $4x^2 - 1 = 0$. (f) $4x^3 + 2x^2 + 3x = 0$.

Problem 2.46. *Simplify each in the form* $a + bi$

(a) $i^2 - 2i^3 + 3i^{29}$.   (b) $\overline{-2 + 3i}$.

(c) $(3 - 7i) - (5 + 4i)$.   (d) $(3i - 2)(1 + 2i)$.

(e) $\frac{3 - 5i}{4 + 3i}$.   (f) $\frac{1 + 2i}{i} + \frac{2 + 2i}{3 + 4i}$.

Problem 2.47. *Solve and graph the solution set*

(a) $5(x - 4) < 3(x - 4)$.   (b) $2(x - 3) + 4 \geq 2x + 7$.

(c) $3(x - 2) - 5(x - 1) \geq -2x - 1$.   (d) $5(x - 1) - 2(x + 3) < 3(x + 1)$.

(e) $-5 \leq 6 - 2x \leq 7$.   (f) $-5 < 3 - 2x \leq 9$.

Problem 2.48. *Solve and graph the solution set*

(a) $(x + 2)(x - 5) \leq 0$.   (b) $x^2 - 3x - 27 > 0$.

(c) $2x^2 - 7x - 3 \geq 0$.   (d) $\frac{x - 4}{(x + 1)(x - 1)} \leq 0$.

(e) $\frac{x - 3}{x + 1} \geq -2$.   (f) $\frac{x - 2}{x - 1} < \frac{x + 4}{x + 3}$.

Problem 2.49. *Solve. Graph the solution where applicable*

(a) $|2x + 3| = 7$.   (b) $|-2x + 1| = 5$.

(c) $|2x - 3| \leq 5$.   (d) $|3x + 2| < 1$.

(e) $|2 - x| > 3$.   (f) $|-x - 3| \geq -2$.

Problem 2.50. *Find 3 consecutive even integers whose sum is 48.*

Problem 2.51. *Jerry borrows $12,000 at a simple annual interest rate of 8% for 3 years. However, he changes his mind at the end of the first year, returns a portion of the amount he had borrowed, and keeps the rest until the end of the three year period. If he pays a total of $1760, how much did Jerry return at the end of the first year?*

Problem 2.52. *George, working together with his assistant, can edit a particular book in 3 days. Working alone, his assistant takes twice as long as George working alone, to complete the job. How long would it take each person working alone to edit the book?*

Chapter 3.

Visualizing, Constructing and Reading Graphs

Introduction

Visualizing data and information is an important technical skill. Computers cannot yet master the ability to see a graph or picture, globally, and extract information from it. One of the major research projects in artificial intelligence has been to "teach" computers to recognize a person just from looking at their face.

This chapter will emphasize methods that will help develop an understanding of how best to visualize, read and construct graphs.

3.1. Relations and Functions

Mathematics connects with the world at-large through sentences.

Definition

Relation

A mathematical **relation** is an imperative sentence (a **command** or **instruction**) which produces one or more numbers from a given number.

Here are some examples of relations:

- Given a positive number, produce the given number plus one.
- Given a number, produce two times the given number.
  The following is an example of a **special type** of **relation**:
- Given a number greater than or equal to zero, produce the positive number or zero which when multiplied by itself yields the given number.
  - ▶ **Note**: This mathematical sentence is known by the name "square root". And exactly one answer is possible for each given number.

This special type of relation, is called a **function**.

Definition

Function

A mathematical **function** is an imperative sentence which produces **exactly one number** from a given number.

Mathematical notation

When referring to **functions** or **relations** mathematical notation is often used. This makes certain concepts and procedures easier to understand.

Example **3.1.** *Give the mathematical notation for the following sentence: Given a number, produce two times the number. Let S be the sentence, and N represent the number. That is:* S : *Given a number, produce two times the given number.*

- **Solution**: $S(N) = 2 \cdot N$

Example **3.2.** *Give the math notation for the following sentences. Let u represent the given number.*

(a) S : *Given a number, produce the given number plus one.*

(b) T : *Given a number, produce the given number times itself.*

(c) H : *Given a number, produce negative one times the given number.*

- **Solution**: (a) $S(u) = u + 1$ (b) $T(u) = u \cdot u = u^2$ (c) $H(u) = (-1)u = -u$

-
> **Definition**
>
> **Independent Variable**
>
> The symbol used to express the given number
>
> is called the **independent variable**.
>
> **Dependent Variable**
>
> The number(s) produced by the relation or function
>
> is called the **dependent variable(s)**.
>
> For $S(N) = 2 \cdot N$, the **independent variable is** N,
>
> while the **dependent variable is** $S(N)$.

Exercise **3.1.** *Identify the dependent and independent variables in the following*:

(a) $f(x) = x^2$

Solution:

(b) $f(x) = x + 3$

Solution:

(c) $H(u) = -u$

Solution:

Variations in notation of relations

Consider the function, $S(u) = u + 1$. If u, the independent variable is equal to 5, the number produced by S is 6. This can be denoted by the notation:

$$5 \overset{S}{\longmapsto} 5 + 1 = 6.$$

Likewise,

$$0 \overset{S}{\longmapsto} 0 + 1 = 1$$

$$3 \overset{S}{\longmapsto} 3 + 1 = 4$$

$$10 \overset{S}{\longmapsto} 10 + 1 = 11$$

$$-1 \overset{S}{\longmapsto} -1 + 1 = 0$$

independent variable $\longmapsto$ dependent variable

Thus the **function** S can be written two different ways:

·_99_·

$$\boxed{\begin{array}{l} S(u) = u + 1 \\ \hline \text{or} \\ \hline u \overset{S}{\longmapsto} u + 1. \end{array}}$$

A third notation (pairs of numbers).

The first number of the pair represents the "input" to the **function** (the **independent variable**) and the second number represents the "output' (or dependent variable) from the function. This is called **ordered pair** notation. Thus:

$0 \overset{S}{\longmapsto} 1$ can be written as $(0,1)$

$3 \overset{S}{\longmapsto} 4$ can be written as $(3,4)$

$5 \overset{S}{\longmapsto} 6$ can be written as $(5,6)$

$-1 \overset{S}{\longmapsto} 0$ can be written as $(-1,0)$

Example **3.3.** *For the function* $R(u) = u^2$, *given* $u = -1/2, 3, 4/3, 6$ *and* -5, *find the dependent variable. Represent the information in the formula notation, the arrow notation and the ordered pair notation.*

- **Solution:**

Formula	Arrow	Pair
$R\left(-\frac{1}{2}\right) = \left(-\frac{1}{2}\right)\left(-\frac{1}{2}\right) = \frac{1}{4}$	$-\frac{1}{2} \overset{R}{\longmapsto} \frac{1}{4}$	$\left(-\frac{1}{2}, \frac{1}{4}\right)$
$R(3) = (3)(3) = 9$	$3 \overset{R}{\longmapsto} 9$	$(3,9)$
$R\left(\frac{4}{3}\right) = \left(\frac{4}{3}\right)\left(\frac{4}{3}\right) = \frac{16}{9}$	$\frac{4}{3} \overset{R}{\longmapsto} \frac{16}{9}$	$\left(\frac{4}{3}, \frac{16}{9}\right)$
$R(6) = (6)(6) = 36$	$6 \overset{R}{\longmapsto} 36$	$(6,36)$
$R(-5) = (-5)(-5) = 25$	$-5 \overset{R}{\longmapsto} 25$	$(-5,25)$
$R(u) = (u)(u) = u^2$	$u \overset{R}{\longmapsto} u^2$	(u,u^2)

Exercise **3.2.** *Let the following sentence define the function* A. *A : Given a number, produce the given number minus three with that whole quantity divided by the given number plus four.*

(*a*) *If* x *is the independent variable name, then give the formula notation of the function.*

Solution:

(b) For the values of $-1/4, 0, 3, 7, 9,$ and -5 give the produced number; represent the information in the formula notation, the arrow notation and the ordered pair notation by completing the chart below:

Formula	Arrow	Pair
$A\left(-\frac{1}{4}\right) = $ _____	$-\frac{1}{4} \overset{A}{\mapsto}$ _____	$\left(-\frac{1}{4}, \underline{\quad}\right)$
$A(0) = $ _____	$0 \overset{A}{\mapsto}$ _____	$(0, \underline{\quad})$
$A(3) = $ _____	$3 \overset{A}{\mapsto}$ _____	$(3, \underline{\quad})$
$A(7) = $ _____	$7 \overset{A}{\mapsto}$ _____	$(7, \underline{\quad})$
$A(-5) = $ _____	$-5 \overset{A}{\mapsto}$ _____	$(-5, \underline{\quad})$
$A(9) = $ _____	$9 \overset{A}{\mapsto}$ _____	$(9, \underline{\quad})$
$A(x) = $ _____	$x \overset{A}{\mapsto}$ _____	$(x, \underline{\quad})$

Exercise **3.3.** *Write the imperative sentences which are determined by the following formulas. Include the name of the sentence with the grammatically correct statement.*

(a) $S(t) = t^2 + 1$

Solution:

(b) $H(x) = 2x + 3$

Solution:

(c) $F(u) = (-1)x + x^2$

Solution:

Exercise **3.4.** *Let the following sentence define the function* A. A : *Given a number, produce the given number provided the given number is zero or positive; produce negative one times the given number if the given number is negative. Denote the independent variable by the name* b. *For the values of* $-1, 6, -6, 2, -4, 0, \frac{1}{2}$ *and* $-\frac{1}{3}$ *give the produced number. Represent the information in the the chart below;*

Formula	Arrow	Pair
$A(\underline{\quad}) = $ _____	_____ $\overset{A}{\mapsto}$ _____	$(\underline{\quad}, \underline{\quad})$
$A(\underline{\quad}) = $ _____	_____ $\overset{A}{\mapsto}$ _____	$(\underline{\quad}, \underline{\quad})$
$A(\underline{\quad}) = $ _____	_____ $\overset{A}{\mapsto}$ _____	$(\underline{\quad}, \underline{\quad})$
$A(\underline{\quad}) = $ _____	_____ $\overset{A}{\mapsto}$ _____	$(\underline{\quad}, \underline{\quad})$
$A(\underline{\quad}) = $ _____	_____ $\overset{A}{\mapsto}$ _____	$(\underline{\quad}, \underline{\quad})$
$A(\underline{\quad}) = $ _____	_____ $\overset{A}{\mapsto}$ _____	$(\underline{\quad}, \underline{\quad})$
$A(\underline{\quad}) = $ _____	_____ $\overset{A}{\mapsto}$ _____	$(\underline{\quad}, \underline{\quad})$
$A(\underline{\quad}) = $ _____	_____ $\overset{A}{\mapsto}$ _____	$(\underline{\quad}, \underline{\quad})$

Exercise **3.5**. Let R be the relation defined by the following sentence: R : Given a number, produce a number so that the given number times itself added to the produced number times itself yields four. Let x be the independent variable. Use the values of $2, 0, -2,$ and $\sqrt{3}$ for the values of the independent variable. Complete the a chart which includes the formula notation, the arrow notation, and the ordered pair notation. | **Note**: | In some cases two outputs may result from one input; you must choose one.

Solution:

Summary 3.1.

- Relations and functions are imperative sentences which take an "input" and produce an "output".
- A **function** produces exactly one output for a given number.
- A **relation** is allowed to produce more than one answer for each given number.
- There are **three common methods of representing functions and/or relation**s:

$S(u) = u + 1$
$u \xrightarrow{S} u + 1$
$(u, u + 1)$

End of Section 3.1. **Problems**

Problem 3.1. Using the English sentence T, complete the chart with the values that are produced by the sentence. Some entries are provided for reference. The variable x is the independent variable. T : Given a number, produce the given number divided by the result of the given number minus 2.

$\underline{\quad} \xrightarrow{T} \underline{\quad}$	$(\underline{\quad}, \underline{\quad})$	$T(\ \ ) =$
$1 \xrightarrow{T} \frac{1}{-1}$	$(1, -1)$	$T(1) = -1$
$0 \xrightarrow{T} \frac{0}{-2}$	$(0, \underline{\quad})$	$T(0) = 0$
$-1 \xrightarrow{T} \frac{-1}{-3}$	$(-1, \underline{\quad})$	$T(-1) =$
$-2 \xrightarrow{T} \underline{\quad}$	$(-2, \underline{\quad})$	$T(-2) =$
$\frac{1}{4} \xrightarrow{T} \frac{\frac{1}{4}}{\frac{1}{4} - 2}$	$(\frac{1}{4}, \underline{\quad})$	$T(\frac{1}{4}) =$
$x \xrightarrow{T}$	$(x, \underline{\quad})$	$T(x) =$

Problem 3.2. Using the English sentence G, complete the chart with the values that are produced by the sentence. Some entries are provided for reference. The variable x is the independent variable. G :Given a number, produce 17.

$\underline{\quad} \xrightarrow{G} \underline{\quad}$	$(\underline{\quad},\underline{\quad})$	$G(\ ) =$
$1 \xrightarrow{G} 17$	$(1,17)$	$G(1) = 17$
$0 \xrightarrow{G} 17$	$(0,\underline{\quad})$	$G(0) =$
$-1 \xrightarrow{G}$	$(-1,\underline{\quad})$	$G(-1) =$
$-2 \xrightarrow{G}$	$(-2,\underline{\quad})$	$G(-2) =$
$\frac{1}{4} \xrightarrow{G}$	$\left(\frac{1}{4},\underline{\quad}\right)$	$G\left(\frac{1}{4}\right) =$
$x \xrightarrow{G}$	$(x,\underline{\quad})$	$G(x) =$

Problem 3.3. Using the English sentence Pro, complete the chart with the values that are produced by the sentence. Some entries are provided for reference. The variable x is the independent variable. Pro: Given a number, produce the given number times itself times negative three minus the given number times two.

$\underline{\quad} \xrightarrow{Pro} \underline{\quad}$	$(\underline{\quad},\underline{\quad})$	$Pro(\ ) =$
$1 \xrightarrow{Pro} -5$	$(1,-5)$	$Pro(1) = -5$
$0 \xrightarrow{Pro} 0$	$(0,\underline{\quad})$	$Pro(0) =$
$-1 \xrightarrow{Pro}$	$(-1,\underline{\quad})$	$Pro(-1) =$
$-2 \xrightarrow{Pro}$	$(-2,\underline{\quad})$	$Pro(-2) =$
$\frac{1}{4} \xrightarrow{Pro}$	$\left(\frac{1}{4},\underline{\quad}\right)$	$Pro\left(\frac{1}{4}\right) =$
$x \xrightarrow{Pro}$	$(x,\underline{\quad})$	$Pro(x) =$

Problem 3.4. Using the English sentence ABS, complete the chart with the values that are produced by the sentence. Some entries are provided for reference. The variable x is the independent variable. ABS: Given a number, produce the given number provided the given number is zero or positive; otherwise, produce negative one times the given number.

$\underline{\quad} \xrightarrow{ABS} \underline{\quad}$	$(\underline{\quad},\underline{\quad})$	$ABS(\ ) =$
$1 \xrightarrow{ABS} 1$	$(1,1)$	$ABS(1) = 1$
$0 \xrightarrow{ABS} 0$	$(0,\underline{\quad})$	$ABS(0) =$
$-1 \xrightarrow{ABS} 1$	$(-1,1)$	$ABS(-1) = 1$
$-2 \xrightarrow{ABS} \underline{\quad}$	$(-2,\underline{\quad})$	$ABS(-2) =$
$\frac{1}{4} \xrightarrow{ABS} \underline{\quad}$	$\left(\frac{1}{4},\underline{\quad}\right)$	$ABS\left(\frac{1}{4}\right) =$
$x\ (x \geq 0) \xrightarrow{ABS} \underline{\quad}$	$(x,\underline{\quad})$	$ABS(x) =$
$x\ (x \leq 0) \xrightarrow{ABS} \underline{\quad}$	$(x,\underline{\quad})$	$ABS(x) =$

Problem 3.5. Using the English sentence R, complete the chart with the values that are produced by the sentence. Some entries are provided for reference. The variable x is the independent variable. R :Given a number, produce any number. NOTE: R is **not** a function.

$\underline{\ \ } \overset{R}{\to} \underline{\ \ }$	$(\underline{\ },\underline{\ })$	$R(\ ) =$
$1 \overset{R}{\to} 17.4$	$(1, 17.4)$	$R(1) = 17.4$
$0 \overset{R}{\to} -3$	$(0, -3)$	$R(0) = -3$
$-1 \overset{R}{\to}$	$(-1, \underline{\ })$	$R(-1) =$
$-2 \overset{R}{\to}$	$(-2, \underline{\ })$	$R(-2) =$
$\frac{1}{4} \overset{R}{\to}$	$(\frac{1}{4}, \underline{\ })$	$R(\frac{1}{4}) =$
$x \overset{R}{\to}$ *any value*	$(x,\ \textit{any value})$	$R(x) = $ *any value*

Problem 3.6. Using the English sentence C, complete the chart with the values that are produced by the sentence. Some entries are provided for reference. The variable x is the independent variable. C: Given a number, produce a number so that the number produced times itself times itself again yields the given number.

$\underline{\ \ } \overset{C}{\to} \underline{\ \ }$	$(\underline{\ },\underline{\ })$	$C(\ ) =$
$8 \overset{C}{\to} 2$	$(8, 2)$	$C(8) = 2$
$1 \overset{C}{\to} 1$	$(1, \underline{\ })$	$C(1) =$
$\frac{-1}{64} \overset{C}{\to}$	$(\frac{-1}{64}, \underline{\ })$	$C(\frac{-1}{64}) =$
$3 \overset{C}{\to}$	$(3, \underline{\ })$	$C(3) =$
$x \overset{C}{\to}$	$(x,\ )$	$C(x) =$

Problems 3.7.- 3.12. Write the English imperative sentence which produces the following chart:

3.7.	3.8.	3.9.	3.10	3.11	3.12.
$0 \overset{S}{\to} 2$	T	$0 \overset{A}{\to} 0$	$f(-4) = 5$	V	$T(6) = \frac{2}{5}$
$1 \overset{S}{\to} 3$	$(-1, 1)$	$-1 \overset{A}{\to} 1$	$f(-3) = 4$	$(0, 3)$	$T(4) = \frac{2}{3}$
$2 \overset{S}{\to} 4$	$(-\frac{1}{2}, \frac{1}{4})$	$-2 \overset{A}{\to} 2$	$f(-2) = 3$	$(2, 4)$	$T(0) = -2$
$3 \overset{S}{\to} 5$	$(-5, 25)$	$\frac{-1}{3} \overset{A}{\to} \frac{1}{3}$	$f(0) = 1$	$(6, 6)$	$T(-1) = -1$
$\frac{1}{2} \overset{S}{\to} \frac{5}{2}$	$(\frac{1}{10}, \frac{1}{100})$	$1 \overset{A}{\to} 1$	$f(\frac{1}{2}) = \frac{3}{2}$	$(-1, \frac{5}{2})$	$T(\frac{1}{2}) = -4$
$\frac{3}{4} \overset{S}{\to} \frac{11}{4}$	$(3, 9)$	$2 \overset{A}{\to} 2$	$f(7) = 8$	$(-2, 2)$	$T(-\frac{1}{3}) = -\frac{3}{2}$

$\frac{2}{x-1}$

$\frac{\frac{2}{3}}{-\frac{1}{3}-1} \cdot \frac{\frac{2}{3}}{\frac{-4}{3}} = \frac{2}{-4} \cdot \frac{3}{-4}$

Problem 3.13. Write an imperative English sentence, begin the sentence with "Given a number, produce..." for:

(a) $T(w) = 2w + 1$ (b) $Sub(x) = 3 - x$ (c) $C(u) = 0u + 10$ (d) $(u, 2u + 1)$

(e) $(w, 2w^2)$ (f) $u \xrightarrow{M} 4u$ (g) $L(u) = 10$ (h) $R(t) = -2t(t) + 2$

(i) $(w, (-1)w)$ (j) $(x, \frac{3}{x})$ (k) $v \xrightarrow{C} v^2 - 1$

Problem 3.14. Given an amount of money, produce the interest that the given amount can earn in one year by multiplying the given amount times the interest rate of $.05$. Let p denote the independent variable and let I be the independent variable. For this function, complete the following chart:

$p \xrightarrow{I} I$	(p, I)	$I(p) =$
$1000 \xrightarrow{I} \underline{\quad}$	$(1000, \underline{\quad})$	$I(1000) = \underline{\quad}$
$500 \xrightarrow{I} \underline{\quad}$	$(500, \underline{\quad})$	$I(500) = \underline{\quad}$
$100 \xrightarrow{I} \underline{\quad}$	$(100, \underline{\quad})$	$I(100) = \underline{\quad}$
$25 \xrightarrow{I} \underline{\quad}$	$(25, \underline{\quad})$	$I(25) = \underline{\quad}$
$0 \xrightarrow{I} \underline{\quad}$	$(0, \underline{\quad})$	$I(0) = \underline{\quad}$

3.2. Reading a Graph

Using the **ordered pair notation** allows us to form a picture of the **function**. The ordered pair notation of a function can correspond to **horizontal** and **vertical points** in a **grid**.

> **Definition**
>
> **Coordinate plane:**
>
> The **coordinate plane** is defined to be all possible ordered pairs (a, b), where a and b can be any real number.

The **coordinate plane** can be drawn in the following manner:

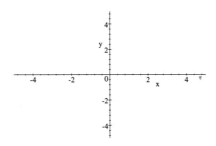

- The **horizontal axis** represents the **independent variable** and is called the **independent axis** (or **abscissa axis**),
- The **vertical axis** represents the **dependent variable** and is called the **dependent axis** (or **ordinate** axis).
- The intersection of the two axes is called the **origin**, denoted by the ordered pair $(0, 0)$.
- A point is denoted by an ordered pair (a, b);
    - ▶ a, denotes a position to the right or left of the origin;
    - ▶ b, denotes a position above or below the origin.

Example **3.4.** *Plot the point* $(2, 3)$.

- **Solution:** 2 is the **independent variable** and therefore determines a distance to the right (since $2 > 0$) of the origin. 3, the dependent variable, determines the distance above (since $3 > 0$) the origin.

    Thus (2,3) is the position in the coordinate plane shown below:

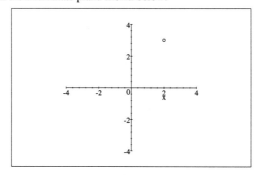

Exercise **3.6**. *Plot the point* $(-4,-2)$.

> **Solution:**

The axes divide the plane into four regions called **quadrants**. The quadrants each have ordered pairs of a distinct form:

Quadrant	Type of ordered pair
I	(+,+)
II	(-,+)
III	(-,-)
IV	(+,-)

The quadrants are numbered in a $\boxed{\text{counter-clockwise}}$ order starting from the top right, as shown in the next example.

Example **3.5**. *Plot the points that are determined by the relation* S *with independent variable* u. *The relation* S *in ordered pair notation consists of exactly the pairs* $(1,3),(2,7),(\frac{3}{4},5),(-1,2),(-\frac{3}{2},0),(\frac{1}{2},-2),(-5,-4),(0,3)$.

• **Solution:**

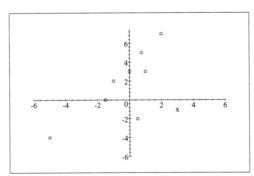

$\boxed{\text{Note:}}$ The point on the **vertical axis** can be labeled $(0,3)$.

$\boxed{\text{Note:}}$ The point on the **horizontal axis** can be labeled $(-\frac{3}{2},0)$.

Remember, for the relation S there are different notations to represent the same information. The following table shows the equivalent notations:

Pair notation	Relation notation	Arrow notation
$(1,3)$	$S(1) = 3$	$1 \xmapsto{S} 3$
$(-\frac{3}{2},0)$	$S(-\frac{3}{2}) = 0$	$-\frac{3}{2} \xmapsto{S} 0$
$(0,3)$	$S(0) = 3$	$0 \xmapsto{S} 3$
$(-5,-4)$	$S(-5) = -4$	$-5 \xmapsto{S} -4$

> **Definition**
>
> **Solve**:
>
> The mathematical term **solve** replaces the phrase:
> "Which value of the independent variable
> forces the statement to be true?"
> For example, "Which values of the independent variable
> force the relation to produce the number zero?" becomes:
> Solve $R(u) = 0$.

Example **3.6**. The points defining the relation $G(x)$, are:

$$(1,3), (2,7), (\tfrac{3}{4},5), (-1,2), (-\tfrac{3}{2},0), (\tfrac{1}{2},-2), (-5,-4), (0,3).$$

The graph is:

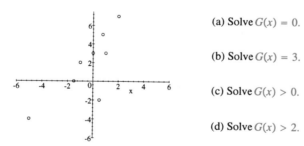

(a) Solve $G(x) = 0$.

(b) Solve $G(x) = 3$.

(c) Solve $G(x) > 0$.

(d) Solve $G(x) > 2$.

- **Solution**:
  (a) $x = -\frac{3}{2}$.
  (b) $x = 1$ or $x = 0$.
  (c) x values of $1, 2, \frac{3}{4}, -1, \frac{1}{2}, 0$.
  (d) x values of $1, 2, \frac{3}{4}$ and 0.

Definition

Let S be a **relation**, let m be a value of the **independent** variable and let h be the number produced by S for the value of m.

Thus, equivalent notations are:

$m \overset{S}{\mapsto} h$, $S(m) = h$, and (m, h).

We will refer to the **second (dependent) value**, h, as the **height** of the relation S produced by the value m.

Note: The concept of height can be positive, negative, or zero. For example, altitude can be above sea level (positive), sea level, or below sea level (negative).

Exercise **3.7**. Let $S(u) = u - 1$. *For independent variable* $= -\frac{1}{2},\ 0,\ 4,\ and\ 6$*, calculate the heights the function S produces?*

Solution:

The use of the word height seems natural in this setting, because it refers to the second value of the ordered pair which, when plotted on a graph, determines the "height" above or below the horizontal axis.

Example **3.7**. *Consider the graph of the following ordered pairs produced by a relation* $R(u)$: $(-5, 4), (0, 2), (2, 3), (1, 0),\ (5, 1),\ (3, -1),\ (-1, -2),\ (4, -3),\ (7, -2),\ (-3, -5)$.

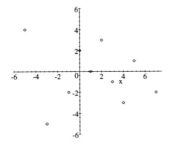

(a) *Which value of u forces R to produce* 1?

- **Solution**: This question is asking which value of the independent variable causes the dependent variable to be 1. This is a question about the location of the point, since it is asking which points of the relation R are at a height of one unit above the origin. We note that $(5, 1)$ is a point on the graph of the relation. So, in arrow notation, $5 \overset{R}{\mapsto} 1$. This means that $u = 5$ is the correct value of the independent variable required to produce 1 for a dependent variable.

(b) *Which value of u forces R to produce* 0?

- **Solution**: The only point of the relation which has a second coordinate of 0 is the point $(1, 0)$. Therefore $1 \overset{R}{\mapsto} 0$ (Read "The relation R takes 1 to 0"). So the correct answer is $u = 1$.

(c) *Which value of u forces R to produce* 3?

- **Solution**: The relevant point is $(2,3)$. The answer is $u = 2$.

(d) *Which value of u forces R to produce* 2? *That is, solve* $R(u) = 2$.

- **Solution**: We need to consider the point (0,2). That is, $0 \overset{R}{\mapsto} 2$. Therefore the answer is $u = 0$.

(e) *Which value of u forces R to produce positive numbers? That is, solve* $R(u) > 0$.

- **Solution**: For this question, the points to consider are all of those points which have a positive (greater than zero) second coordinate. In this example, these are the points (-5,4), (0,2), (2,3), and (5,1). Thus the answer is $u = -5$, $u = 0$, $u = 2$, or $u = 5$.

(f) *Solve* $R(u) = 4$.

- **Solution**: Has a solution of $u = -5$ since the point (-5,4) is part of the relation R.

(g) *Solve* $R(u) = -2$.

- **Solution**: The answer to this question provides the proof that R is a relation and not a function. To solve $R(u) = -2$, we must consider the points (-1,-2) and (7,-2). These points imply that $R(-1) = -2$ and $R(7) = -2$. Thus, the correct values of u which yield -2 as a relation value are $u = -1$ and $u = 7$.

(h) *Solve* $R(u) < 0$. $R(u) = 4$.

- **Solution**: In this case we are looking for values of the relation which are negative. The ordered pairs are $(-3,-5)$, $(-1,-2)$, $(3,-1)$, $(4,-3)$, and $(7,-2)$. Therefore the answer to the question are the values of $u = -3$, $u = -1$, $u = 3$, $u = 4$, and $u = 7$.

(i) *Solve* $R(u) > 1$.

- **Solution** This question allows us to look at a solution in relation to a number other than 0. The required ordered pairs are those with a second coordinate which is strictly greater than 1: (-5,4), (0,2), and (2,3). Thus the values of -5, 0, and 2 produce values of R which are greater than 1. The relevant point is (2,3). So the correct answer is $u = -5, 0$ and 2.

Exercise **3.8**. *Consider the graph of the following ordered pairs produced by a relation*
$R(u)$: $(-4,3), (0,-2), (2,-3), (-1,0), (-3,3), (1,1), (3,7), (3,-5)$.

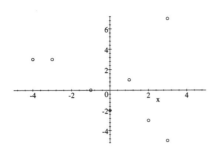

Answer the following questions:

(a) Which value of u forces R to produce 1?

Solution:

(b) Which value of u forces R to produce 0?

Solution:

(c) Which value of u forces R to produce 3?

Solution:

(d) Which value of u forces R to produce 7? *That is, solve* $R(u) = 7$.

Solution:

(e) Which value of u forces R to produce positive numbers? i.e. solve $R(u) > 0$.

Solution:

(f) Solve $R(u) = -5$.

Solution:

(g) Solve $R(u) = -2$.

Solution:

(h) Solve $R(u) < 0$.

Solution:

(i) Solve $R(u) > 1$.

Solution:

Summary 3.2.

- A **graph** allows information to be recorded in a form that is more visual —and therefore more readily analyzed.
- The **independent** variable is recorded on the horizontal axis.
- The **dependent** variable is recorded on the vertical axis.
- We can use the graph to help us quickly answer questions about which independent values produce a given answer.
  - ▶ This often involves the analysis of points in terms of height above or below the horizontal axis.

End of Section 3.2. Problems

Problem 3.15. *On the grid provided, plot the following points:*

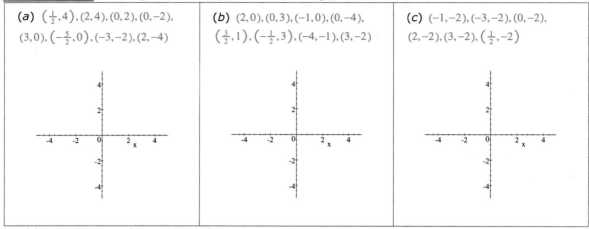

(a) $\left(\frac{1}{2},4\right),(2,4),(0,2),(0,-2),$
$(3,0),\left(-\frac{5}{2},0\right),(-3,-2),(2,-4)$

(b) $(2,0),(0,3),(-1,0),(0,-4),$
$\left(\frac{3}{2},1\right),\left(-\frac{1}{2},3\right),(-4,-1),(3,-2)$

(c) $(-1,-2),(-3,-2),(0,-2),$
$(2,-2),(3,-2),\left(\frac{1}{2},-2\right)$

Problem 3.16. *For the two graphs below answer each of the following questions:*

   i. Find the value(s) of u for all points positioned at $(u,0)$.

   ii. Find the value(s) of u for all points positioned at $(u,2)$.

  iii. Find the value(s) of u for all points positioned at (u,w), where w is strictly a negative height.

  iv. Find the value(s) of the height b, so that the point $(-1,b)$ is a position on the graph.

   v. Find the value(s) of the height b, so that the point $(0,b)$ is a position on the vertical axis.

  vi. Which values of u have the property that two or more heights b are plotted using that particular value of u?

  vii. Find all the values of u so that $(u,-2)$ is a point on the graph.

 viii. Find all the values of u so that $(u,4)$ is a point on the graph.

  ix. Find all the values of u so that (u,b) is a point on the graph and $b > 0$.

   x. Find all the values of u so that (u,b) is a point on the graph.

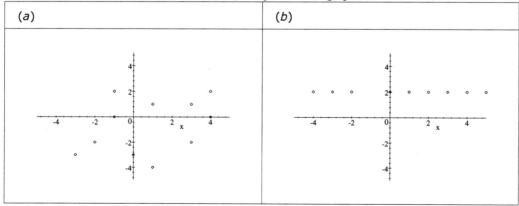

(a)

(b)

Problem 3.17. *Identify the independent variable for the notation:*

  (a) $T(u)$ **(b)** $F(z)$ **(c)** $(k,\_\_\_)$ **(d)** $b \xrightarrow{S} \_\_\_\_$

Problem 3.18. *Write the name of the independent variable for the following sentence:*

  (a) $(u,2u+1)$ **(b)** $f(x) = x^2 + 1$ **(c)** $\left(x,\frac{2x-1}{x-2}\right)$ **(d)** $g(w) = \frac{w-1}{w+1}$

Problem 3.19. For the next four problems, choose an independent variable for each equation; then solve the equation for the dependent variable.

(a) $w + u = 3$ (b) $2p + q - 4 = 0$ (c) $\sqrt{u} + q - 4 = 0$ (d) $t + 3p = 7$

Problem 3.20. The relation R is defined by the list of points $(-1,1), (-2,-6), \left(\frac{1}{2}, \frac{17}{8}\right), (0,2), \left(-\sqrt[3]{2}, 0\right)$. Answer the following questions for the relation R :. The graph of the relation is given by

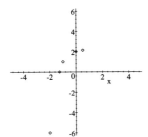

(a) Solve $R(u) = 0$.

(b) Solve $R(u) = -6$.

(c) Solve $R(u) > 0$.

(d) Solve $R(u) \leq 0$.

(e) Solve $R(u) = -2$.

Problem 3.21. For the three graphs below, answer the following questions.

i. Solve $T(x) = 0$.   vi. Solve $T(x) \leq 1$.

ii. Solve $T(x) > 0$.   vii. Solve $T(x) > 2$.

iii. Solve $T(x) < 0$.   viii. Solve $T(x) = \frac{1}{2}$.

iv. Solve $T(x) \leq 0$.   ix. Solve $T(x) = -4$.

v. Solve $T(x) \geq 0$.   x. Solve $-2 < T(x) < 3$

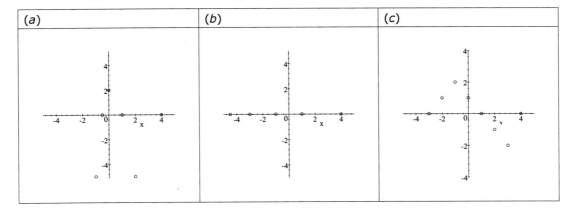

3.3. Relations, Functions and Graphs

A **function** on the real numbers can take on values at any point which will generate a **dependent value**. To obtain a good idea of what a **function** is doing, one will need to plot a very large number of ordered pairs on the grid, also known as the **coordinate plane** or **Cartesian plane**. For example, the hourly price of a stock, charted over a seven year period is given below.

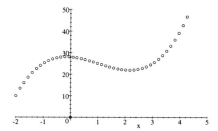

The vertical axis records dollars while, the horizontal axis records time.

Additionally, we denote time 0 as a particular historical date. Negative numbers on the horizontal axis represent the number of years **before** that set date, while positive numbers represent years **after** that date. To plot the graph for just one year requires plotting 6,248 ordered pairs. After plotting sufficient numbers of points, the graph begins to resemble a "continuous" curve.

- If you are to analyze the characteristics of a particular mathematical expression, then each ordered pair is determined by choosing a value of the **independent variable** and then determining the **height** (**dependent variable**) which it produces.

Consider the following example of the graph of a curve:

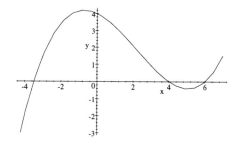

- Any point on the curve can be described by an ordered pair.
  - ▶ A point which is directly above the independent value $x = -2$, is positioned at the ordered pair $(-2, 3.5)$.
  - ▶ We describe an ordered pair as (a, b).
  - ▶ The **independent variable**, a, which corresponds to a distance left or right from the axes crossing point.
  - ▶ The value b (**dependent variable**) is the height which is produced by the function from the independent value a.

The easiest ordered pairs to locate on a graph, and often of the most interest to a scientist or economist, are those points which have a height of exactly zero. In terms of our pair notation, these are the points of the form $(a, 0)$.

> **Definition**
>
> **Roots** or **zeros**:
>
> For a function $f(x)$, the points for which $f(x) = 0$, are known as the **roots** or **zeros** of the function.

Zeros on the graph of a function.

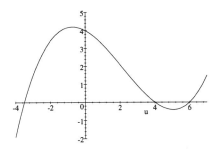

The points resting exactly on the independent axis are $\left(-3\frac{1}{2}, 0\right)$, $(4,0)$, and $(6,0)$. Thus for the function $t(u)$ we know the following information:

$t\left(-\frac{7}{2}\right) = 0,$	$t(4) = 0,$	$t(6) = 0;$
or $-\frac{7}{2} \overset{t}{\mapsto} 0,$	$-4 \overset{t}{\mapsto} 0,$	$-6 \overset{t}{\mapsto} 0.$

It is important to remember that we are estimating the values of the independent variable from the graph; instead of -4, the value might truly be -3.95. One of the goals of mathematics is to exactly pinpoint the independent values which produce **zero**.

Example **3.8**. *Consider this graph*:

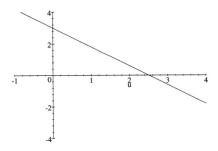

(a) *Which value of u forces the height of the graph to be zero?*

- **Solution**: There is one point which is neither above nor below the axis–and the coordinates of that point appear to be $\left(\frac{5}{2}, 0\right)$. Therefore the value of u which produces a height of zero is $u = \frac{5}{2}$.

(b) *How many values of u produce height zero?*

- **Solution**: There is only one position where the graph of the function has height zero.

(c) *What is an estimate of the value $f(1)$.*

- **Solution**: The value $f(1)$ is actually another way of requesting information about the point on the graph $(1, ?)$; that is, what is the second coordinate (height) for the point on the graph above the value $u = 1$? The value $u = 1$ produces a height $\frac{9}{5}$.

(d) *What is the height* $f(0)$*?*

- **Solution**: We need to realize that we are looking for the point $(0, f(0))$. So we estimate that $f(0) = 3$.

(e) *What is an estimate of the height* $f(4)$*?*

- **Solution**: Again, locate the point $(4, f(4))$ on the graph. In this case, the point on the graph is directly below the value $u = 4$. We estimate the height of this point to be -2. That is, we determine that $4 \overset{f}{\mapsto} -2$, or $f(4) = -2$, or $(4, -2)$ is a point on the graph of f.

(f) *Solve* $f(u) > 0$*.*

- **Solution**: Written as a sentence, this is requesting which value(s) of u force the height, $f(u)$, to be strictly positive (> 0). Scan the graph from left to right, listing the one or more points that correspond to where the function crosses the independent variable's axis. There is only one point, $\left(\frac{5}{2}, 0\right)$ Any value of u which is located to the left of $\frac{5}{2}$ on the u-axis produces a point on the graph which lies above the u-axis. So our answer is $u < \frac{5}{2}$, or u is in the interval $\left(-\infty, \frac{5}{2}\right)$.

  | **Note:** | The value of $f\left(\frac{5}{2}\right)$ is exactly 0 and therefore neither positive or negative, and therefore not part of the answer.

(g) *Solve* $f(x) < 0$*.*

- **Solution**: We see that the function must have heights which are negative and therefore the plotted point lies below the u-axis. Once again the point $\left(\frac{5}{2}, 0\right)$ provides a key position. For those values of u which lie to the right of $\frac{5}{2}$, $f(u)$ will be negative.

  Thus the answer to this question is $\frac{5}{2} < u$, or u is in the interval$(5, \infty)$.

Exercise 3.9. *Consider the graph of a second function,* $f(z)$:

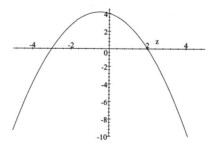

Answer the following questions:

(a) *What is the height* $f(0)$?

Solution:

(b) *Approximate* $f(1)$.

Solution:

(c) *Approximate* $f(3)$.

Solution:

(d) *Solve* $f(z) = 0$.

Solution:

(e) *Solve* $f(z) > 0$.

Solution:

(f) *Solve* $f(z) < 0$.

Solution:

Exercise **3.10. Consider the graph** $g(w)$.

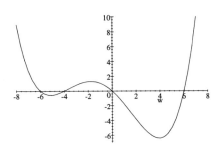

(a) *Solve* $g(w) = 0$.

Solution:

(b) *Solve* $g(w) > 0$.

Solution:

(c) *Solve* $g(w) < 0$.

Solution:

(d) *Solve* $g(w) \geq 0$.

Solution:

(e) *Solve* $g(w) \leq 0$.

Solution:

Exercise 3.11. **For the following exercises consider the graph of** $Q(x)$.

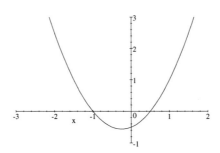

(a) Solve $Q(x) = 0$.

Solution:

(b) Solve $Q(x) > 0$.

Solution:

(c) Solve $Q(x) < 0$.

Solution:

(d) Solve $Q(x) \geq 0$.

Solution:

(e) Solve $Q(x) \leq 0$.

Solution:

(f) Approximate $Q(1)$.

Solution:

(g) Approximate $Q(-1)$.

Solution:

(h) Approximate $Q(-2)$.

Solution:

(i) Approximate $Q(2)$.

Solution:

Exercise 3.12. **For the following exercises consider the graph of** $h(x)$.

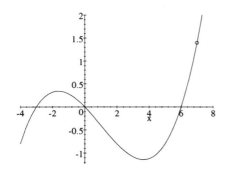

Note: The circle denotes the point $\left(7, \frac{7}{5}\right)$, and means that point is deleted from the graph in this instance. Therefore, all of your answers when written as intervals must exclude the value of 7.

Answer the following questions:

(a) *Solve* $h(x) = 0$.

Solution:

(b) *Solve* $h(x) > 0$.

Solution:

(c) *Solve* $h(x) < 0$.

Solution:

(d) *Solve* $h(x) \geq 0$.

Solution:

(e) *Solve* $h(x) \leq 0$.

Solution:

(f) *Which independent value has no definition in the graph?.*

Solution:

(g) *Which independent values are used in the graph?*

Solution:

(h) *Approximate* $h(1)$.

Solution:

(i) *Approximate* $h(-1)$.

Solution:

(j) *Approximate* $h(-2)$.

Solution:

(k) *Approximate* $h(5)$.

Solution:

(l) *Approximate* $h(0)$.

Solution:

Summary 3.3.

- The graph of a function is obtained by selecting many values for the **independent variable** and then plotting the corresponding **ordered pairs** generated by the function.
- One of the most often asked questions is where the function crosses the axis of the independent variable–that is where it has **height zero**.
- It is important to remember that the **value** of the function refers to the **height** of the graph of the function above or below the independent axis.

End of Section 3.3. *Problems*

Problem 3.22. *Identify the independent and the dependent variable for the mathematical expressions*:

(a) $w(u) = 2u^{-2} + u + 1$ (b) $y = \frac{z+1}{3z-1}$ (c) $x = 6y^2 + 1$

(d) $u \xrightarrow{S} 2u + 1$ (e) $v \xrightarrow{T} \frac{v-1}{\sqrt{v}}$ (f) $t \xrightarrow{L} -\frac{1}{2}t^2 + 10t$

Problem 3.23. Identify the independent and the dependent variable for the mathematical expressions whose graph is shown:

(*a*) $j(x)$ (*b*) $w(u)$ (*c*) $P(x) = 7$

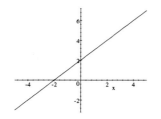 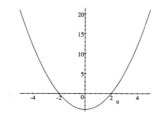 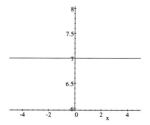

Problem 3.24. The input into an English sentence is the independent variable. The product of the sentence is noted by the dependent variable. For the following processes, determine the independent and the dependent variables.

(*a*) Given the temperature in degrees centigrade C, produce the temperature in degrees Fahrenheit, F.

(*b*) Given the speed in miles per hour v, produce the speed in feet per second, w.

(*c*) Given a length in inches i, produce the length in centimeters, c.

Problem 3.25. Find the roots of the functions, given the following graphs:

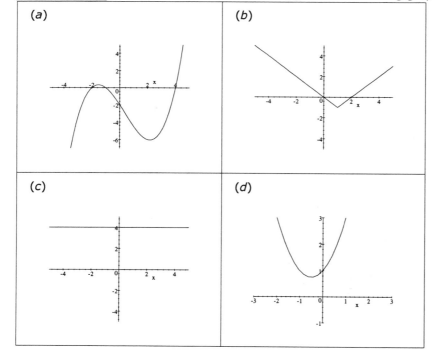

Problem 3.26. Using the graph of $f(x)$ below, answer the following questions:

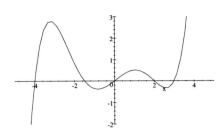

(a) What is the height $f(0)$?

(b) What is the height $f(3)$?

(c) What is the height $f(1)$?

(d) Approximate the height $f\left(\frac{5}{2}\right)$?

(e) Find all the values of x which force $f(x) = 0$.

(f) Find all the values of x which force $f(x) > 0$.

(g) Find all the values of x which force $f(x) \geq 0$.

(h) Find all the values of x which force $f(x) < 0$.

(i) Find all the values of x which force $f(x) \leq 0$.

(j) Approximate $f(-2)$.

(k) Is $f\left(\frac{1}{2}\right) < f(-3)$?

(l) Is $f(-3) < f(6)$?

Problem 3.27. Use the graph of $y = f(x) = -(x-2)^2 - 1$, provided below, to answer the following questions

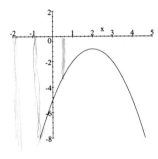

(a) What is the height $f(0)$? (b) What is the height $f(3)$?
(c) Find all values x which force $f(x) > 0$.
(d) Find all values x which force $f(x) \geq 0$.
(e) Find all values x which force $f(x) = 0$.
(f) Find all values x which force $f(x) < 0$.
(g) Find all values x which force $f(x) \leq 0$.
(h) Approximate $f(-2)$. (i) Is $f\left(\frac{1}{2}\right) < f(-3)$?
(j) Is $f(-3) < f(6)$?

Problem 3.28. Use the graph of $y = f(x) = \frac{1}{2}(x^3 - 4x)$, provided below, to answer the following questions

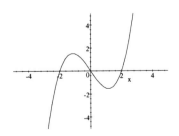

(a) What is the height $f(0)$?.
(b) What is the height $f(3)$?
(c) Approximate the height $f(1)$
(d) Find all values x which force $f(x) > 0$.
(e) Find all values x which force $f(x) \geq 0$.
(f) Find all values x which force $f(x) < 0$.
(g) Find all values x which force $f(x) \leq 0$.

3.4. Graphs of Standard Functions

■ There are **five basic types** of functions:

▶ polynomials

▶ rational functions

▶ algebraic functions

▶ trigonometric functions

▶ exponential and logarithmic functions

In this chapter we will examine only the first type of function: **Polynomial functions**. These are the most basic functions.

3.4.i. Straight lines

The most frequently used straight line is the **oblique** (diagonal) **straight line**. The graph of this line passes diagonally across the grid:

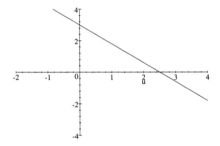

Consider the following function:

S : Given a number, produce two times the given number plus 3; that is, $S(u) = 2u + 3$ where u is the independent variable.

If we select a few values for the independent variable, we can find several ordered pairs. For example:

$S\left(\frac{1}{2}\right) = 4$	or:	$\left(\frac{1}{2}, 4\right)$
$S\left(-\frac{3}{2}\right) = 0$	or:	$\left(-\frac{3}{2}, 0\right)$
$S(1) = 5$	or:	$(1, 5)$
$S(4) = 11$	or:	$(4, 11)$

We can plot these four points on a grid to see the relationship between them.

It appears that a straight line would pass through all four of the points. In fact, if you were to plot many points, they would all lie in a straight line. If you plot sufficient points, they appear to coalesce and form a straight line:

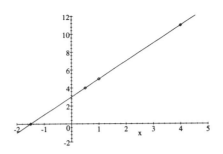

For example, the formula for the statement: "Given a number, produce the given number", would be $I(x) = x$. This is of the correct form to be a straight line since we can write $I(x) = 1 \cdot x + 0$

In the ordered pair notation, $I(x) = x$ is represented as (x,x). The graph of the function is:

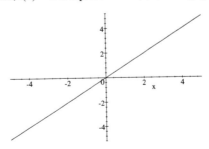

The graph of the function crosses the x-axis at the point $(0,0)$. Hence: $I(x) > 0$, when $x > 0$; and $I(x) < 0$, when $x < 0$.

■ Lines that **rise** as you move from **left to right** are said to have **positive slope** (as in the above example).

■ Lines that **fall** as you move **left to right** are said to have a **negative slope**. For example, the following line has negative slope:

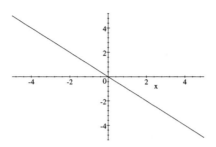

There are two special types of lines which do not give an oblique (diagonal) graph.

I. The first is the constant function

Consider the line illustrated by the sentence: S : Given a number, S produces the number 2. Notationally, we have: $S(u) = 2$, or $u \overset{S}{\longmapsto} 2$, or $(u, 2)$.

No matter which values are given, the output is the same constant number, in this case 2.

▶ This is known as a **constant function**.

▶ The plotted points all have the same height.

▶ S above is a constant function.

The graph of the function is:

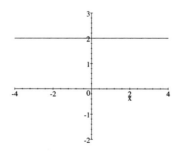

▶ The graph of the function is clearly a **straight, horizontal line**.

▶ It is of the correct form, since $S(u) = 2$ can be rewritten as $S(u) = 0 \bullet u + 2$.

Constant function:

Any function of the form $S(u) = c$,

where c is a **constant** will graph as a **horizontal** line.

These lines are said to have **zero slope**.

Note: The graph of any **zero degree polynomial** will be of this type.

Exercise **3.13**. *Sketch the graph of* $f(x) = -2$.

Solution:

II. **The second is a vertical line**.

All of the points on the line have the same first coordinate, for example 3, and a different second coordinate.

This is **not a function**, since there is no unique height produced by 3. These lines are said to have <u>no</u> **slope (or infinite slope)**. This type of line is **not** a <u>function</u> and therefore is **not a** <u>polynomial</u> (since all polynomials are functions), and hence is not considered further here.

The graph of $x = 3$ is shown below.

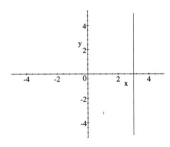

3.4.ii. Second degree polynomials

Second degree polynomials have graphs with some curvature.
For example, the graph of $f(x) = x^2$ is:

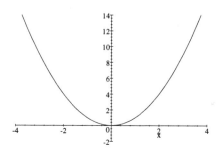

Notice that the graph of the function **touches** the **horizontal axis** at **one** point.

If you adjust the height of the function by moving it up   units, you get the equation $g(x) = x^2 + 6$, which has the following graph:

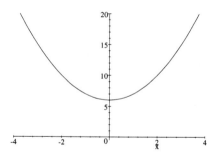

Note: This graph does not cross the horizontal axis, therefore the equation has no **zeros**. This tells us that the **function** is **prime** and in **factored form**. The sign of the discriminant $\boxed{b^2 - 4ac}$ (strictly negative in this case) also tells us the function is prime.

If you shift the original function down 6 units, you actually get a graph which crosses the horizontal axis twice:

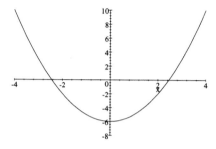

The equation for this polynomial is $h(x) = x^2 - 6$. There are two **roots** (**zeros**) for this polynomial. If you use the **quadratic formula**, or the **difference of squares method**, you will find that the roots are $\pm \sqrt{6}$.

> **Definition**
>
> • **Graphs of second degree polynomials**:
>
> **All** of the graphs of **second degree polynomials** are of the shape in the previous graph and are called **parabolas**.

Parabolas can open up, as in the previous examples; <u>or</u>, they can open down:

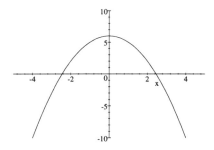

Exercise **3.14.** *Sketch the graph of* $-x^2$.

Solution:

3.4.iii.   Third degree polynomials

The simplest **third degree polynomial** is given by $p(x) = x^3$. The graph of this polynomial is:

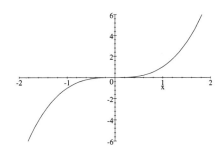

When the <u>second</u> and <u>first</u> degree terms are <u>non-zero</u>, you may get additional bends in the graph. For example, the graph of $y = x(x-1)(x+2) = x^3 + x^2 - 2x$:

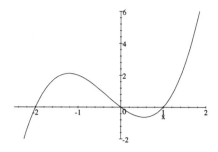

A <u>negative</u> coefficient in front of the <u>third</u> degree term will switch the portions of the graph above and below the axis. The graph of $y = -x(x-1)(x+2) = -x^3 - x^2 + 2x$ is shown here:

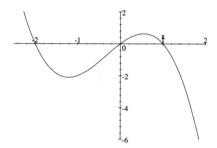

> | Definition |
>
> **Cubics**:
>
> All <u>third degree polynomials</u> are referred to as
>
> **cubics**, and will have the same general shapes
>
> as displayed in these examples.

Determining the graph of a third degree polynomial is greatly assisted by **factoring**, as you will see in the next section. If you cannot factor the polynomial, you are reduced to drawing its graph through plotting a large number of points.

Exercise 3.15. *Sketch the graph of the following*:

(*a*) *A first degree polynomial.*

Solution:

(*b*) *A constant polynomial.*

Solution:

(c) A cubic.

Solution:

(d) $x^2 - 4$.

Solution:

(e) $4 - x^2$.

Solution:

(f) $x^3 - x^2 + 2$.

Solution:

Summary 3.4.

- The graphs of polynomials of degree zero, one, two, and three are initially found through plotting a large number of points.
- It is important to be able to use the equation of the polynomial to predict the general shape of the graph for polynomials of degree zero, one, two, and three.

End of Section 3.4. Problems

Problem 3.29. On the grid, graph the following zero degree polynomials. Determine whether the slope of each polynomial is zero, negative or positive.

(a) $f(x) = 0x + 3$.　(b) $g(x) = 0x - 2$　(c) $h(x) = 0x + 0$

Problem 3.30. For the following sets of points

i. Plot the points on the grid and graph the line determined by the points.

ii. Most lines can be represented by a polynomial equation of either degree one or two. State the degree of the polynomial for this line, if possible.

iii. State whether the slope of the line is zero, negative, or positive.

(a) $(0,-2)$ and $(4,0)$.

(b) $\left(0,\frac{1}{2}\right)$, $\left(4,\frac{1}{2}\right)$ and $\left(-5,\frac{1}{2}\right)$.

(c) $(1,-2)$, $(-1,1)$ and $\left(0,-\frac{1}{2}\right)$.

(d) $(1,2)$, $(1,4)$ and $(1,-3)$

Problem 3.31. $W(u) = -2u + 3$ is a polynomial of degree one. If we choose $u = 2$, the polynomial produces -1. That is $2 \xrightarrow{W} -1$. Similarly, $-1 \xrightarrow{W} 5$.

(a) Plot these two points and graph the polynomial on the grid

(b) What is the slope of the line?

Using the graph, solve:

(c) $-2u + 3 = 0$ and plot the point $(u, 0)$. (d) $-2u + 3 > 0$. (e) $-2u + 3 < 0$.

Problem 3.32. Consider the following graph of the function $y = mx + c$.

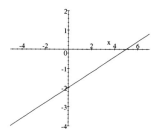

(a) What is the degree of the polynomial producing this line?

(b) Is the slope positive, negative, or zero?

Using the graph, solve:

(c) $mx + c = 0$. (d) $mx + c > 0$.

(e) $mx + c < 0$. (f) $mx + c \geq 0$.

Problem 3.33. Consider the following graph of the function $y = mx + c$.

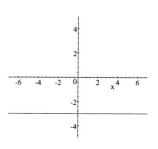

(a) What is the degree of the polynomial producing this line?

(b) What is the slope of this polynomial?

(c) Find and label two points on this line.

(d) What is the value of m?

(e) What is the value of c?

(f) Using the graph, solve $mx + c = 0$.

Problem 3.34.

i. Solve the following equations for w.

ii. Determine the degree of the resulting expression for w (w is a polynomial in x).

iii. State whether it represents a line.

(a) $2w - 2x = 0$.   (b) $3w - x = 1$.   (c) $w - 0x - 3 = 0$.   (d) $w - x^2 + 3 = 2x$.

(e) $w - x^3 + 1 = 0$.   (f) $w - 14 = 0$.   (g) $w = 3x + 1$.   (h) $-2w = 2x - 4$.

Problem 3.35. Consider the polynomial $P(x) = -x^2 + 5x - 6 = (-1)(x - 3)(x - 5)$

(a) What is the degree of $P(x)$?   (d) Graph $P(x)$ on the grid.

(b) Is the graph of $P(x)$ a line?   (e) Solve $(-1)(x - 3)(x - 5) = 0$.

(c) Which values of x force $P(x)$ to be zero?   (f) Solve $-x^2 + 5x - 6 > 0$.

3.5. Graphing and Factoring

The topic of this section is the **linkage** between **graphing** and the **factorization of polynomials**. For polynomials of **degree zero, one, two**, or **three**, we will be able to reproduce the **factored form** of the polynomial **from the graph**.

3.5.i. Using graphs to simplify polynomials

- In order to use graphing of polynomials to simplify factoring, we must first,
    - ▶ "read" the graph of a polynomial,
    - ▶ find the zeros, and then,
    - ▶ factor the polynomial.

Consider the graph of $f(x) = -8x^2 - 31x + 4$.

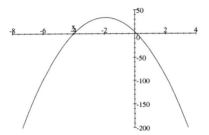

The **zeros** for this polynomial are at $\frac{1}{4}$ and -4.

From the second part of the quadratic formula, we know that the polynomial $f(x)$ factors as:

$$\boxed{a(x - r)(x - s)}$$

where r and s are the **roots** of the polynomial and a is the **coefficient of the second degree term** ($ax^2 + bx + c$). Consequently, using the **graph** of the function and **without using the quadratic equation** for finding the roots, we can see that the factorization of $f(x) = -8x^2 - 31x + 4$, from $a(x - r)(x - s)$, is:

$$-8\left(x - \frac{1}{4}\right)(x + 4).$$

- This can be verified using "FOIL" to multiply.

Example **3.9**. *State the function that gives the following graph*:

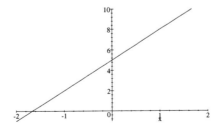

- **Solution**: Remember the graph of a straight line is produced by a first degree polynomial of the form $p(x) = ax + b$. From the graph, we can see that the points $\left(-\frac{5}{3}, 0\right)$ and $(0, 5)$ are on the graph. We know from forcing a factorization that $ax + b = a\left(x + \frac{b}{a}\right)$, and $-\frac{b}{a}$ is the **zero** of the polynomial.

Therefore $-\frac{b}{a} = -\frac{5}{3}$. So the equation of $p(x) = ax + b = a\left(x + \frac{5}{3}\right)$.

When you substitute 0 for the value of x, you get $a\left(0 + \frac{5}{3}\right) = 5$. So the value of a must be 3. Thus the equation is $p(x) = 3x + 5$.

Exercise 3.16. *State the function that gives the following graph:*

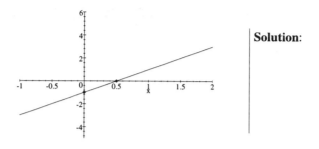

Solution:

Consider the following graph of a <u>third degree</u> polynomial:

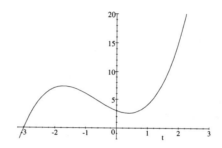

The graph implies that there is only **one root** to the **cubic polynomial**. Since the polynomial is of **degree three (cubic)**, we know that **either** it is the **product of three first degree** polynomials (in which case there would be **three roots**), <u>or</u> it is the **product of a first degree** polynomial and a **prime second degree** polynomial (in which case there would only be **one root**, since **prime polynomials have no roots**)

Since there is in fact only **one root**, we know that when we factor this polynomial we are going to have a prime second degree component, and that the first degree portion will be $t + 3$ (since the root is -3).

In fact the polynomial which generated this graph is $P(t) = t^3 + 2t^2 - 2t + 3$. If you use long division to divide $t + 3$ into $t^3 + 2t^2 - 2t + 3$, you get $P(t) = (t+3)(t^2 - t + 1)$. The discriminant $\boxed{b^2 - 4ac}$ of the second degree polynomial $(t^2 - t + 1)$ is -3, so it is in fact **prime**. This agrees with our deductions from the graph.

Consider the graph of $P(x) = \frac{1}{10}(2x^5 - 15x^4 + 27x^3 - 11x^2 + 7x - 10)$:

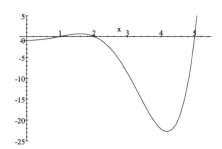

This polynomial is of degree five. The graph indicates that there are zeros at $1, 2,$ and 5; thus $P(x) = \frac{1}{10}(x-1)(x-2)(x-5)Q(x)$, where $Q(x)$ is of **degree two and prime** (if it were not prime, there would be additional roots.) If needed, you can use long division to determine the exact nature of $Q(x)$.

Exercise **3.17.** *Use the graph of* $P(x) = 2x^2 + x + 1$ *to determine whether* $P(x)$ *is prime or factorable.*

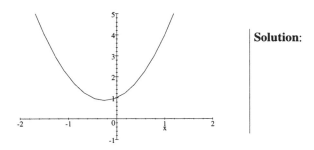

Solution:

3.5.ii. Graphs from factored form polynomials

In the instance of a <u>first degree</u> polynomial, $y = mx + b$, we know from the previous section that this is straight a line. It is a horizontal line if $m = 0$, and otherwise oblique or vertical.

■ To graph the line it is simply a matter of finding two points, one of which is easily found through factoring.

Example **3.10.** *Graph* $T(x) = 2x - 1$.

● **Solution:** Force a factorization of the coefficient of x. This gives us $T(x) = 2\left(x - \frac{1}{2}\right)$. Thus we know that the zero of the function is at $\frac{1}{2}$ and therefore the point $\left(\frac{1}{2}, 0\right)$ is on the graph.

Find the point corresponding to the independent variable value of $0 : (0, -1)$.

Plotting both of these points and then connecting them with a straight line gives us the graph.

■ A straight line because this is the "shape" of <u>all</u> first degree polynomials.

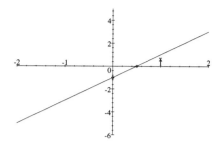

■ | Note: | These two points are identified as the **intercepts** of the function, since they are the points at which the function intercepts the <u>independent</u> and <u>dependent</u> axes.

In graphing a **function**, the most visually identifiable points are the **intercepts**. The relationship between the **roots** of the function and the intercepts is exploitable because of the following property of the real numbers.

| Definition |

<u>Integral Domain Property</u>:

If two numbers, say a and b, when multiplied yield zero, then one of the original numbers must be zero.

In terms of equations, $a \cdot b = 0$ implies either $a = 0$ or $b = 0$.

This property is true if more that two numbers are multiplied together.

Example 3.11. Consider $f(x) = (2x + 1)(x - 3)(x + 2)$. *Find the intercepts for this function and its graph.*

Solution: $f(x) = (2x + 1)(x - 3)(x + 2) = 0$.

From the <u>Integral Domain Property</u>, we know that either

$$2x + 1 = 0, x - 3 = 0, \text{ or } x + 2 = 0.$$

Solving each of these, we see that either

$$x = -\frac{1}{2}, \ x = 3, \text{ or } x = -2.$$

This gives us three roots to the equation.

To find the y-intercept, substitute $x = 0$:

$$y = (2 \cdot 0 + 1)(0 - 3)(0 + 2)$$
$$= -6$$

Knowing that $f(x)$ is a third degree equation gives us a standard shape for $f(x)$, and knowing the roots and the y-intercept allows us to sketch a fairly accurate graph of $f(x)$. We may also need to **check a point between the roots** to determine if the graph of the function is **above** or **below** the axes on that side of the intercept.

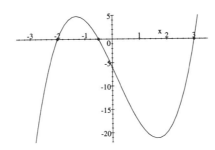

Exercise **3.18**. *Consider* $f(x) = (-6x + 1)(2x + 1)$. *Find the intercepts for this function and its graph.*

Solution:

Exercise **3.19**. *Use the graph of* $P(x) = -3x^2 - 8x + 35$ *to determine whether* $P(x)$ *is prime or factorable. Estimate the factorization from the graph if possible and then verify using previous techniques.*

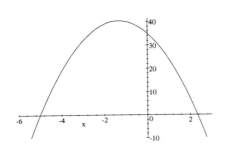

Solution:

Exercise **3.20**. *Factor and graph* $f(x) = -12x^2 - 4x + 1$.

Solution:

Exercise **3.21.** *Use the graph of* $P(x) = x^2 - 16$ *to determine whether* $P(x)$ *is prime or factorable. Estimate the factorization from the graph if possible and then verify using previous techniques.*

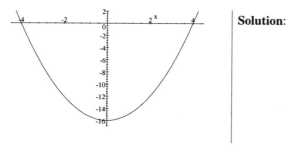

Solution:

Exercise **3.22.** *Use the graph of* $P(x) = x^5 - 2x^4 - 24x^3 - x^2 + 2x + 24$ *to determine whether* $P(x)$ *is prime or factorable. Estimate the factorization from the graph if possible and then verify using previous techniques. In this instance, you will need to use long division of polynomials.*

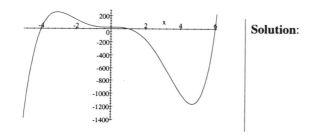

Solution:

Exercise **3.23. Determine the graph of:** $f(x) = (-4x + 7)(x - 3)(x + 5)$.

Solution:

Summary 3.5.

- Factoring and graphing are closely interrelated.
- You can use the graph of a function to predict the number of zeros, which tells you the number of times the graph crosses the independent axis, and thus aids in the factoring of a function.
- To move from the factorization of a function to its graph.
  - Recognizing the function as one of the standard graphs. (If you cannot do this, then your only recourse is to plot points and then add the result to your list of standard graphs.)
  - Factoring the function to identify the intercepts with the independent axis.
  - Plotting the intercepts and use the standard graph to complete the sketch.

> Noting that all values of the independent variable other than the zeros yield points with either **positive or negative heights**. Thus we may need to **check a point between the intercepts** to determine if the graph of the function is **above or below** the axes on that side of the intercept.

End of Section 3.5. Problems

<u>Remember</u> that the **degree two polynomial** is **prime** provided the degree two polynomial has **no zeros**. Conversely, if a degree two polynomial has zeros, then the polynomial has two factors.

<u>Problem</u> 3.36. Examine the graph of $f(x)$ provided below and: **i.** Find the zeros, if any. **ii.** Using the zeros, determine if the polynomial is prime or if it factors. **iii.** If it factors, give the factorization.

(a) $f(x) = ax^2 + bx + c;\ a = 2.$ **(b)** $f(x) = -\frac{1}{2}x^2 + 4x - 8$ **(c)** $f(z) = -2z^2 + z - 1.$

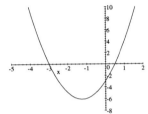 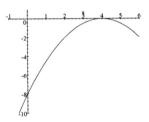 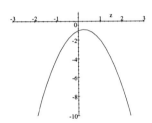

(d) $f(z) = -4z^2 + 1$ **(e)** $f(u) = \frac{1}{2}u^2 + \frac{3}{2}u + \frac{3}{2}$

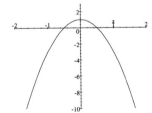 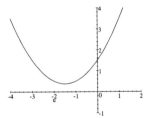

<u>Problem</u> 3.37. The polynomials below are degree one polynomials. **i.** Factor the polynomial into a degree one polynomial multiplied by a constant. **ii.** Find the zero of the polynomial.

(a) $g(u) = -\frac{1}{2}u - 3$ **(b)** $h(z) = 2z - 6$

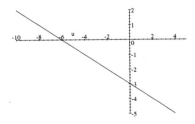 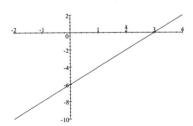

Problem 3.38. The debt of an individual is given by the degree one polynomial $D(t) = 100t - 1500$, where the variable t is interpreted as time measured in years and $D(t)$ is measured in dollars. The graph of $D(t)$ is given below.

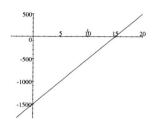

(a) Using the graph, what is the debt at time zero?

(b) At what time is the debt $\$0$?

(c) Using the graph, factor $100t - 1500$.

Problem 3.39. Consider the graphs below. Remember that every polynomial of degree greater than two will factor. **i.** Find the zeros of the following polynomials. **ii.** Factor the polynomials.

(a) $f(x) = x^4 - 10x^2 + 9$

(b) $g(x) = -2x^3 + 5x^2 + 4x - 3$

(c) $f(x) = 4x^5 - 12x^4 - 17x^3 + 51x^2 + 4x - 12$

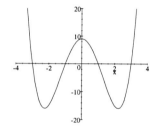
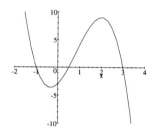
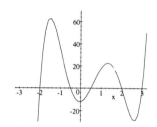

(d) $h(u) = 9u^3 - 42u^2 + 49u - 12$

(e) $f(x) = x^3 + 3x^2 + 3x + 1$.

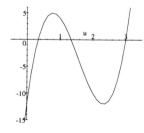
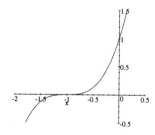

Problem 3.40. Use the **integral domain property** to solve the following polynomial equations:

(a) $(3x - 6)(x - 1) = 0$

(b) $(3x - 6)(x - 1) = 0$

(c) $-3(x - 1)(x - 1)(x - 1) = 0$

(d) $-2(x - 1)\left(x - \frac{1}{3}\right)\left(x + \frac{1}{4}\right)(x - 5) = 0$

3.6. Applications of Graphs to Absolute Value, Domain and Inequalities

i. Absolute value

The absolute value, as a function, must at all times be **positive**. Thus the graph of the function will always lie above the independent axis. One of the characteristics of an absolute value graph is that it will have sharp points, also known as **cusps**.

- ■ Strategy
    - ▶ First graph the inner function (without the absolute value)
    - ▶ reflect all points below the independent axis up to the corresponding point above the axis.
    - ▶ | **Note:** | $|x|$ means the absolute value of x.

Example **3.12.** *Develop the graph of* $g(t) = |2t - 1|$.

- • **Solution:** First, consider the graph of $f(t) = 2t - 1$.

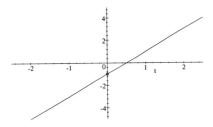

- • Every point on the graph which lies below the t-axis must be moved to a point above the axis. For example, the point $(0, -1)$ (marked with a diamond) must become the point $(0, 1)$:

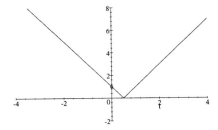

Exercise **3.24.** *Develop the graph of* $g(x) = |19x - 5|$.

Solution:

Example **3.13.** *Construct the graph of* $h(x) = |x^2 - 4|$.

- **Solution**: The graph of the function is a parabola with zeros of -2 and 2; the zeros come from the factorization of $x^2 - 4 = (x - 2)(x + 2)$. The value of the function $x^2 - 4$ at 0 is -4, therefore the parabola is below the axis between -2 and 2. Using this and the standard shape of a parabola, we get the graph of $x^2 - 4$.

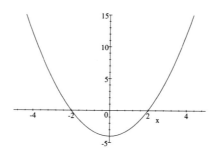

- Applying absolute values to $x^2 - 4$ will cause the middle of the graph to move from below the axis to above the axis. As an example, the point $(0, -4)$ will become the point $(0, 4)$. This gives the graph of $h(x) = |x^2 - 4|$. The graph has two cusps.

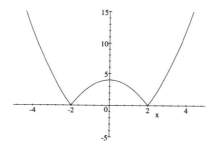

ii. Domain

> Definition
>
> **Domain**:
>
> Domain is a set of numbers that may replace the independent variable of a function.

There are functions which have a restricted domain, for example the function $\sqrt{x}$. This function can not accept an input which is a negative number. For any **root** with an even index, that is, for roots of the form $\sqrt[N]{}$ where N is an **integer**, the numbers substituted into the expression must be **positive or zero**.

Example **3.14.** *Find the numbers which may be used for* x *in the function* $f(x) = \sqrt{x - 1}$

- **Solution**: We must determine for which values of x, is $x - 1 \geq 0$. Using algebra, we would solve the inequality by adding 1 to both sides and determining that $x \geq 1$. Using **graphing**, we construct the graph of $x - 1$ and see that once again the graph lies on or above the independent axis for those values of x **greater than or equal to** 1.

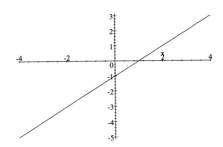

- **Remember:** To find the domain of a function, we look for specific heights and then the x values (the domain) which generated them.
  - **There are fairly complex examples which can require an analysis of the domain:**
    - ▶ $f(x) = \sqrt{x^2 - 3x + 1}$, which requires that we know where $x^2 - 3x + 1 \geq 0$.
    - ▶ $f(x) = \sqrt{\frac{x-1}{x+3}}$, which requires that we know where $\frac{x-1}{x+3} \geq 0$.
    - ▶ $f(x) = \sqrt[4]{x^3 - 4x + 1}$, which requires that we know where $x^3 - 4x + 1 \geq 0$.
    - ▶ $f(x) = \sqrt[10]{\frac{x^3 - 4x + 1}{x^2 + x - 3}}$, which requires that we know where $\frac{x^3 - 4x + 1}{x^2 + x - 3} \geq 0$.

  While each of these problems may be solved algebraically, it becomes more and more complicated to do so. A **combination** of **factoring and graphing** can lead to a faster solution.

Example 3.15. For $f(x) = \sqrt[4]{3x^2 + 4x + 1}$, find the domain.

- **Solution**: We must determine where $3x^2 + 4x + 1 \geq 0$. Factoring, we see that
$$g(x) = 3x^2 + 4x + 1$$
$$= (3x + 1)(x + 1)$$
$$= 3\left(x + \tfrac{1}{3}\right)(x + 1)$$

Thus the zeros of the function are $-\frac{1}{3}$ and -1. This is a **second degree** polynomial, so we know from the standard graphs that it is a parabola. Since $g\left(-\frac{1}{2}\right) = -\frac{1}{4}$, the "center" portion of the graph is below the axis. The graph of $g(x)$ is:

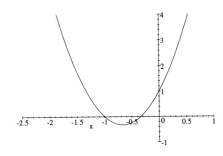

- That portion of the function $g(x)$ which is unusable, that is, negative, corresponds to that portion of the graph which lies below the independent axis. Consequently all remaining points which yield heights of zero and above are usable. Therefore the domain of $f(x) = \sqrt[4]{3x^2 + 4x + 1}$ is $(-\infty, -1]$ and $[-\frac{1}{3}, \infty)$.

iii. Inequalities

Inequalities play an important role in **graphing** and **finding the domain** of functions.

With the advent of new technologies that allow a quick production of the graph of a function, we can begin to use a graph to motivate the understanding and quick solution of inequalities. This can also motivate the traditional algebraic methods of solving an inequality and help to better understand the solutions to the algebraic process.

Example **3.16.** *Solve* $-2x + 3 < 0$.

- **Solution:** The graph of $-2x + 3$ is given by

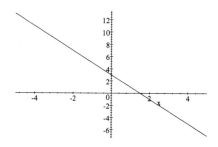

- All values of the independent variable which generate points of negative height lie to the right of the graph. In particular, since the x-intercept is $\frac{3}{2}$, the solution to $-2x + 3 < 0$ is $x > \frac{3}{2}$.

Example **3.17.** *Solve the inequality* $x^2 - 3x + 2 < 0$.

- **Solution:** The graph of $f(x) = x^2 - 3x + 2$, is:

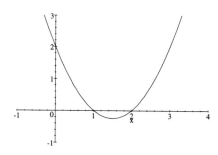

- Those points of the graph which lie strictly below the independent axis, that is those points for which the height is less than zero, correspond to the solution of $x^2 - 3x + 2 < 0$. This gives a solution of $(-1, 2)$.

> **Note:** The numbers -1 and 2 come from the factorization of $x^2 - 3x + 2$. The construction of the graph uses information about the shape of the graph of all second degree polynomials.

Exercise **3.25.** *Solve the inequality* $x^2 - 12x < 0$.

> Solution:

iv. Solving inequalities by comparing graphs of functions

Example **3.18.** *Solve the inequality* $2x + 2 > 3$.

- **Solution:** First we graph the two functions $f(x) = 2x + 2$ and $g(x) = 3$.

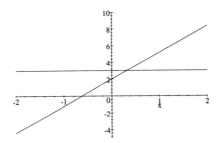

- It is clear that the functions cross to the right of zero; to the right of that point, $2x + 2 > 3$. We need to exactly locate the value of the point where the functions intersect.

  So we solve the equality
  $$2x + 2 = 3$$
  $$\text{to get } 2x = 1$$
  $$x = \frac{1}{2}$$

  Therefore the solution to $2x + 2 > 3$ is that $x > \frac{1}{2}$.

Exercise **3.26.** *Solve the inequality* $|x - 2| > 3$.

> Solution:

Example **3.19.** *Solve the inequality* $x^2 - 2x > 2x$.

- **Solution**: The function on the left side of the inequality is a **parabola** with **zeros** at 0 and 2. The function on the right is a line through the origin. Graphing these two functions gives the following:

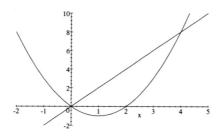

- We are concerned with those independent values which allow the parabola to lie above the line. As before, we must find the intersection points:

$$x^2 - 2x = 2x$$
$$x^2 - 4x = 0$$
$$x(x - 4) = 0$$

So the intersection points correspond to $x = 0$ and $x = 4$. This means the solution to $x^2 - 2x > 2x$ is found to the left and to the right of the two intersection points, respectively (see graph); so the solution is $x < 0$ or $x > 4$.

Exercise **3.27. Draw the graph of**:

(a) $f(x) = |(x - 1)(x - 3)(x - 5)|$

Solution:

(b) $\sqrt{(x - 2)(x + 1)}$; *and, find the domain.*

Solution:

Exercise **3.28. Solve the following inequalities, for the function** $P(x) = -3x^2 + 16x - 5$, **the graph of** $P(x)$ **is**:

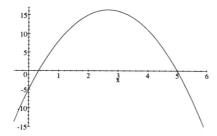

(a) $P(x) > 0$

Solution:

(b) $P(x) \le 0$

Solution:

Exercise 3.29. *Draw the graph of* $f(x) = |-(x+2)(x-1)(x-5)|.$ *The graph of* $-(x+2)(x-1)(x-5)$ *is:*

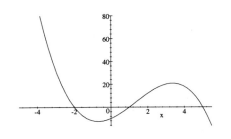

Solution:

Exercise 3.30. *Draw the graph of* $f(x) = |5(x-1)(x-3)|.$ *The graph of* $5(x-1)(x-3)$ *is:*

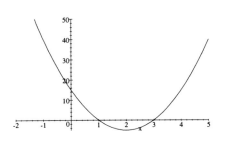

Solution:

Exercise 3.31. *Find the domain of the function* $\sqrt{-15x^2 - x + 1}$. *The graph of* $-15x^2 - x + 1$ *is:*

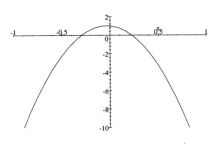

Solution:

Exercise 3.32. *Find the domain of the function* $\sqrt{\dfrac{(x-1)}{(x+2)}}$. *The graph of* $\dfrac{(x-1)}{(x+2)}$ *is:*

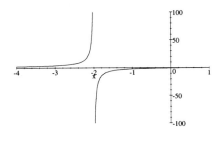

Solution:

Exercise **3.33.i.** *Construct the graph of* $P(x) = (2x-1)(x-2)(x+3)(x-4)$.

Solution:

Exercise **3.33. ii.** *Use the graph in* **3.33.i** *to answer the following inequalities*:

(a) $(2x-1)(x-2)(x+3)(x-4) > 0$.

(b) $(2x-1)(x-2)(x+3)(x-4) \leq 0$.

Solution:

Solution:

Exercise **3.34.** *Given the graph of* $P(x) = \frac{1}{2}x^3 - \frac{3}{2}x^2 - 5x + 12$ *use it to answer the inequalities*:

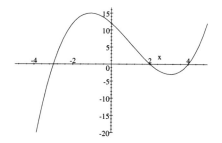

(b) $\frac{1}{2}x^3 - \frac{3}{2}x^2 - 5x + 12 \geq 0$.

Solution:

(c) $\frac{1}{2}x^3 - \frac{3}{2}x^2 - 5x + 12 < 0$.

Solution:

Exercise **3.35.** *Solve the inequality* $(2x-1)(x-4) > -\frac{4}{3}x^2 + \frac{43}{3}x - 16$, *given the graph,* (**Note**: $(2x-1)(x-4)$ *is the thick lick on the graph*):

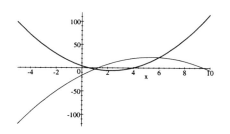

Solution:

Summary 3.6.

■ **Graphing** and **factoring** techniques can be combined to provide a powerful tool for determining the **domain** of a function and for **solving inequalities**.

■ This section also illustrates how to **graph absolute values**.

■ Graphing can be used to **compare a function to a fixed value**; also, it can be used in the comparison of two functions.

End of Section 3.6. Problems

Problem 3.41. *The first degree polynomial* $f(x) = -x + 3$ *has a graph as shown:*

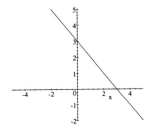

(a) *Graph* $h(x) = |-x + 3|$
(b) *Solve* $|-x + 3| = 0$

Problem 3.42. *The graph of* $y = -x^2 + 9$ *is shown along with the function* $L(x) = 2$:

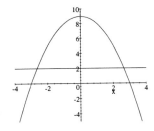

(a) *Graph* $|-x^2 + 9|$ *and* $L(x) = 2$.
(b) *How many solutions are there to* $|-x^2 + 9| = 2$?

Problem 3.43. $y = 2x - 3$ *has a graph that is a straight line. Describe the graph of* $|2x - 3|$

Problem 3.44. *The graphs of* $g(x) = 2x - 1$ *and* $h(x) = -1(2x - 1)$ *are as given.* (a) *Graph* $|2x - 1|$ *and* $|-1(2x - 1)|$. (b) *What do the graphs indicate is true for* $|2x - 1|$ *and* $|-1(2x - 1)|$?

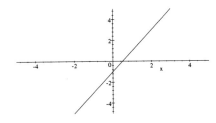

$g(x) = 2x - 1$

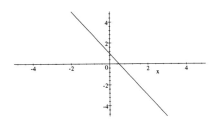

$h(x) = -1(2x - 1)$

Problem 3.45. Remember that $\sqrt{g(x)}$, $\sqrt[4]{g(x)}$, $\sqrt[6]{g(x)}$, etc. may use only those values of x for which the function $g(x)$ is strictly positive or zero. That is, $g(x) \geq 0$.

(a) Graph $g(x) = x^2 - 3x + 2 = (x-2)(x-1)$.

(b) Solve $(x-2)(x-1) \geq 0$.

(c) Find the domain of $\sqrt[4]{x^2 - 3x + 2}$.

Problem 3.46. The graph of $K(u) = |(-1)(u^2 - 4)|$ is given below as well as the graph of the horizontal line $f(u) = 5$. The points of intersection are $(-3, 5)$ and $(3, 5)$.

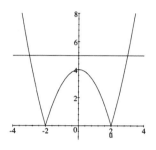

(a) For which values of u is the graph of $K(u) = |(-1)(u^2 - 4)|$ strictly below the graph of $f(u) = 5$?

(b) Solve $|(-1)(u^2 - 4)| < 5$

Problem 3.47. The graph of $f(x) = x^4 + 1$ is:

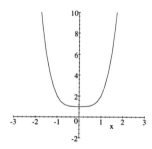

(a) Solve $x^4 + 1 > 0$

(b) Solve $x^4 + 1 < 0$

(c) Solve $x^4 + 1 = 0$

Problem 3.48. Use the graph of $K(w) = .02(w+4)(w-3)(w-2)\left(w+\frac{3}{2}\right)w$, to solve:

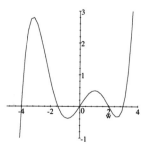

(a) $K(w) = 0$.

(b) $K(w) > 0$.

(c) $K(w) \geq 0$.

(d) $K(w) < 0$.

(e) $K(w) \leq 0$.

._148_.

End of Chapter 3. Problems

Problem 3.49. *Using the English sentence E, complete the chart with the values that are produced by the sentence. Some entries are provided for reference. The variable x is the independent variable.*
Given a number, produce two times the given number minus four.

$\xrightarrow{E}$ ___	$(\_,\_)$	$E(\ ) =$
$8 \xrightarrow{E} 12$	$(8,12)$	$E(8) = 12$
$1 \xrightarrow{E} -2$	$(1,\_)$	$E(1) = \_$
$\frac{-1}{64} \xrightarrow{E}$	$(\frac{-1}{64},\_)$	$E(\frac{-1}{64}) = \_$
$3 \xrightarrow{E}$	$(3,\_)$	$E(3) = \_$
$x \xrightarrow{E}$	$(x,\ \_)$	$E(x) = \_$

Problem 3.50. *Using the English sentence Jump, complete the chart with the values that are produced by the sentence. Some entries produce twice the given number plus one provided the given number is strictly greater than one; otherwise, produce negative one times the given number (when the given number is less than or equal to one.)*

$\xrightarrow{\text{Jump}}$ ___	$(\_,\_)$	$\textbf{Jump}(\ ) =$
$-1 \xrightarrow{\text{Jump}} 1$	$(-1,1)$	$\textbf{Jump}(-1) = 1$
$0 \xrightarrow{\text{Jump}} 0$	$(0,0)$	$\textbf{Jump}(0) = \_$
$\frac{3}{4} \xrightarrow{\text{Jump}}$	$(\frac{3}{4},\_)$	$\textbf{Jump}(\frac{3}{4}) = \_$
$1 \xrightarrow{\text{Jump}}$	$(1,\_)$	$\textbf{Jump}(1) = \_$
$x\ (x > 1) \xrightarrow{\text{Jump}}$ ___	$(x,\_)$	$\textbf{Jump}(x) = \_$
$x\ (x \le 1) \xrightarrow{\text{Jump}}$ ___	$(x,\ \_)$	$\textbf{Jump}(x) = \_$

Problem 3.51. *Write the English imperative sentence which produces the following charts:*

(a)	(b)
	H
$R(1) = 1$	$(0,0)$
$R(\frac{1}{2}) = 2$	$(\frac{1}{4},\frac{1}{2})$
$R(2) = \frac{1}{2}$	$(16,4)$
$R(-3) = \frac{-1}{3}$	$(4,2)$
$R(\frac{1}{6}) = 6$	

Problem 3.52. *Write an imperative English sentence for the following . Begin the sentence with "Given a number, produce...".*

(a) $Q(w) = \dfrac{(x+1)}{(x^2+1)}$. (b) $S(u) = (u, u-1)$. (c) $t \xrightarrow{M} -16t^2 + 5$. (d) $x \xrightarrow{E} 2^x$.

Problem 3.53. *Given a time measured in seconds, produce the height of a ball which is thrown vertically upward at* 100 *ft./sec. The height is produced by multiplying* -16 *times the time in the air squared and then adding* 100 *times the time in the air. If we call this calculation* h, *evaluate the following:*

(a) $h(0) =$ (b) $1 \overset{h}{\to}$ ____   (c) $h(2) =$ (d) $\left(\tfrac{1}{2}sec, \text{____}\right)$ (e) $h(t) =$

Problem 3.54. *The 1996 tax information packet for the 1040 Tax forms explains the calculation of an individual's federal income tax for an individual who is married, filing jointly, by: Given a taxable amount from line 37 which is between \$40,100 and \$96,900, pay taxes of \$6,015 plus .28 times the difference of the given taxable amount and \$40,100. Let* I *be the taxable amount on line 37 and let* FIT *(Federal Income Tax)be the name of the relation that calculates the tax. Complete the following table:*

$I \overset{FIT}{\to}$ ____	$(I, \text{__})$	$FIT(I) =$
$\$40,100 \overset{FIT}{\to} \$6,015$	$(\$40,100, \$6,015)$	$FIT(\$40,100) = \$6,015$
$\$41,100 \overset{FIT}{\to} \$6,295$	$\$41,100, \$6,295)$	$FIT(\$41,100) = \$6,295$
$\$52,100 \overset{FIT}{\to}$	$(\$52,100, \text{_____})$	$FIT(\$52,100) = \text{_____}$
$\$62,100 \overset{FIT}{\to}$	$(\$62,100, \text{_____})$	$FIT(\$62,100) = \text{_____}$
$\$96,100 \overset{FIT}{\to}$	$\left(\$96,100, \text{_____}\right)$	$FIT(\$96,100) = \text{_____}$
$\$96,900 \overset{FIT}{\to} \$21,919$	$(\$96,900, \$21,919)$	$FIT(\$96,900) = \$21,919$

Problem 3.55. *On the grid plot the following points:*

$(-3,0), \left(-\tfrac{1}{2},-0\right), (0,0), (1,0), \left(\tfrac{3}{2},0\right), (4,0), (5,0), (6,0).$

Problem 3.56. *For the following graph:*

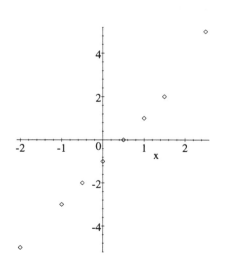

(a) *Find value(s) of* u *for all points positioned at* $(u,0)$.

(b) *Find value(s) of* u *for all points positioned at* $(u,2)$.

(c) *Find value(s) of* u *for all points positioned at* (u,w), *where* w *is strictly a negative height.*

(d) *Find value(s) of the height* b, *so that the point* $(-1,b)$ *is a position on the graph.*

(e) *Find value(s) of the height* b, *so that the point* $(0,b)$ *is a position on the vertical axis.*

(f) *Which values of* u *have the property that two or more heights* b *are plotted using that value of* u?

(g) *Find all values of* u *so that* $(u,-2)$ *is a point on the graph.*

(h) *Find all the values of* u *so that* $(u,4)$ *is a point.*

(i) *Find all the values of* u *so that* (u,b) *is a point. and* $b > 0$.

(j) *Find all the values of* u *so that* (u,b) *is a point.*

Problem 3.57. *Identify the independent variable for the following*

(a) (v,w).   (b) $y = x^2 + 1$.   (c) $T = w^2 + 4$.

Problem 3.58. *Choose an independent variable for each of the following equations; then solve the equation for the dependent variable.*

(a) $zw = 8$. (b) $zh - z = -4$. (c) $\Delta^2 - 3\square = 0$.

Problem 3.59. *For the following graph $T(x)$:*

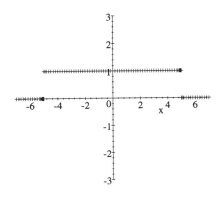

(a) *Solve* $T(x) = 0$.
(b) *Solve* $T(x) > 0$.
(c) *Solve* $T(x) < 0$.
(d) *Solve* $T(x) \le 0$.
(e) *Solve* $T(x) \ge 0$.
(f) *Solve* $T(x) \le 1$.
(g) *Solve* $T(x) > 2$.
(h) *Solve* $T(x) = \frac{1}{2}$.
(i) *Solve* $T(x) = -4$.
(j) *Solve* $-2 < T(x) < 3$.

Problem 3.60. *Let $T(x)$ be defined by the points:*

$(2,4), (2,3), (2,1), (2,0), (2,-1), (2,-2), (2,-3), (2,-4)$.

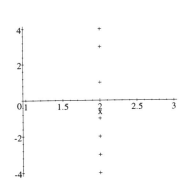

(a) *Solve* $T(x) = 0$.
(b) *Solve* $T(x) > 0$.
(c) *Solve* $T(x) < 0$.
(d) *Solve* $T(x) \le 0$.
(e) *Solve* $T(x) \ge 0$.
(f) *Solve* $T(x) \le 1$.
(g) *Solve* $T(x) > 2$.
(h) *Solve* $T(x) = \frac{1}{2}$.
(i) *Solve* $T(x) = -4$.
(j) *Solve* $-2 < T(x) < 3$.

Problem 3.61. *Identify the independent (input) and dependent variable (output) for the following:*

(a) $w = -2u + 3u + 1$ (b) $y = 3x^2 + 1$.

(c) *The federal income tax tables which, when given a taxable income I, produces a federal tax T.*

(d) *A bank which, when given a deposit A, for a savings account then produces interest I, on this deposit.*

(e) *A power company which, when given a customer's energy use K, produces the customer cost C.*

Problem 3.62. *For the next three graphs answer the following questions:*

(a) *What is* $f(0)$?   (b) *What is* $f(-1)$?   (c) *Solve* $f(x) = 0$ (d) *Solve* $f(x) > 0$

(e) *Solve* $f(x) \geq 0$ (f) *Solve* $f(x) < 0$ (g) *Solve* $f(x) \geq 0$ (h) *Is* $f(2) < f(3)$?

i. $f(x) = -2x + 1$	**ii.** $f(x) = x^3$	**iii.** $f(x) = -1(x-2)(x-3)$
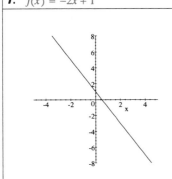	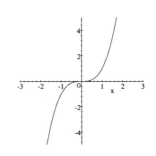	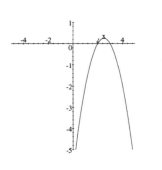

Problem 3.63. *Consider the points* $(2,0)$, $(1,4)$ *and* $(0,8)$.

(a) *Plot the points on the grid and graph the polynomial* $P(x)$ *determined by these points.*

(b) *What is the degree of the polynomial determined by these points.*

(c) *If the graph is a line, is the slope of the line positive, negative, or zero?*

(d) *Solve* $P(x) = 0$ (e) *Solve* $P(x) > 0$ (f) *Solve* $P(x) < 0$

Problem 3.64. *Consider the polynomial* $P(u) = -\frac{1}{2}u + 1$

(a) *Graph the polynomial* $P(u)$ *on the grid provided.*

(b) *What is the slope of* $P(u)$?

(c) *Solve* $P(u) = 0$; *that is , when is* $-\frac{1}{2}u + 1 = 0$? *Plot this point on your graph above.*

(d) *Solve* $-\frac{1}{2}u + 1 > 0$ (e) *Solve* $-\frac{1}{2}u + 1 \geq 0$ (f) *Solve* $-\frac{1}{2}u + 1 \leq 0$ (g) *Solve* $-\frac{1}{2}u + 1 < 0$

Problem 3.65. *Let* $3y - 2x + 3 = 2y + (2 - 2x)$.

(a) *Solve for the dependent variable* y.

(b) *Is* y *a polynomial?*

(c) *If* y *is a polynomial, what is the degree of* y?

(d) *Graph* y.

(e) *If* y *graphs as a line, what is the slope of the line?*

(f) *Solve* $y(x) = 0$

(g) *Solve* $y(x) > 0$

Problem 3.66. For each of the following expressions, fill out the table. Polynomial refers to the expression being a polynomial in u.

Expression	Polynomial	Horiz. line	Oblique line	Slope	Quadratic	Deg.
$y = 2u + 1$	Yes	No	Yes	2	No	1
$y = 0u + 3$						
$y = 8u^3$						
$y = -2u^2$						
$y = 3u^2 + 4u + 1$						
$y = (-u + 1)(-u + 1)$						
$y - u = 1 - u$						
$y = 6(u - 1)^3$						
$y = 2u^2 - 4\left(\frac{1}{2} + \frac{1}{2}u^2\right)$						
$0y + u = 2$						

Problem 3.67. Label each of the following true or false.

(a) If $P(x)$ is a degree one polynomial, then the graph of $P(x)$ is a line.

(b) If $P(x)$ is a degree one polynomial, then the graph of $P(x)$ is an oblique line.

(c) If $P(x)$ is a degree zero polynomial, then $P(x) = k$ where k is a constant.

(d) If $P(x)$ is a degree zero polynomial, then the graph of $P(x)$ is a horizontal line.

(e) If $P(x)$ has a graph that is a horizontal straight line, then $P(x)$ has degree zero.

(f) If $P(x)$ has a value of x that $P(x)$ cannot use, then $P(x)$ is not a polynomial.

(g) If $P(x) = ax^2 + bx + c$, then the graph of $P(x)$ is concave up.

(h) If $P(x) = x^3$, then the graph of $P(x)$ is a line.

(i) If $P(x) = x^4 + 1$, then the graph of $P(x)$ is a line.

(j) If $P(x)$ has degree one, then $P(x) = mx + b$ and $m \neq 0$.

Problem 3.68. Consider the graph of $y(x) = 12x^2 - 5x - 3$.

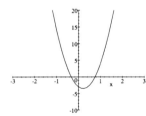

From the graph,

(a) Find the zeros of $y(x)$.

(b) Does $y(x)$ factor?

(c) Provide the factorization if the answer is yes.

Problem 3.69. Consider the graph of $F(u) = au^2 + bu + c$:

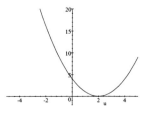

From the graph,

(a) Why does $F(u)$ factor?

(b) From the graph, find the two zeros of $F(u)$

(c) What is true about the factorization of $au^2 + bu + c$?

Problem 3.70. *A ball is shot vertically upward from a tower that is 64 feet tall.* $h(t) = -16t^2 + 64$ *determines the height of the ball at any time t (measured in seconds) is . The height $h(t)$ is measured in feet. Consider the graph of $h(t) = -16t^2 + 64$:*

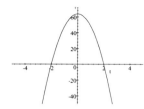

(a) *When does the ball hit the ground?*

(b) *What is the maximum height the ball attains?*

(c) *Factor* $h(t) = -16t^2 + 64$.

Problem 3.71. *The first degree polynomial that produces temperature measured in degrees Fahrenheit from a given temperature measured in degrees Centigrade is* $f(C) = \frac{9}{5}C + 32$. *(a) What temperature in Centigrade produces zero degrees Fahrenheit? (b) Using the graph below for reference, factor* $\frac{9}{5}C + 32$:

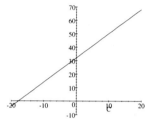

Problem 3.72. *Let* $f(x) = 3x - 12$. *The graph of* $f(x)$ *is*:

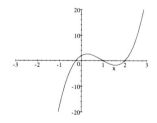

From the graph,

(a) *Which values of* x *force* $f(x) = 0$?

(b) *Factor* $3x - 12$.

Problem 3.73. *Consider the graph of* $f(x) = 4x^3 - 11x^2 + 5x + 2$:

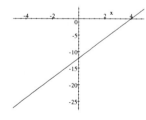

From the graph,

(a) *Find the solution to* $4x^3 - 11x^2 + 5x + 2 = 0$.

(b) *Factor* $4x^3 - 11x^2 + 5x + 2$.

Problem 3.74. *The profi of a company is given by* $V(t) = t^3 - 5t^2 + 2t + 8,$ *where* t *is measured in days from today.* $V(t)$ *is measured in dollars. The graph of* $V(t)$ *is :*

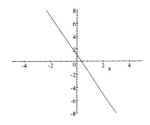

From the graph,

(a) On which days, from today, is the value of the stock zero?

(b) Factor $V(t)$.

Problem 3.75. *Solve*:

$(a) 2(x + 1) = 0$. $(b) 2\left(x - \sqrt{2}\right)(x + 4)\left(x - \sqrt{3}\right) = 0$.

Problem 3.76. *Let* $f(x) = -3x + 1$. *The graph of* $f(x)$ *is as shown*:

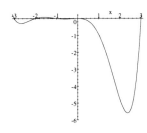

From the graph,

(a) Solve $-3x + 1 = 0$.

(b) Graph $g(x) = |-3x + 1|$.

(c) Solve $|-3x + 1| = 0$.

Problem 3.77. *Let* $g(x) = 2x^2 + x - 6 = (2x - 3)(x + 2)$.

(a) Graph $g(x) = (2x - 3)(x + 2)$.

(b) Graph $h(x) = |(2x - 3)(x + 2)|$.

Problem 3.78. *Let* $h(x) = \frac{2}{100}(x - 3)(x + 3)(x + 2)(x + 1)x^2$. *The graph of* $h(x)$ *is shown here*:

From the graph,

(a) Solve $h(x) = 0$.

(b) Solve $h(x) < 0$.

(c) Solve $h(x) > 0$.

Chapter 4.

Functions

4.1. Functions and Our Daily Life

A basic building block of mathematics is given by the concept of a function. Functions often operate naturally, and invisibly, in our every-day lives.

For example, a graph used by pediatricians to determine the normal growth of children is shown below:

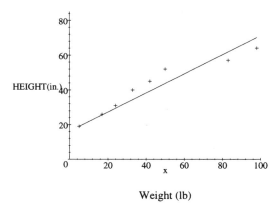

Weight (lb)

This graph uses gathered data, represented in the following table:

Weight (lbs.)	5	17	24	33	42	50	83	98
Height (inches)	19	26	31	40	45	52	57	64

The plotted data reveals a pattern: a straight line that passes through most of the points. This tells us that there is a special relationship between the height of a child and their weight.

- This **relationship** associates a given **weight** (of a child) to **exactly one height** on the graph.
- This relationship is called a **function**.

Definition

Functions

Informally, **a function** is an **association between two sets of values**.

This association can be expressed by words, by graphs, by tables, or by equations.

Example **4.1.** *Write an* **equation** *to express a relationship between the* **total cost** $(f(x))$ *of a plumbing job, the* **hours**(x) *the plumber spent working, plus the* **cost of materials** (c).

- **Solution:** $f(x) = 25x + c$, where $f(x)$ = the total cost, x = number of hours spent working, c = cost of materials, and 25 = the hourly wage in $'s.

Our discussion of a function can be summarized formally by the following definition:

> **Definition**
>
> **Function, domain and range**
>
> A **function** is a relationship between two sets of values, namely D (**domain**) and R (**range**), such that each value d in D is related (associated) to **exactly one** value r in R.
>
> This relationship can be described by $\boxed{r = f(d)}$
>
> -where r is known to be the **output** of the function f for the **input** d.

Example **4.2.** Use the above symbolic notation to describe the following function: For every normal cat (c) there are four legs (l).

- Solution: $l = f(c) = 4c$. That is, of the form $\boxed{r = f(d)}$. Thus, when $c = 2$, $f(c) = 4c = 8 = l = $ legs.

Exercise **4.1.** Assume that $y = 9$ for all possible real values of x.

(a) Is y a function of x ?

Solution:

(b) Is x a function of y ?

Solution:

Exercise **4.2.** Assume that $x = 8$ for all possible real values of y.

(a) Is y a function of x ?

Solution:

(b) Is x a function of y ?

Solution:

Exercise **4.3.** Determine which of the given situations always defines a function from set D to set R. Support your answers.

(**a**) For a final examination in College Algebra, the following relationship is defined:

Set D is defined to be the set of possible scores on the examination. Set R is defined to be the set of numbers of students who had the same score. Each possible score in A is paired with the number of students who had that score.

Solution:

(**b**) Set^D is defined to be the set of all possible gross taxable incomes in the United States of America the 1996 year. Set^R is defined to be the set of all amounts paid in Federal Income Tax for the year 1996. Each gross taxable income from the set ^{D} is paired with the corresponding amount of Tax paid for that income in the year 1996 in ^{R}.

Solution:

Exercise **4.4.**There are 400 students in a given history class. For each calendar date, you can associate every student in the class who has that birth date.

(a) State why this association does not define a function from the set of all calendar dates to the set of all students in the history class.

Solution:

(b) Assume that you are reversing the association, meaning that you consider the set of students in the history class to be the domain and the set of birth dates to be the range. Show that this reverse association defines a function.

Solution:

4.2. Numerical or Tabular Representation of Functions

The following table contains some data from the 1996 IRS Income Tax Table:

Income ($)	10,000	11,000	12,000	13,000	14,000
Tax ($)	1504	1654	1804	1954	2104

This table shows that each $1,000 increase in income results in an increase in tax of $150. This means that the **tax** amount **varies directly** with the **income** amount.

Recalling the definition of a function (a relationship between two sets of values or quantities, namely **Domain** and **Range**, such that every value in the domain is associated with exactly one value in the Range) we can easily verify that this table represents a function.

This function can also be represented by the set of ordered pairs: $(10000, 1504), (11000, 1604), (12000, 1804), (13000, 1954), (14000, 2104)$.

Exercise **4.5**. *A car, costing $30,000, depreciates at a rate of $2,500 per year for the first six years.*

(a) Complete the following table:

Years owned	0	1	2	3	4	5
Value of House	$30,000	$27,500				

(c) Give a graphical representation for this function.

Solution:

(b) Show that this table represents a function.

Solution:

(d) Give a symbolical representation for this function.

Solution:

(e) Identify a set of ordered pairs that defines the above table as a function.

Solution:

Exercise **4.6**. *A house, costing $85,000, appreciates at a rate of 4 percent per year for the first 5 years.*

(a) Complete the following table:

Years owned	0	1	2	3	4	5
Value of House	$85,000	$88,400				

(c) Give a graphical representation for this function.

Solution:

(b) Show that this table represents a function.

Solution:

(d) Give a symbolical representation for this function.

Solution:

(e) Identify a set of ordered pairs that defines the above table as a function.

Solution:

Exercise 4.7. *The following table gives the height of an object, at various times after being dropped from the top of a building* $1,250$ *feet high.*

(a) Complete the following table:

Time(in seconds)	0	2	4	5	8
Height (in feet)	1250	1186	994	850	

(c) Give a graphical representation for this function.

Solution:

(b) Show that this table represents a function.

Solution:

(d) Identify a set of ordered pairs that defines the above table as a function.

Solution:

4.3. Graphical or Visual Representation of Functions

There are several effective ways of expressing functions graphically: bar charts, pie charts, and scatter plotting. The graph shown below displays the amount of federal income tax for certain incomes.

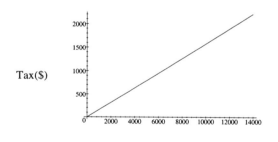

Tax($)

Income ($)

Symbolically we represent this function as

$f(Income) = Tax$ or $Tax = f(Income)$.

That is, every domain value (income) has exactly one range value (tax). That is, tax is a function of income.

Similarly, the following graph shows the performance of Dow-Jones stocks for the period of 1980-1996.

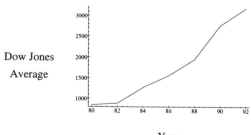

Dow Jones Average

Year

Graphical representations of functions are valuable because they provide us with a clear overview of the behavior of the involved quantities. The graphical form of a function can be converted into a table, or a numerical form of the function. For example, the following table is a representation of the previous graph.

Year	80	82	84	86	88	90	92
Dow-Jones	839	875	1259	1547	1939	2753	3169

When using graphs to define functions, see Chapter 3, the following terms are often used:

Definition

Independent Variable (Domain)

The symbol used to express the given number

is called the **independent variable**.

Dependent Variable (Range)

The number(s) produced by the function

are called the **dependent variable(s)**.

For $f(x) = 2x$, the **independent variable** is x,

while the **dependent variable** is $f(x)$.

Example **4.3. i**. *Identify which of the following graphs defines a function*

ii. *Determine their* **domains** *and their* **ranges**.

(a)   (b)

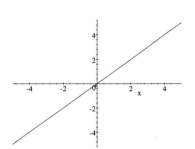 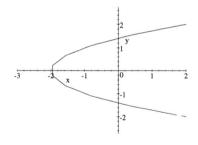

Solution: **(a) i**. This is a function, since every x-value (domain) has only one y-value (range).

ii. Domain: $\{x \mid x \text{ is real}\}$, Range: $\{y \mid y \text{ is real}\}$.

(b) i. This is not a function, since the domain value has more than one corresponding range value.

ii. Domain: $\{x \mid x \geq -2\}$, Range: $\{y \mid y \text{ is real}\}$.

Exercise **4.8**. *Identify which of the following graphs define a function; those that* **do**, *determine their* **domains** *and their* **ranges**.

(a)

(b)

(c)

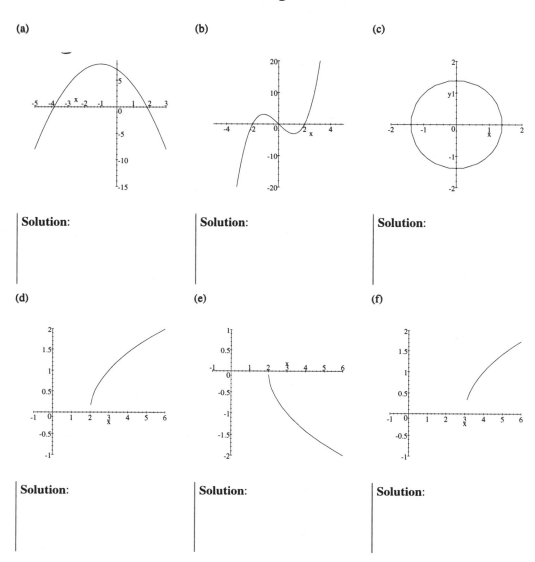

Solution:

Solution:

Solution:

(d)

(e)

(f)

Solution:

Solution:

Solution:

Exercise **4.9**. *Identify the intervals over which the function is constant, decreasing, increasing, strictly decreasing or strictly increasing.*

(a)

(b)

(c)

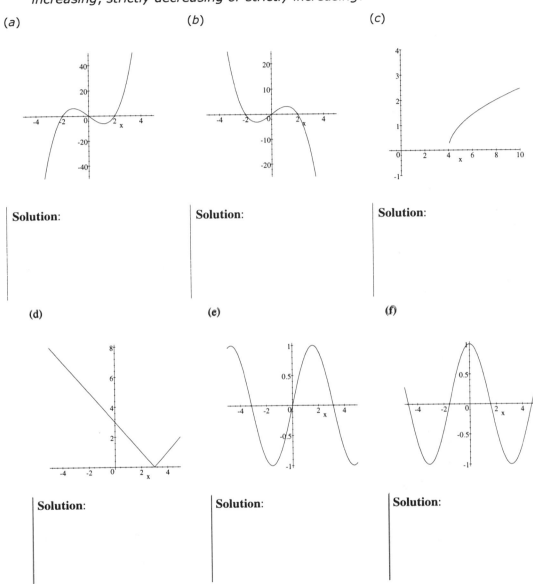

Solution:

Solution:

Solution:

(d)

(e)

(f)

Solution:

Solution:

Solution:

Exercise *4.10. An uncovered piece of roast beef with a temperature of* $65°F$ *is placed into a hot oven at a temperature of* $400°F$ *and left to cook.*

(a) Sketch a reasonable graph for the temperature of the beef as a function of time.

(b) Identify appropriate values for the independent and dependent variables of this function.

Solution:

Solution:

4.4. Symbolic or Algebraic Representation of Functions

Let's consider the following expression: $y = 5(x + 4)$.

This expression describes a rule for calculating the value of y for a given value of x. The value of y for $x = 2$ will be computed first by adding 4 to 2 and then by multiplying the result by 5, which is 30. If you choose 3 for x, then the computed value for y will be 35. It is clear then that the value of y depends on the chosen value of x. Each chosen value of x will produce a unique value for y through the expression $y = 5(x + 4)$.

■ Hence, this expression represents a **function**.

▶ Naming this function $f(x)$, we produce a standard notation that easily identifies the entity as a function. From now on, we can refer to this function as:

◀ $y = f(x)$ $= 5(x + 4)$.

Similarly, the expression $\sqrt{y} + 5$, means: take the square root of a given non-negative real number y and add 5 to the computed square root.

▶ Naming this function $g(y)$, we have: $x = g(y)$ $= \sqrt{y} + 5$.

The value of x for $y = 4$, will be computed by taking the square root of 4, which is 2, and adding this value to 5, which is 7.

◀ $5(x + 4)$ and $\sqrt{y} + 5$ are symbolic or algebraic forms of the function $f(x)$ and $g(y)$, respectively.

Example *4.4. Determine the* **domain** *and* **range** *of the following equation* $x - y^2 = -2$.

• **Solution**: Solve for x : $x = -2 + y^2$. Range $\{x | x \geq -2\}$.

Solve for y : $y = \sqrt{x + 2}$. Domain $\{x | x \geq -2\}$.

⏐Note⏐ This equation does **not** represent a function, since certain values of x in $\pm\sqrt{x + 2}$ give two solutions.

Exercise 4.11. *Identify which of the following equations define* y *as a function of* x:

(a) $5x + 3y = 1$.

Solution:

(b) $y + |x| = 4$.

Solution:

(c) $4x^2 - y^2 = 1$.

Solution:

(d) $y - x^3 = 5$.

Solution:

(e) $x^4 - 2x + y = 0$.

Solution:

(f) $y^3 + 8 - x = 16$.

Solution:

Exercise 4.12. *Complete the following exercises.*

(a) For the function $f(x) = |x| + 4$, compute the values that corresponding with $x = -2, 2, -3, 4$

Solution:

(b) For the function $r(x) = \frac{x}{x+4}$, compute the values corresponding with $x = 4, 0, -6$

Solution:

(c) Given $s(x) = \frac{x}{|x|}$, find the values corresponding with $x = -1$, $x = 1$, $x = 5$, $x = -5$

Solution:

Exercise 4.13. *Determine the domain of the function represented by the given equation.*

(a) $f(x) = \sqrt{7 - x}$.

Solution:

(b) $f(x) = \sqrt{x^2 - 9}$.

Solution:

(c) $f(x) = \frac{1}{\sqrt{x-4}}$.

Solution:

(d) $f(x) = 1 - 2x$

Solution:

(e) $f(x) = \frac{\sqrt{7-x}}{5}$.

Solution:

(f) $f(x) = 2x$

Solution:

(g) $f(x) = \sqrt{x}$.

Solution:

(h) $f(x) = 2 - x^2$.

Solution:

4.5. Inverse Functions

One To One Functions

A function can be represented by a set of ordered pairs.

For example, the function $y = f(x) = x + 3$, defined over the set $D_f = \{3, 4, 5, 6\}$, can be represented by the following set of ordered pairs:

$$\Rightarrow \{(3, f(3)), (4, f(4)), (5, f(5)), (6, f(6))\}$$
$$= \{(3, 6), (4, 7), (5, 8), (6, 9)\}$$

For this function (f) you can start at any **range** value y in R_f and trace it back to exactly one **domain** value x in D_f .

■   Functions with this property are said to be **one-to-one functions**.

Conversely, the function defined over the set $D_g \{-2, 1, 2, 3\}$ by $g(x) = x^2$, can be represented by the following set of ordered pairs:

$$\Rightarrow \{(-2, g(-2)), (2, g(2)), (1, g(1)), (3, g(3))\}$$
$$= \{(-2, 4), (2, 4), (1, 1), (3, 9)\}$$

For the function (g), you can **not** start at any range value y in R_f and trace it back to exactly one domain value x in D_g . This means that for the function g, there exists at least one value in the range R_g which is associated with more than one value from the domain D_g . For example, 4 from R_g is associated with -2 **and** 2, from D_g .

■   Clearly $g(x) = x^2$ is **not** a one-to-one function.

| Definition |

One-to-one Functions

A function with domain D and range R is said to be **one-to-one** if for x_1 and x_2 in D, $f(x_1) = f(x_2)$ implies that $x_1 = x_2$.

Example **4.5.** *Show that the function* $f(x) = 2x + 5$ *is one-to-one.*

●   **Solution:** Assume that $f(x_1) = f(x_2)$. Then,

$$2x_1 + 5 = 2x_2 + 5 \ (Subtracting \ 5 \ from \ both \ sides)$$
$$2x_1 = 2x_2$$
$$x_1 = x_2 .$$

Example **4.6.** *Show that the function* $f(x) = x^4$ *is not one-to-one.*

●   **Solution:** $f(-1) = (-1)^4 = 1$ and $f(1) = (1)^4 = 1$ while $1 \neq -1$.

Inverse Functions

By exchanging the first and the second coordinates of each of the ordered pairs of f and g, you can construct the **inverse** of f and g as follows:

$$f = \{(3,6),(4,7),(5,8),(6,9)\}$$
$$\text{Inverse of } f = \{(6,3),(7,4),(8,5),(9,6)\}$$

$$g = \{(-2,4),(2,4),(1,1),(3,9)\}$$
$$\text{Inverse of } g = \{(4,-2),(1,-1),(4,2),(9,3)\}$$

The inverse of f is a relationship defined from the set $Rf = \{6,7,8,9\}$ to the set $D_f = \{3,4,5,6\}$.

■ This relationship is a function and is called the **inverse function of** f

▶ It can be written as $f^{-1}(x)$

The inverse of g is also a relationship defined from R_g to D_g. But this relationship is **not a function**.

> We **conclude** that the **inverse of a one-to-one** function **is a function**,
> and the **inverse of a non one-to-one** function is **not** a function.

Definition

Inverse functions

• Two functions f and g are said to be **inverse functions** of each other if:

$f(g(x)) = x$ for all values of x in the domain of g and
$g(f(x)) = x$ for all values of x in the domain of f

Example 4.7. Show that $f(x) = 2x+3$ and $g(x) = \dfrac{(x-3)}{2}$ are inverses of each other.

• **Solution:**

$$f(g(x)) = f(\tfrac{x-3}{2})$$
$$= 2(\tfrac{x-3}{2}) + 3$$
$$= x - 3 + 3$$
$$= x$$

$$\text{and } g(f(x)) = g(2x+3)$$
$$= \frac{2x+3-3}{2}$$
$$= \frac{2x}{2}$$
$$= x$$

Example **4.8**. Show that $f(x) = 2x + 3$ and $g(x) = \frac{(x+3)}{2}$ are not inverses of each other.

- **Solution**:

$$f(g(x)) = f(\frac{x+3}{2})$$
$$= 2(\frac{x+3}{2}) + 3$$
$$= x + 3 + 3$$
$$= x + 6$$
$$\text{and } g(f(x)) = g(2x + 3)$$
$$= \frac{2x + 3 + 3}{2}$$
$$= \frac{2x + 6}{2}$$
$$= \frac{2x}{2} + \frac{6}{2}$$
$$= x + 3$$

Exercise **4.14**. In the following exercises use the given graph to determine if the represented function is a one-to-one function.

(a)

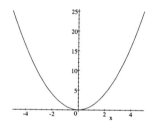

Solution:

(b)

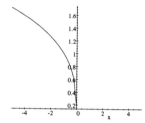

Solution:

(c)

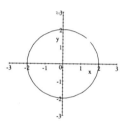

Solution:

(d)

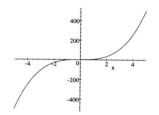

Solution:

(e)

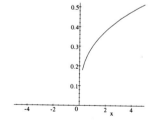

Solution:

(f)

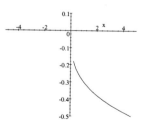

Solution:

Exercise 4.15. *In the following exercises show that the functions f and g are inverses of each other.*

(a) $f(x) = x + 3$; $g(x) = x - 3$.

Solution:

(b) $f(x) = 2x + 5$; $g(x) = \frac{x-5}{2}$.

Solution:

(c) $f(x) = \frac{3}{x}$; $g(x) = \frac{3}{x}$.

Solution:

(d)) $f(x) = x^3$; $g(x) = \sqrt[3]{x}$.

Solution:

(e) $f(x) = \frac{1}{x-5}$; $g(x) = 5 + \frac{1}{x}$.

Solution:

(f) $f(x) = \sqrt[3]{x+2}$; $g(x) = x^5 - 2$.

Solution:

Exercise 4.16. *Pair the graph with the graph of its inverse.*

(a)

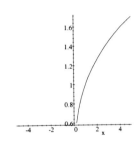

i.

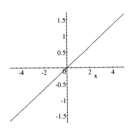

ii.

iii.

Solution:

(b)

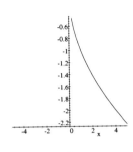

i

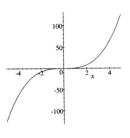

ii

iii

Solution:

Exercise **4.17**. *In the following exercises, verify graphically that the function $f(x)$ has an inverse. Sketch the graph of $f^{-1}(x)$ for these exercises.*

(a)

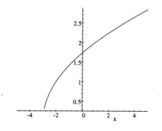

Solution:

(b)

Solution:

(c)

Solution:

Exercise **4.18**. *In the following exercises, verify graphically that the function f does not have an inverse.*

(a)

Solution:

(b)

Plot

Solution:

(c)

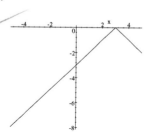

Solution:

4.6. Power and Polynomial Functions

Power Functions

Power functions are important functions. These functions are used extensively in geometry, chemistry and physics.

Examples are:

- The area A of a square with side a is $A(a) = a^2$.
- The volume V, of a sphere with radius r is $V(r) = \frac{4}{3}\pi r^3$.
- The area of a circle with radius r is $A(r) = \pi r^2$.
- The positive square root r, of a number n is $r(n) = \sqrt{n} = n^{(1/2)}$.

| Definition |

Power Functions

A **power function** of x is any function of the form

$$f(x) = cx^p$$

where c and p are any constants.

Example **4.9.** *Which of the following functions are power functions? For each power function, identify the value of c and p in the formula $f(x) = cx^p$.*

(a) $p(x) = 1000(1.06)^x$.

(b) $q(x) = 5x^2$.

(c) $k(x) = x^{\frac{-3}{2}}$.

(d) $g(x) = 25(x-3)^5$.

(e) $h(x) = (f)g(x) = 3x$.

- **Solution:**

  (a) The function p is not a power function.

  (b) The function q is a power function. $c = 5$ and $p = 2$.

  (c) The function k is a power function. $c = 1$ and $p = -\frac{-3}{2}$.

  (d) The function g is not a power function because g is proportional to the fifth power of $(x\text{-}2)$, it is not proportional to a power of x.

  (e) The function h is a power function. $c = 1$ and $p = 1/3$.

Polynomial Functions

Another important and interesting family of functions is the **polynomial family**. They can be constructed by a combination of sums and/or differences of constant multipliers of non-negative power functions.

- The **highest power** of the variable present in a polynomial is called the **degree of the polynomial**.
- The **coefficients** of a polynomial are the constant multiples present.

For example, the following are all polynomials

$3x^4 + 2x^3 - 3x^2 + 6$	4	$3, 2, -3, 6$
$8x^{10} - 10x^8 + 5x^2 + 3.5 - 4$	10	$8, -10, 5, 3.5, -4$
$3x^2 + 5$	2	$3, 5$
$5x^9$	9	5
x	1	1 (and not 0)

Definition

Polynomials in Standard Form

Let n be a non-negative integer and let $a_n, a_{n-1}, \ldots a_1,$ and a_0 be real numbers. A polynomial of degree n in variable x in standard form is an expression of the form $a_n x^{n-1} + a_{n-1}x^{n-1} + \ldots + a_1 x^1 + a_0$, where $a_n \neq 0$.

a_n is called the **leading coefficient**, and

a_0 is called the **constant term**.

Polynomials with **one**, **two**, or **three terms** are called **monomials**, **binomials** or **trinomials**, respectively.

For example:

- $3x^4, 2x^3, -3x^2$, and 6 and are monomials.
- $3x^4 + 2x^3$, and $-3x^2 + 6$ are examples of binomials.
- $3x^4 + 2x^3 - 3x^2$, and $2x^3 - 3x^2 + 6$ are examples of trinomials.
- A polynomial with **all zero coefficients** is called the **zero polynomial**, denoted by 0.

4.6.i.   Operations with Polynomials

It is possible to add and subtract polynomials in the same way that you add and subtract real numbers, as long as you add or subtract like terms (terms having the same exponent on the variable; that is, the terms have the same degree) by adding or subtracting their coefficients.

For example

$$3y^2 + 5y^2 = (3 + 5)y^2$$
$$= 8y^2$$

and

$$3y^2 - 5y^2 = (3 - 5)y^2$$
$$= -2y^2$$

Example **4.10.** *Solve* $(5y^3 - 7y^2 + 4) + (y^3 + 2y^2 - y + 10)$.

Solution:

$$\Rightarrow (5y^3 - 7y^2 + 4) + (y^3 + 2y^2 - y + 10)$$
$$= (5y^3 + y^3) + (-7y^2 + 2y^2)4) - y + (10 + 4)$$
$$= (5 + 1)y^3 + (-7 + 2)y^2 - y + 14$$
$$= 6y^3 - 5y^2 - y + 14$$

Example **4.11.** *Solve* $(7y^4 - y^3 + 4y - 5) - (3y^4 + 4y^2 + 3y - 7)$.

- **Solution:**

$$\Rightarrow (7y^4 - y^3 + 4y - 5) - (3y^4 + 4y^2 + 3y - 7)$$
$$= (7y^4 - 3y^4) - y^3 - 4y^2 + (4y - 3y)(7 - 5)$$
$$= (7 - 3)y^4 - y^3 - 4y^2 + (4 - 3)y + 2$$
$$= 4y^4 - y^3 - 4y^2 + y + 2$$

You can multiply two polynomials by using the left and the right distributive properties.

For example, if you treat $(3y + 4)$ as a single term, you can multiply $(2y - 3)$ by $(3y + 4)$, as follows:
$$(2y - 3)(3y + 4) = 2y(3y + 4) - 3(3y + 4)$$
right distributive property
$$= (2y)(3y) + (2y)(4) - (3)(3y) - (3)(4)$$
left distributive property
$$= 6y^2 + 8y - 9y - 12$$

- The **FOIL Method** can be applied for speed.

Example **4.12.** *Compute the product of* $(ax + b)$ *and* $(cx + d)$ *by using the FOIL Method.*

- **Solution:**

$$(ax + b)(cx + d)$$

First terms
Outer terms
Inside terms
Last terms

$$= acx^2 + (ad + bc)x + bd$$

Note: The left and the right distributive properties can be extended to multiply polynomials with three terms (trinomials) or more.

Example **4.13.** *Compute the product of* $(x^2 + 2x - 2)$ *and* $(x^2 - 2x + 3)$.

- **Solution:** $(x^2 - 2x + 3)(x^2 + 2x - 2)$

$$= \underbrace{(x^2(x^2 + 2x - 2))} + \underbrace{(-2x(x^2 + 2x - 2))} + \underbrace{(3(x^2 + 2x - 2))}$$
$$= (x^4 + 2x^3 - 2x^2) + (-2x^3 - 4x^2 + 4x) + (3x^2 + 6x - 6)$$
$$= x^4 + 0x^3 - 3x^2 + 10x - 6$$

Exercise 4.19. *In the following exercises, identify which of the algebraic expressions are polynomials. For each polynomial, identify:*

The degree and the nonzero coefficients.
Whether the polynomial is a monomial, binomial, or trinomial.

(a) $x^3 + 0.3x - 4$.

Solution:

(b) $x^5 - 5x^2 - 4$.

Solution:

(c) $x^{0.5} + 0.3x - 4$.

Solution:

(d) $x^3 - 4$.

Solution:

(e) $4x^{\frac{1}{4}}$.

Solution:

(f) $12x^9$.

Solution:

Exercise 4.20. *Sketch the graphs of the following equations; identify any similarities or differences in the graphs.*

(a) $y = x^2$.

Solution:

(b) $y = x^4$.

Solution:

(c) $y = x^6$.

Solution:

(d) $y = x^8$.

Solution:

Exercise 4.21. *Sketch the graphs of the following expressions; identify any similarities or differences in the graphs.*

(a) $y = x^1$.

Solution:

(b) $y = x^3$.

Solution:

(c) $y = x^5$.

Solution:

(d) $y = x^7$.

Solution:

Exercise 4.22. *Sketch the graphs of the following expressions; identify any similarities or differences in the graphs.*

(a) $y = x^{-1}$.

Solution:

(b) $y = x^{-2}$.

Solution:

Exercise 4.23. *Sketch the graphs of the following expressions; identify any similarities or differences in the graphs.*

(a) $y = x^{\frac{1}{2}}$.

Solution:

(b) $y = x^{\frac{1}{3}}$.

Solution:

Exercise 4.24. *In the following exercises, identify which of the algebraic expressions are polynomials. For each polynomial, identify:*

i. *The degree and the nonzero coefficients.*

ii. *Whether the polynomial is a monomial, binomial, or trinomial.*

(a) $x^3 + 0\ 3x - 4$.

Solution:

(b) $x^5 - 5x^2 - 4$.

Solution:

(c) $x^{0.5} + 0\ 3x - 4$.

Solution:

(d) $x^3 - 4$.

Solution:

(e) $4x^{\frac{1}{4}}$.

Solution:

(f) $12x^9$.

Solution:

(g) x .

Solution:

(h) -4 .

Solution:

Exercise 4.25. *In the following exercises, perform the indicated operations and simplify the results.*

(a) $(3x^4 - 5x^2 + 13x - 4)$
$+(x^4 - 5x^2 + 20)$.

Solution:

(b) $(3x^4 - 5x^2 + 13x - 4)$
$-(x^4 - 5x^2 + 20)$.

Solution:

(c) $(2x + 1)(2x - 1)(4x^2 - 1)$
$-(16x^4 - 2)$.

Solution:

(d) $(3x^3 - 5x^2 + 13x - 4)(5x + 20)$.

Solution:

(e) $(x^3 - x^2 + 13x)(x^2 - 5x + 4)$.

Solution:

(f) $(2y + 3)(y^3 - 5y + 6)$.

Solution:

4.7. Rational Functions (see also Chapter 5.3)

The **quotient** of two **polynomials** is called a **rational function**. For example, the quotient of $f(x) = x + 1$ and $g(x) = x - 2$, is a rational function, and can be written as $q(x) = \frac{x+1}{x-2}$.

The **domain** of a rational function of x includes all real numbers except values which make its denominator zero.

For example:

- $h(x) = \frac{x-2}{x+1}$ the domain of h excludes $x = -1$ and $x = 0$.
- $q(x) = \frac{(x+1)(x-2)}{x(x-3)}$ the domain of q excludes $x = -3$.
- $l(x) = \frac{x^2+1}{x(x-3)}$ the domain of l excludes $x = 0$ and $x = 3$.
- $k(x) = \frac{3x+4}{x^2+1}$ the domain of k is the set of all real numbers.

Example **4.14**. *Find the domain, and analyze the behavior, of the function f near any value that makes its denominator zero.*

- **Solution**: Since the denominator is zero when $x = 3$, the domain of f is the set of all real numbers except $x = 3$.

4.7.i. Asymptotes

Observe the following table and associated graphs:

x	2.9	2.99	2.999	3	3.001	3.01	3.1
$f(x)$	−59	−599	−5999	$-\infty$ $+\infty$	6111	611	61

As x approaches 3 from the left of the number line, $f(x)$ decreases without bound.
As x approaches 3 from the right of the number line, $f(x)$ increases without bound.

- The graph of $f(x)$ shown below demonstrates this behavior visually:

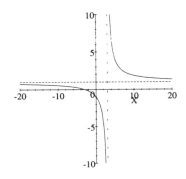

▶ We call the line $x = 3$ a **vertical asymptote** of the graph of f.
As x decreases without bound or increases without bound, the values of $f(x)$ approach $y = 1$.
▶ We call the line $y = 1$ the **horizontal asymptote**.

Vertical and Horizontal Asymptotes

The line $x = a$ is a **vertical asymptote** of the graph of a
function f if $f(x) \to -\infty$ or $f(x) \to \infty$ as x approaches a
from either the left or the right.
The line $y = b$ is a **horizontal asymptote** of the graph of a
function $f(x)$ if $f(x) \to b$ as $x \to -\infty$ or $x \to \infty$.

Example **4.15**. *Find horizontal asymptotes of the rational function* $f(x) = \frac{2x+1}{x-1}$.

- **Solution**: The line $\boxed{y = 2 \text{ is a } \textbf{horizontal asymptote}}$ for the graph of $f(x)$, as shown below.
- $\boxed{\text{Note}}$ the **degree** of the **numerator** is the **same** as the **degree of the denominator**.

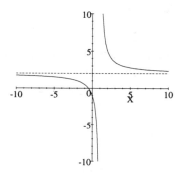

Example **4.16**. *Find horizontal asymptotes of the rational function* $g(x) = \frac{1}{x+2}$.

- **Solution**: The line $\boxed{y = 0 \text{ is a } \textbf{horizontal asymptote}}$ for the graph of g, as shown below.
- $\boxed{\text{Note}}$ that the **degree of numerator** is **less than** the **degree of the denominator**.

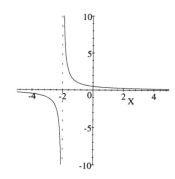

Example **4.17.** *Find horizontal asymptotes of the rational function* $h(x) = \frac{x^2 - 4}{x - 3}$ *.*

- **Solution**: The graph of $\boxed{h \text{ has } \underline{\textbf{no horizontal asymptote}}}$ because as x decreases or increases without bound the values of f decreases or increases without bound too.

- $\boxed{\text{Note}}$ the **degree** of the **numerator** is **greater than** the **degree** of the **denominator**.

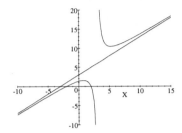

4.7.i.a. Finding asymptotes of a rational function.

Use the following procedure:

Let $q(x) = \dfrac{f(x)}{g(x)} = \dfrac{a_n x^n + a_{n-1} + \ldots + a_0}{b_m x^m + b_{m-1} x^{m-1} + \ldots + b_0}$, and assume that $f(x)$ and $g(x)$ have no common factors.

Then:

- ▪ If $g(u) = 0$, then $x = u$ is a **vertical asymptote**.
- ▪ If $n < m$, then $y = 0$ is a **horizontal asymptote** (numerator degree $<$ denominator degree).
- ▪ If $n = m$, then $y = \frac{a_n}{b_m}$ is a **horizontal asymptote** (numerator degree $=$ denominator degree).
- ▪ If $n > m$, then the graph of q has **no horizontal asymptote** (numerator degree $>$ denominator degree).

Example **4.18.** *Analyze the behavior of the graph of the rational function* $r(x) = \frac{x - 1}{x^2 - 4}$ *.*

- **Solution**:

  $x - 1 \Rightarrow x = 1$. Therefore $r(x)$ has a zero at $x = 1$.

  $x^2 - 4 = 0 \Rightarrow (x - 2)(x + 2) = 0$. Thus the denominator of $r(x)$ has two zeroes at $x = 2$ and $x = -2$. This means that the lines $x = 2$ and $x = -2$ are vertical asymptotes of $r(x)$.

  Since the **degree** of the **numerator is less than** the **degree** of the **denominator**, the line $y = 0$ is a **horizontal asymptote** for the graph of $r(x)$.

  Knowing when the graph of r is **above or below** the x-axis is a useful aid to analyze its behavior. Since $r(x) = 0$ if and only if $x = 1$, we need to consider other possible values of x. To find the signs of $r(x)$, we select two convenient test values (one greater than 1 and one less than 1) for the numerator and three convenient test values (different form 2 and -2) in the intervals: $(-\infty, -2)$, $(-2, +2)$ and $(2, \infty)$ as follows:

- $(-\infty, 1)$: Test value : $x = 0$: Then, $x - 1 = 0 - 1 = -1$.
- $(1, \infty)$: Test value: $x = 3$ Then, $x - 1 = 3 - 1 = 2$.
- $(-\infty, -2)$: Test value: $x = -3$ Then, $x^2 - 4 = (-3)^2 - 4 = 5$.
- $(-2, 2)$: Test value: $x = 0$ Then, $x^2 - 4 = (0)^2 - 4 = -4$.
- $(2, \infty)$: Test value: $x = 3$ Then, $x^2 - 4 = (3)^2 - 4 = 5$.

The results of these steps are summarized in the following sign table:

Test point	Sign $(x-1)$	Sign(x^2-4)	Sign $r(x) = \dfrac{(x-1)}{(x^2-4)}$	Graph relative to x-axis
$-\infty$	$-\infty$	$-\infty$	0	Below
-3	$-$	$+$	$-$	Below
-2	$-$	$+$	$-$	Below
0	$-$	$-$	$+$	Above
1	$-$	$-$	$+$	Above
2	$+$	$-$	$-$	Below
3	$+$	$+$	$+$	Above
∞	$+$	$+$	$+$	Above

■ $y = 0$ is a horizontal asymptote.

■ $x = 2$ and $x = -2$ are vertical asymptotes.

Using the gathered information in the above table gives us the following graph for $r(x)$.

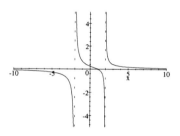

Note: The graph of a function may intersect its horizontal asymptote.

Exercise 4.26. Use a sign table to sketch the graph of:

(a) $y = f(x) = \dfrac{x+2}{x-2}$.

Solution:

(b) $y = f(x) = \dfrac{2x^2}{x^2-9}$.

Solution:

(c) $y = f(x) = \dfrac{3}{x^2 + 1}$.

Solution:

(d) $y = f(x) = \dfrac{3}{x^2 - 1}$.

Solution:

Exercise 4.27. *In the following exercises*

- **i.** Identify the domain of the function. **ii.** Identify any horizontal or vertical asymptotes. **iii.** Sketch the graph of the function.

(a) $f(x) = \dfrac{1}{x + 3}$.

Solution:

(b) $f(x) = \dfrac{x^3}{x^2 - 1}$.

Solution:

(c)) $f(x) = \dfrac{x - 3}{x + 3}$.

Solution:

(d) $f(x) = \dfrac{4}{x^2 - 4}$.

Solution:

4.8. Translation of Functions

Graphs of many functions are simple transformations of the graphs of a **few common functions**. The graphs of these __fundamental functions__ are shown below:

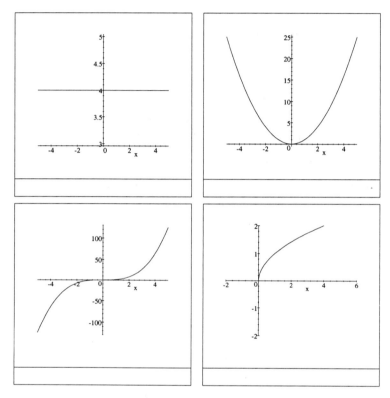

For example, you can use the graph of $f(x) = x^2$ as a guide to obtain the graph of $h(x) = x^2 - 2$. You need to shift the graph of $f(x)$ down two units as shown below:

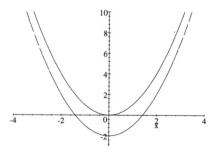

Similarly, you can get the graph of $g(x) = (x + 2)^2$ by shifting the graph of $f(x) = x^2$ to the left two units, as shown below:

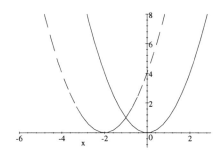

In the above examples, the functions h and g have the following relationship with the function f:

$$h(x) = x^2 - 2 = f(x) - 2$$
$$g(x) = (x+2)^2 = f(x+2)$$

Example 4.19. Use the graph of $f(x) = x^3$, to graph

    (a) $g(x) = x^3 + 3$.

    (b) $h(x) = (x-3)^3$.

• **Solution:**

    (a) $g(x) = f(x) + 3$ **(b)** $h(x) = (x-3)^3 = f(x-3)$

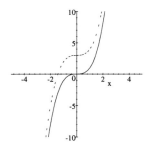 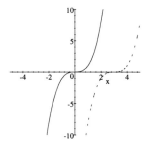

Exercise 4.28. Use the graph of $y = f(x)$, shown below, to sketch the graph of each of the following:

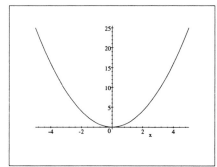

(a) $y = f(x - 2) + 3$.

Solution:

(b) $y = -2f(x) + 3$.

Solution:

(c) $y = f(x) + 4$.

Solution:

(d)) $y = -f(x + 2)$.

Solution:

Exercise **4.29**. *Use the graph of* $f(x) = -x^2 + 8$ *to sketch the graph of each of the following*:

Plot $f(x) = -x^2 + 8$ first.

(a) $y = f(x) + 2$.

Solution:

(b) $y = f(x - 2)$.

Solution:

(c) $y = f(x - 2) + 3$.

Solution:

(d)) $y = 3f(x - 2) - 3$.

Solution:

Chapter Summary

- Function.
  - ▶ A Function f is a rule or an association between two sets of values, namely **domain** and **range**, such that each value d in the domain is associated exactly with one value r in the range. The association can be expressed by words, by graphs, by tables, by a set of ordered pairs or by equations or formulas.
- Vertical line test.
  - ▶ If any vertical line intersects a given graph at more than one point, the given graph does not represent a function. If any vertical line intersects a given graph at one point only, then the given graph represents a function.
- Horizontal line test.
  - ▶ If any horizontal line intersects a given graph at more than one point, then the inverse of the function, represented by the graph, is not a function.
- Composition of two functions
  - ▶ If f and g are functions, the composition function $f \circ g$ is defined by $(f \circ g)(x) = f(g(x))$.
- Undoing functions
  - ▶ A function with domain D and range R is one-to-one if, for x_1 and x_2 in D, $f(x_1) = f(x_2)$, and implies that $x_1 = x_2$. Those functions that are one-to-one have undoing functions called **inverse functions**. The inverse of f, denoted by f^{-1} satisfies $(f^{-1} \circ f)(x) = f^{-1}(f(x)) = x$.

End of Chapter 4. Problems

Problem 4.1. *Find the domain of each of the following functions:*

(a) $f(x) = 5x^2 + 10$. (b) $f(x) = \sqrt{x^2 - 4}$. (c) $f(x) = \dfrac{x^2}{x^2 - 2}$. (d) $f(x) = \sqrt{2x + 10}$.

(e) $f(x) = \dfrac{\sqrt{6x - 12}}{3x - 15}$. (f) $f(x) = \dfrac{5x + 4}{x^2}$. (g) $f(x) = x^3 + 2x^2 + 1$. (h) $f(x) = \sqrt{x} + 1$.

Problem 4.2. *In the following tables is y a function of x.*

(a)

x	4	0	-4	4
y	8	0	-8	16

(b)

x	2	3	0	1
y	5	7	1	3

(c)

x	5	1	9	-3
y	2	0	4	-2

(d)

x	4	9	4	16
y	2	3	-2	4

Problem 4.3. *Graph each of the following functions by hand. You may check your graph by using a graphing calculator.*

(a) $f(x) = 2x^2$. (b) $f(x) = 2(x - 3)^2$. (c) $f(x) = 2(x - 3)^2 - 4$. (d) $f(x) = 2(x + 5)^2 + 3$.

Problem 4.4. *Find the inverse of each of the following functions*

(a) $f(x) = \dfrac{5x - 2}{2x - 3}$. (b) $f(x) = 3x + 5$. (c) $f(x) = \dfrac{5}{3x + 5}$. (d) $f(x) = x^{\frac{1}{3}}$.

(e) $f(x) = \sqrt{x + 2}$. (f) $f(x) = \dfrac{2x - 3}{5x + 2}$. (g) $f(x) = \dfrac{5x + 2}{2x - 3}$. (h) $f(x) = x^3$.

Problem 4.5. *For the following functions*

(a) $f(x) = \frac{x+1}{x-1}$. (b) $f(x) = \frac{x^2-4}{x^2+4}$. (c) $f(x) = \frac{x^2+5x-6}{x^2-1}$. (d) $f(x) = \frac{x^2+5x-6}{x^2+5x+6}$.

Find

i. x intercepts of $f(x)$, if any.

ii. y intercepts of $f(x)$, if any.

iii. The equations of the vertical asymptotes, if any.

iv. The equations of the horizontal asymptotes, if any.

v. A sign table.

vi. A graph of $f(x)$.

Problem 4.6. *Find a tabular or numerical representation for the following functions.*

(a)   (b)   (c)

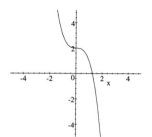 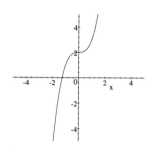

Problem 4.7. *Find a tabular or numerical representation for the following functions:*

(a) $f(x) = 3x^4 + 2x^3 + 5$. (b) $f(x) = 9$. (c) $f(x) = 3x^2 + \sqrt{x}$. (d) $f(x) = \frac{x^2-4}{x^2-1}$.

(e) $f(x) = 5x^5 + 4x^{\frac{1}{4}} + 10$. (f) $f(x) = \frac{1}{5}x^6 + \frac{3}{4}x + 10$. (g) $f(x) = 9x$. (h) $f(x) = \sqrt{x}$.

Problem 4.8. *Graph each of the following polynomials by hand. You may verify your answer by using a graphing calculator.*

(a) $f(x) = x^2 - 4$
$= (x-2)(x+2)$.

(b) $f(x) = x^2 - x - 6$
$= (x-3)(x+2)$.

(c) $f(x) = x^3 - 6x^2 - 4x + 24$
$= (x-6)(x^2-4)$.

(d) $f(x) = x^3 + x^2 - 5x + 3$
$= (x-1)^2(x+3)$.

(e) $f(x) = x^3 + 2x^2$
$= x^2(x+2)$.

(f) $f(x) = x^4 + 3x^3 - 4x^2$
$= x^2(x+4)(x-1)$.

(g) $f(x) = x^2 + x - 2$
$= (x-1)(x+2)$.

(h) $f(x) = x^3 + 2x^2 - x - 2$
$= (x^2-1)(x+2)$.

Problem 4.9. *Perform the following indicated operations:*

(a) $(x^2 + 2x + 4)(x-2)$. (b) $(y^2 - 2y + 4)(y+2)$.

(c) $(x^2 + 2x - 5)(x^2 - 2x + 5)$. (d) $(z^3 - 3z + 5)(z^3 + 3z - 5)$.

Chapter 5.

Polynomials and rational functions

5.1. Quadratic functions

> **Definition**
>
> **Polynomial function**:
>
> If a_n is a real number with n a whole number, $n \in W$, and $a_n \neq 0$,
> then the n-th degree function of x, denoted by
> $$P_n(x) = f(x) = y = a_n x^n + a_{n-1} x^{n-1} + a_{n-2} x^{n-2} + ... + a_1 x + a_0,$$
> is defined to be a **polynomial function** of the n-th degree.

Types of Polynomial Functions

- Constant function (0 degree)
  - ▶ $P_0(x) = 7$

- Linear function (1st degree)
  - ▶ $P_1(x) = 2x + 3$

- Quadratic function (2nd degree)
  - ▶ $P_2(x) = \frac{1}{3}x^2 + 3x - 5$

- Cubic function (3rd degree)
  - ▶ $P_3(x) = x^3 + 2x^2 - 5x + 2.34$

- Quartic function (4th degree)
  - ▶ $P_4(x) = x^4 - 7x^3 + 5x^2 - x + 4$

- n-th degree function
  - ▶ $P_n(x) = f(x) = y = a_n x^n + a_{n-1} x^{n-1} + a_{n-2} x^{n-2} + ... + a_1 x + a_0$

These examples illustrate that the coefficients of x are real numbers.

In this section polynomials of **degree two**, called **quadratic functions**, will be discussed in some detail.

> **Definition**
>
> **Quadratic function**:
>
> If a, b, and c are real number constants and a is not equal to zero $(a \neq 0)$,
> then $f(x)$, defined by, $y = f(x) = ax^2 + bx + c$, is a **quadratic function**.
> The **highest** degree **power** of any **term** is **two**.

The graph of a quadratic function is a parabola.

If a is **positive** $(a > 0)$ then, $y \geq \frac{4ac - b^2}{4a}$, and the graph, $y = f(x)$, has a minimum value of $\frac{4ac - b^2}{4a}$, at $x = \frac{-b}{2a}$, and the parabola will **open upwards**.

If a is **negative** $(a < 0)$ then, $y \leq \frac{4ac - b^2}{4a}$, and the graph, $y = f(x)$, has a maximum value of $\frac{4ac - b^2}{4a}$, $x = \frac{-b}{2a}$, and the parabola will **open downwards**.

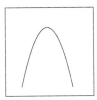

Definition

Vertex:

The **maximum** or **minimum** point on the graph is called the **vertex**, and will be the point:

$$\left[\frac{-b}{2a}, \frac{4ac - b^2}{4a} \right] = \left[\frac{-b}{2a}, f\left(\frac{-b}{2a} \right) \right]$$

Note: $\frac{4ac - b^2}{4a}$ is simply the y-value of the function at $x = \frac{-b}{2a}$ and therefore need not be memorized.

The simplest quadratic function is $y = f(x) = x^2$ where, $a = 1$, $b = 0$, $c = 0$.

Since $a = 1 > 0$, the parabola opens upward and the function has a minimum value. The vertex is $(0,0)$. The **minimum** occurs at $x = 0$, and the **minimum** value is $y = 0$. The graph is given below:

x	-3	-2	-1	0	1	2	3
y	9	4	1	0	1	4	9

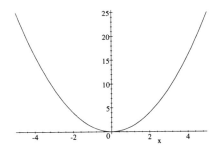

5.1.i. Graphing Quadratics

■ To graph a quadratic function, which is a parabola, follow these steps:
- ◄ Determine the **vertex**.
- ◄ Determine whether the graph **opens upward** or **downward**.
- ◄ Plot the vertex and sketch the graph.

Example **5.1**. *Graph* $y = x^2 - 6x + 5$.

- **Solution**:
  Determine the vertex for $y = x^2 - 6x + 5$, $a = 1$, $b = -6$, and $c = 5$.

- **Method 1**:
  Substituting these values into the formula $\left(\frac{-b}{2a}, \frac{4ac - b^2}{4a} \right)$

  $$= \left(\frac{-(-6)}{2(1)}, \frac{4(1)(5) - (-6)^2}{4(1)} \right)$$
  $$= \left(\frac{6}{2}, \frac{20 - 36}{4} \right)$$
  $$= \left(3, \frac{-16}{4} \right)$$
  $$= (3, -4)$$

  Thus $(3, -4)$ is the vertex of the parabola.

- **Method 2**:
  Find the value of x for the vertex using $x = \frac{-b}{2a}$.

  $$x = \frac{-(-6)}{2(1)}$$
  $$= 3$$

  Substitute this value for x into the original equation.

  $$y = f(3) = 3^2 - 6(3) + 5$$
  $$= 9 - 18 + 5$$
  $$= -4$$

The vertex is $(3, -4)$.

Determine whether the graph opens upward or downward. Since $a = 1$ $(a > 0)$ the graph opens upward, and the graph has a minimum value. The minimum value is -4 when $x = 3$. Plot the graph:

x	-2	-1	0	1	2	3	4	5	6
y	21	12	5	0	-3	-4	-3	0	5

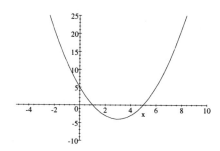

Exercise **5.1**. *For* $f(x) = 2x^2 - 3x + 4$, *find the vertex, the maximum or minimum, value and graph the function.*

Solution:

5.1.ii. Completing the square

Another method for finding the **vertex** is by completing the square. The quadratic equation in the form

$$y = f(x) = ax^2 + bx + c$$

is put into the form

$$y = f(x) = \boxed{a(x-h)^2 + k}$$

where (h, k) is the **vertex**.

■ Steps for completing the square:

▶ Group the x^2 and x terms together.

$$y = (ax^2 + bx) + c$$

▶ If $a \neq 1$, factor a from these two terms.

$$y = a\left(x^2 + \frac{b}{a}\right) + c$$

▶ Make the **grouped** terms into a perfect square trinomial.

◀ This is accomplished by adding $\left[\frac{1}{2}\left(\frac{b}{a}\right)\right]^2$ to the grouped terms.

▶ Add to c, $\left[-a\left[\frac{1}{2}\left(\frac{b}{a}\right)\right]^2\right]$ to cancel the value added in the previous step.

$$y = \left(x^2 + \frac{b}{a}x + \left[\frac{1}{2}\left(\frac{b}{a}\right)\right]^2\right) + c - a\left[\frac{1}{2}\left(\frac{b}{a}\right)\right]^2$$

$$y = a\left(x + \left[\frac{1}{2}\left(\frac{b}{a}\right)\right]^2\right) + c - a\left[\frac{1}{2}\left(\frac{b}{a}\right)\right]^2$$

▶ Factor the trinomial into a binomial squared. Thus $\boxed{y = a(x-h)^2 + k}$, where

$$h = -\frac{1}{2}\left(\frac{b}{a}\right) \text{ and } k = c - a\left[\frac{1}{2}\left(\frac{b}{a}\right)\right]^2.$$

This method is demonstrated below.

Example **5.2**. *Given the function* $y = x^2 - 6x + 5$, *find the vertex.*

• **Solution:**

$a = 1, b = -6$ and $c = 5$.

Group the x^2 and x terms together.

$$y = (x^2 - 6x) + 5$$

$a = 1$, hence there is no need to factor the grouped terms.

Make the grouped terms into a perfect square trinomial.

$$(x^2 - 6x) + 5$$

$$= (x^2 - 6x + \left[\tfrac{1}{2}\left(-\tfrac{6}{1}\right)\right]^2) + 5 + \left[\left[\tfrac{1}{2}\left(-\tfrac{6}{1}\right)\right]^2\right]$$

$$= (x^2 - 6x + 9) + 5 + [-1(9)]$$

$$= (x - 3)(x - 3) + 5 + [-1(9)]$$

$$= (x - 3)^2 - 4$$

$$c.f. \quad a(x - h)^2 + k$$

The vertex is given by (h, k), hence the vertex is $(3, -4)$.

■ ☐ **Note** It is important to notice the sign of h and k:
The vertex for $y = 3(x + 6)^2 - 2$ is, $(-6, -2)$.

The vertex for $y = 2(x - 1)^2 + 6$ is $(1, 6)$.

The vertex for $y = (x - 5)^2 - 8$, is $(5, -8)$.

The vertex for $y = 4(x + 2)^2 + 7$ is, $(-2, 7)$.

Example 5.3. *Find the vertex and graph the function* $y = -2x^2 + 8x + 1$.

Solution:

$a = -2, b = 8$ and $c = 1$.

Group x^2 and x terms together

$$y = (-2x^2 + 8x) + 1$$

$a \neq 1$, hence factor grouped terms

$$-2(x^2 - 4x) + 1$$

$$= -2(x^2 - 4x + \left[\tfrac{1}{2}\left(\tfrac{8}{-2}\right)\right]^2) + 1 + \left[-(-2)\left[\tfrac{1}{2}\left(\tfrac{8}{-2}\right)\right]^2\right]$$

$$= -2(x^2 - 4x + 4) + 1 + 8$$

$$= -2(x^2 - 4x + 4) + 9$$

$$c.f. \quad a(x - h)^2 + k$$

The vertex is given by (h, k), hence the vertex is $(2, 9)$.
The graph turns downward since $a = -2 < 0$.

x	-2	-1	0	1	2	3	4	5	6
y	-23	-9	1	7	9	7	1	-9	-23

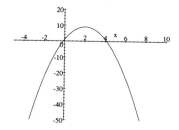

Exercise 5.2. *For the following , find the vertex, the maximum or minimum, value and graph the function:*

(a) $y = 4 + 3x - x^2$.

Solution:

(b) $y = x^2 + 2x - 6$.

Solution:

Summary 5.1.

Quadratic Function

■ $y = f(x) = ax^2 + bx + c$,

▶ where a, b and c are real numbers, and $a \neq 1$, is a quadratic function.

■ The vertex is the point $\left[\frac{-b}{2a}, \frac{4ac - b^2}{4a} \right]$ or $\left[\frac{-b}{2a}, f\left(\frac{-b}{2a} \right) \right]$

■ If $a > 0$, the function has a **minimum** value at the vertex, and the function **opens upward**.

■ If $a < 0$, the function has a **maximum** value at the vertex, and the function **opens downward**.

End of Section 5.1. *Problems*

Problem 5.1. *For each of the following quadratic functions, give the vertex, the maximum or minimum value, the opening direction, and the graph.*

(a) $y = 5x^2 + 3$.   (b) $y = -x^2 + 8x - 4$.   (c) $y = -3x^2 - 2x + 5$.

(d) $y = 2x^2 - 5x + 6$.   (e) $y = -(x + 7)^2 - 3$.   (f) $y = x^2 + 5x + 1$.

(g) $y = 4x^2 - 2x + 3$.   (h) $y = -x^2 - x - 1$.   (i) $y = -4x^2 - x + 5$.

Problem 5.2. *For each of the following quadratic functions, give the vertex, the maximum or minimum value, the opening direction, and the graph.*

(a) $y = 3x^2 + x - 1$.   (b) $y = 6x^2 - 14x + 10$. (c) $y = 4x^2 + 12 + 1$.

(d) $y = -x^2 + 12x + 1$. (e) $y = -x^2 + 3x - 4$.   (f) $y = 2x^2 + 4x + 1$.

(g) $y = -4x^2 + 8x$.   (h) $y = x^2 + 5x + 3$.   (i) $y = -x^2 + x + 2$.

5.2. Polynomial Functions

Polynomial functions of degree two, quadratic functions, have already been discussed. Polynomial functions, in general, are explored now.

When

$$P_n(x) = f(x) = y = a_n x^n + a_{n-1} x^{n-1} + a_{n-2} x^{n-2} + \dots + a_1 x + a_0,$$

and a_n is a real number, n is a whole number, and $a_n \neq 0$, then $P_n(x)$ is an n-th degree polynomial function of x.

Odd degree functions (1st, 3rd, 5th ...)

If the polynomial is an **odd degree function** (first degree or linear, third degree or cubic, fifth degree, etc.) the graph will **begin "up"** on the **left** and **end "down"** on the **right**, or the graph will **begin "down"** on the **left** and **end "up"** on the **right**. Which way depends on the value of a_n.

■ If $a_n > 0$ the graph will begin **down** on the **left** and end **up** on the **right**.

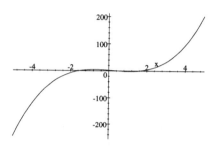

■ If $a_n < 0$ the graph will begin **up** on the **left** and **end down** on the **right**.

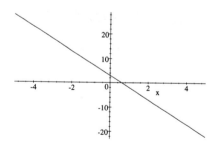

Examples are given below:

■ $P_1(x) = 2x + 3$

This is a linear function and $a_1 = 2 > 0$. The graph will look as follows:

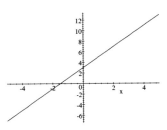

■ $P_3(x) = 2x^3 - x^2 - 5x + 2$

This is a third degree function and $a_3 = 2 > 0$. The graph will look as follows:

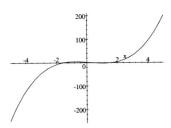

■ $P_1(x) = -5x + 3$

This is a linear function and $a_1 = -5 < 0$. The graph will look as follows:

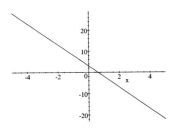

■ $P_3(x) = -2x^3 + x^2 + 5x - 2$

This is a third degree function and $a_3 = -2 < 0$. The graph will look as follows:

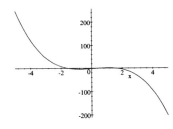

Even degree functions (2nd, 4th, 6th...)

Now lets consider an **even** degree function (second degree or quadratic, fourth degree, etc.) polynomial. The graph will either **go up on both ends**, or it will **go down on both ends**. Again, which way depends on the value of a_n.

- If $a_n > 0$ the graph will turn **up on both ends**.

Think of something <u>positive (> 0)</u> happening. You are happy and you smile, the corners of your mouth go up.

For example, $P_2(x) = 4x^2 + x + 7$, is a second degree (quadratic) function and $a_2 = 4 > 0$. The graph will look as follows:

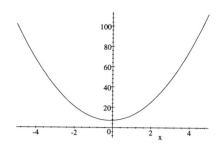

- If $a_n < 0$ the graph will turn **down on both ends**.

Think of something <u>negative (< 0)</u> happening. You are sad and the corners of your mouth go down.

For example, $P_4(x) = -x^4 - 2x^3 + 7x^2 - 8x + 12$, is a fourth degree function and $a_4 = -1 < 0$. The graph will look as follows:

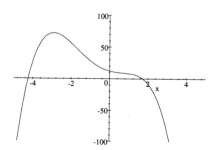

5.2.i. Zeros of a polynomial

The **Fundamental Theorem of Algebra** states that every polynomial function of degree $n > 0$ has at least one real or imaginary zero.

- The real number zeros of a function indicate where the graph will cross or touch the x-axis.
  - They equate to the points $(x_n, 0)$, the x-intercept(s).
    - The real **zeros** can also be thought of as the x values that produce a **zero** height.

A polynomial function $P_n(x)$ of degree $n > 0$ has exactly n real and/or imaginary **zeros**, some of which may be repeated.

The approaches discussed below demonstrate how to find the **real number zeros** of any polynomial.

- ■ When a polynomial function is in **FACTORED FORM, OR CAN BE FACTORED**, the **zeros** can easily be found by setting each factor equal to 0 and solving.

$Example$ 5.4. *If $P_5(x) = (2x - 1)(x + 3)(x - 1)(x + 1)(x + 2)$, find the zeros and graph.*

- • **Solution:**

$2x - 1 = 0$ $2x = 1$ $x = \frac{1}{2}$	$x + 3 = 0$ $x = -3$	$x - 1 = 0$ $x = 1$	$x + 1 = 0$ $x = -1$	$x + 2 = 0$ $x = -2$

- • The zeros of the function are $\frac{1}{2}, -3, 1, -1, -2$. Plot the points $(1/2, 0), (-3, 0), (1, 0), (-1, 0), (-2, 0)$.

  The y-intercept is found by letting $x = 0$. $P_5(0) = (2(0) - 1)(0 + 3)(0 - 1)(0 + 1)(0 + 2) = 6$. Thus $(0, 6)$ is the y-intercept.

  The graph will start down on the left and end up on the right, since all the coefficients of x in the factors are positive and there is no coefficient in front of these, $a_5 > 0$, and it is a fifth degree function. In other words, $2x \bullet x \bullet x \bullet x \bullet x = 2x^5$.

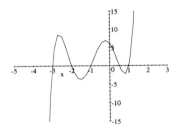

$Example$ 5.5. *If $f(x) = 3x^4 + x^3 - 10x^2$, find the zeros and graph the function.*

- • **Solution:**

$$f(x) = 3x^4 + x^3 - 10x^2$$
$$= x^2(3x^2 + x - 10)$$
$$= x^2(3x - 5)(x + 2)$$

$x^2 = 0$ $x = 0$	$3x - 5 = 0$ $3x = 5$ $x = \frac{5}{3}$	$x + 2 = 0$ $x = -2$

- • The **zeros** are $x = -2$, 0, $5/3$. ☐ **Note:** ☐ The zero, $x = 0$ is actually repeated, since $x^2 = x \bullet x = 0$.

  Therefore, the graph touches the x-axis at $x = 0$ but does not cross.

  The x-intercepts are $(-2, 0)$, $(0, 0)$ and $(5/3, 0)$.

  Since the function is an even degree (quartic or fourth degree) and $a_4 = 3$, a positive number, the graph will turn up on both ends.

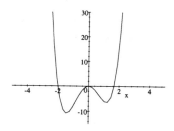

■ When the polynomial function is **NOT IN FACTORED FORM**, a $\boxed{\text{graphing calculator}}$ or a table of values can be used to find the real number zeros.

▶ Using a graphing calculator the function $y = P_n(x)$ is entered using the $\boxed{\text{Y=}}$ key of the calculator and then graphed. Using the $\boxed{\text{ZOOM}}$ and $\boxed{\text{TRACE}}$ functions the zeros are found.

▶ The $\boxed{\text{TRACE}}$ function will also help in identifying the intervals where the function is increasing and decreasing.

Exercise **5.3**. *Given the following polynomial functions , find the zeros and graph the functions*:

(a)$f(x) = x^5 + 3x^4 + x^2 + x - 8$.

Solution:

(b)$f(x) = 2x^3 + 5x^2 - x - 1$.

Solution:

Example **5.6**. *Graph* $f(x) = 3x(2x - 7)(x + 3)(x - 1)$.

• **Solution**: Multiply the factors,

$$3x \cdot 2x \cdot x \cdot x = 6x^4 = a_n x^n$$

We find that this is an even degree, or fourth degree, function, and $a_n = 6 > 0$. Hence a_n is positive (smile) The shape of the graph will be:

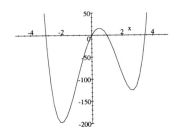

• The zeros of the function will be obtained by setting each factor equal to zero.

	$2x - 7 = 0$		
$3x = 0$	$2x = 7$	$x + 3 = 0$	$x - 1 = 0$
$x = 0$	$x = \dfrac{7}{2}$	$x = -3$	$x = 1$

This gives the points $(0,0)$, $(7/2,0)$, $(-3,0)$, and $(1,0)$. Thus,

Exercise **5.4**. *Graph giving x-intercept, y-intercept, starting position (up or down) and ending position.*

$(a)\, y = -x(x + 1)(x - 3)$.

Solution:

$(b)\, y = 2(x - 1)^2(x + 3)(2x - 3)$.

Solution:

Example **5.7**. Graph $y = -x^5 - 3x^4 + 2x^3 + x^2 - 5x + 1$.

- **Solution**: This is an odd degree, or fifth degree, function.
  $a_n = -1 < 0$. Hence, a_n is negative (sad), and the shape of the graph will be:

- The zeros of the function are not easily determined, therefore either use a graphing calculator or use a table of values.

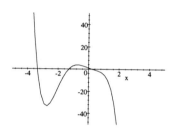

Exercise **5.5**. Graph giving x-intercept, y-intercept, starting position (up or down) and ending position.

(a) $y = -2x(x-1)(x+2)(2x-1)$.

Solution:

(b) $y = x^3 - x^2 - 6x$.

Solution:

Summary 5.2.

To graph a polynomial function:

- Determine the degree of the polynomial
  - ▶ even or odd (first, second, third, fourth, fifth, etc.).
- Check the value of a_n.
  - ▶ $a_n > 0$ or $a_n < 0$.
- Determine, roughly, the shape of the graph.
- Find the zeros of the polynomial, if possible.
- Find the y-intercept, if possible.
- Graph the polynomial.
- Ask yourself, "Does the graph look reasonable?"

End of Section 5.2. Problems

Problem 5.3. *For the following polynomials, give the degree of the zeros and the graph.*

(a) $y = x(x+3)(x-2)$.

(b) $y = (x+1)(x-1)(x+2)(x-2)$.

(c) $y = -x(x+1)(2x-1)$.

(d) $y = (x+4)(2x-3)(x-5)$.

(e) $y = -(x-2)^2(x+1)^2$.

Problem 5.4. *For the following polynomials, give the degree of the zeros and the graph.*

(a) $y = (x+1)^3(x-2)^2$.

(b) $y = 2x^3 - 4x^2 + 3x - 1$.

(c) $y = x^4 - 3x^2 - 4$.

(d) $y = -x^5 + 3x - 2$.

(e) $y = 3x^3 - 4x$.

5.3. Rational functions

> **Definition**
>
> **Rational function**:
>
> When $N(x)$ and $D(x)$ are polynomial functions, and $D(x) \neq 0$,
> the function, $f(x) = \dfrac{N(x)}{D(x)}$, is called a **rational function**.

Asymptotes occurs when x and/or y approach a value, getting closer and closer to the value from both sides.

There are **three types of asymptotes**: **vertical, horizontal**, and **slant**.

- **Vertical asymptotes** can be found by setting the denominator of the rational function equal to zero, $D(x) = 0$, and solving for a constant k. The line(s) $x = k$ is the vertical asymptote, as long as the value of k does not make the numerator zero.

For example: $y = \dfrac{1}{x-1}$,
Set the denominator equal to zero

$$x - 1 = 0$$
$$x = 1$$

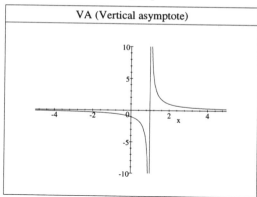

VA (Vertical asymptote)

- **Horizontal asymptotes** occur:
  a) When the **degree of the numerator**, $N(x)$, is **less than** the **degree of the denominator**, $D(x)$. Here, the asymptote is $y = 0$.

For example:

$$y = \frac{1}{x-1} = \frac{N(x)}{D(x)}$$

Degree of $N(x) = 0$ < **Degree of** $D(x) = 1$

Thus, $y = 0$

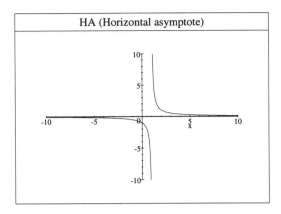

HA (Horizontal asymptote)

b) When the **degree of the numerator**, $N(x)$, and the **degree of the denominator**, $D(x)$, are **equal**. Here, the asymptote is the **ratio of the coefficient** of the **highest degree term in the numerator and in the denominator**.

For example,

$$y = \frac{3x - 1}{2x + 5}$$

The degree of the numerator and the degree of the denominator are both one, therefore

$$y = \frac{3}{2}$$

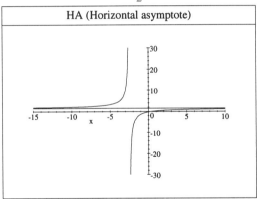

HA (Horizontal asymptote)

- **Slant asymptotes** occur when the **degree of the numerator**, $N(x)$, is **one more than the degree of the denominator**, $D(x)$. The slant asymptote is the equation of the line obtained by dividing the numerator by the denominator and disregarding the remainder, if there is one.

For example:

$$y = \frac{x^2 - 3x + 4}{x - 4}$$

Degree of Numerator $2 > 1$ **Degree of Denominator**

$$\Rightarrow y = x + 1 \text{ remainder } 8$$

The slant asymptote is

$$y = x + 1$$

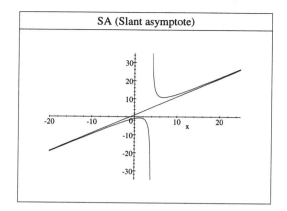

SA (Slant asymptote)

5.3.i.   Graphing

- ■ **To graph a rational function**:
  - ▶ Identify the asymptotes.
  - ▶ Find the x and y intercepts.
  - ▶ Construct a table of values, using values for x for points on either side of the asymptotes.

Example **5.8**. *Graph* $y = \dfrac{4}{2x - 3}$

Solution: To find the **vertical asymptote**, set the denominator equal to zero.

$$2x - 3 = 0$$
$$2x = 3$$
$$x = 3/2 \text{ (Vertical asymptote)}$$

Since the degree of the numerator is zero and the degree of the denominator is one, the **horizontal asymptote** is $y = 0$.

- To find the x-intercept, let $y = 0$ and solve the equation or in a rational function, set the numerator equal to zero and solve.

$$y = \frac{4}{2x - 3}$$
$$0 = \frac{4}{2x - 3} \quad \text{Multiply both sides by } (2x - 3)$$
$$(2x - 3)0 = 4 \quad \text{Multiplication property of zero}$$
$$0 \ne 4 \quad \text{In this case, there is NO } x\text{-intercept}$$

- To find the y-intercept, let $x = 0$.

$$y = \frac{4}{2(0) - 3}$$
$$= \frac{4}{3}$$

- Thus the y-intercept is $\left(0, \dfrac{4}{-3}\right)$.

- Use a table of values to include values to the left and to the right of the vertical asymptote.

x	-2	-1	0	1	2	3	4	5
y	0.6	0.8	1.33	4	-4	-1.3	-0.8	-0.6

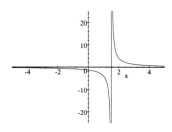

Exercise 5.6. *Graph* $y = \dfrac{3x - 4}{5x + 1}$

> **Solution:**

$$f(x) = \dfrac{2x^2 - 3x + 6}{x - 4}$$

-

$$x - 4 = 0$$
$$x = 4$$

$$y = \dfrac{2x^2 - 3x + 6}{x - 4}$$
$$= 2x + 5 \ \textbf{plus remainder}$$

$$= \textbf{slant asymptote}$$

- x-intercept:

$2x^2 - 3x + 6 = 0$ is in the form $ax^2 + bx + c$

$\Rightarrow a = 2, b = -3$ and $c = 6$. Apply

$$x = \dfrac{-b \pm \sqrt{b^2 - 4ac}}{2a}$$
$$= \dfrac{-(-30) \pm \sqrt{(-3)^2 - 4(2)(6)}}{2(2)}$$
$$= \dfrac{30 \pm \sqrt{-39}}{4}$$

Since $\sqrt{-39} = i\sqrt{39}$, an imaginary number, there is no x-intercept.

y-intercept:

$$y = \frac{2(0)^2 - 3(0) + 6}{(0) - 4}$$

$$= \frac{6}{-4}$$

$$\Rightarrow (0, -\tfrac{3}{2})$$

Table of values

x	0	1	2	3	5	6	7
y	-1.5	-1.7	-4	-15	41	30	27.7

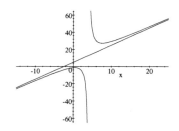

5.3.ii.   Missing points

A vertical asymptote was found by setting the denominator of a rational function equal to zero and solving for $x = k$. The line(s) $x = k$ is a <u>vertical</u> **asymptote** as long as the value k **DOES NOT** make the numerator zero as well.

To see what happens when k makes the numerator as well as the denominator zero, we will investigate the following function

$$y = \frac{x^2 - 4}{x - 2}$$

To test for a vertical asymptote let

$$x - 2 = 0$$

$$x = 2$$

However, $x = 2$ makes the numerator zero. When this occurs there is not an asymptote at $x = 2$, but a **missing point**. To understand this, look at the following: $y_1 = \frac{x^2 - 4}{x - 2} = \frac{(x - 2)(x + 2)}{x - 2}$ which is almost equivalent to $y_2 = x + 2$.

The only difference between y_1 and y_2 is the point $(2, 4)$. The point $(2, 4)$ is a point on y_2 (the simplified function), but is <u>not a point</u> on y_1. Since $y_2 = x + 2$ is easier to graph, graph $y = x + 2$ and then remove the extra point $(2, 4)$.

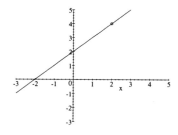

Example **5.10.** *Graph* $y = \dfrac{(x^2 - 1)(x + 3)}{(x - 2)(x + 3)}$.

- **Solution:** Notice that this simplifies to $y = \dfrac{(x^2 - 1)}{(x - 2)}$.

  x-intercept: Set the numerator equal to 0 and solve.
  $$x^2 - 1 = 0$$
  $$(x + 1)(x - 1) = 0$$
  $$x = -1, x = 1$$

- y-intercept: Set $x = 0$.
  $$y = \frac{(0^2 - 1)}{(0 - 2)} = \frac{-1}{-2} = \frac{1}{2}$$

- Vertical asymptote: $(x - 2)(x + 3) = 0$
  $$x - 2 = 0 \qquad x + 3 = 0$$
  $$x = 2 \qquad x = -3$$

- $x = 2$ is a vertical asymptote, but $x = -3$ is a missing point.

  If $x = -3$. $D(-3) = 0$ and also $N(-3) = 0$.
  $$D(-3) = ((-3) - 2)((-3) + 3)$$
  $$= (-5) \times 0$$
  $$= 0$$
  $$N(-3) = ((-3)2 - 1)((-3) + 3)$$
  $$= (8) \times 0$$
  $$= 0$$
  $$x = -3$$
  $$y = \frac{(-3)^2 - 1}{(-3) - 2} = \frac{8}{-5}$$

  $(-3, 8/-5)$

  $$y = \frac{x^2 + 0 - 1}{x - 2}$$
  $$= x + 2 + \frac{3}{x - 2}$$
  $$y = x + 2$$

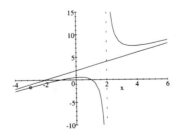

- Notice the lines, $x = 2$ and $y = x + 2$ are asymptotes. These are NOT part of the graph. The missing point is $(-3, -8/5)$.

Exercise 5.7. *Graph the following*:

(a) $y = \dfrac{3x - 5}{x + 1}$.

Solution:

(b) $y = \dfrac{4x - 1}{x + 5}$.

Solution:

(c) $f(x) = \dfrac{x^2 - 16}{(x + 4)(x - 3)}$.

Solution:

(d) $f(x) = \dfrac{x^2 - 4x + 3}{x + 2}$.

Solution:

Summary 5.3.

- A rational function, $F(x) = \dfrac{N(x)}{D(x)}$, may have the following features:
- **Vertical Asymptote**: Occurs when for some point x_1
  - ▶ $D(x_1) = 0$, but $N(x_1) \neq 0$.
- **Missing Point**: Occurs when $D(x_1) = 0$ and $N(x_1) = 0$.
- **Horizontal Asymptote**: Occurs when the degree of $D(x)$ is greater than the degree of $N(x)$.
- **Slant Asymptote**: Occurs when the degree of $D(x)$ is ONE less than the degree of $N(x)$.
- **To Graph Rational Functions**:
  - ▶ Determine any vertical, horizontal, slant asymptotes or missing points.
  - ▶ Determine the x and y intercepts.
  - ▶ Determine other points, if necessary, using a table.

End of Section 5.3. Problems

Problem 5.5. *Graph each of the following:*

(a) $y = \frac{x-1}{2x+3}$. (b) $y = \frac{3}{x-2}$. (c) $y = \frac{2x}{x-1}$. (d) $y = \frac{x^2+4}{x-2}$.

(e) $y = \frac{x^2}{2x+3}$. (f) $y = \frac{x^3}{x^2+x-1}$. (g) $y = \frac{x^3-1}{x-1}$. (h) $y = \frac{x^2+2x+1}{x^2-1}$.

Problem 5.6. *Graph each of the following:*

(a) $y = \frac{3x-1}{5x+7}$. (b) $y = \frac{6x}{6x-1}$. (c) $y = \frac{7x-1}{5x}$. (d) $y = \frac{3x^2}{x+2}$.

(e) $y = \frac{2x^3-1}{x^2-1}$. (f) $y = \frac{13}{x-1}$. (g) $y = \frac{2x-5}{7x+9}$.

End of Chapter 5. Problems

Problem 5.7. *Graph the following quadratic functions, giving the vertex and opening direction.*

(a) $y = 2x^2 - 1$. (b) $y = 3x^2 + 1$. (c) $y = -x^2 + 3$. (d) $y = -2x^2 - 5$.

(e) $y = x^2 - 4x + 1$. (f) $y = -x^2 - 3x + 1$.

Problem 5.8. *Graph the following quadratic functions, giving the vertex and opening direction.*

(a) $y = 2x^2 - 4x - 3$. (b) $y = -3x^2 + x + 3$. (c) $y = -3(x+1)^2 - 5$. (d) $y = 4(x-3)^2 + 4$.

Problem 5.9. *Graph the following polynomial functions giving y-intercepts, x-intercepts, starting position (up or down) and ending position (up or down).*

(a) $y = (x+3)(x-2)(2x+1)$. (b) $y = -2(x-1)(x+1)(3x-2)$. (c) $y = 2x(x-1)(x+1)(x-2)$.

(d) $y = -x(2x-1)(x+3)(x-1)$. (e) $y = -x^3 - -3x^2 + 5x + 3$.

Problem 5.10. *Graph the following polynomial functions giving y-intercepts, x-intercepts, starting position (up or down) and ending position (up or down).*

(a) $y = 3(x-1)(x-3)(x+2)(x-2)(2x-3)$. (b) $y = x^3 - x^2 - 6x$. (c) $y = -4x^3 - 10x^2 + 6x$.

(d) $y = -2x(x-1)(x+1)(x-2)^2$. (e) $y = x^4 - 3x^2 - 1$.

Problem 5.11. *Graph the following rational functions, giving any vertical asymptotes, horizontal asymptotes, slant asymptotes, and/or missing points. Also, give y-intercepts and x-intercepts, if possible.*

(a) $y = \frac{1}{x+1}$.   (b) $y = \frac{-2}{x+3}$.   (c) $y = \frac{x+1}{2x-1}$.   (d) $y = \frac{3x-2}{x+2}$.

(e) $y = \frac{x^2-1}{x+1}$.   (f) $y = \frac{2x^2-x-1}{2x^2+7x+3}$.

Problem 5.12. *Graph the following rational functions, giving any vertical asymptotes, horizontal asymptotes, slant asymptotes, and/or missing points. Also, give y-intercepts and x-intercepts, if possible.*

(a) $y = \frac{2+x-x^2}{4+4x+x^2}$.   (b) $y = \frac{x^2+3x+1}{x+4}$.   (c) $y = \frac{-x^2-3x+1}{x-1}$.   (d) $y = \frac{x^3+2x^2-5x-6}{x^2+2x-3}$.

Problem 5.13. *Find the maximum value (if one exists) of the function* $y = -3x^2 + 6x - 5$ *and tell where it exists.*

Problem 5.14. *Find the minimum value (if one exists) of the function* $y = 3x^2 + 6x - 5$ *and tell where it exists.*

Problem 5.15. *Is the function* $y = 2(x-2)^2 + 3$ *ever negative? Why or why not?*

Problem 5.16. *If a football is kicked straight up with an initial velocity of* 64 *feet per second from a height of* 3 *feet, then its height above the earth is a function of time given by* $h = -16t^2 + 64t + 3$. *What is the maximum height the ball will reach?*

Chapter 6.

Exponential and logarithmic functions

6.1. Exponential functions

> **Definition**
>
> **Exponential function**:
>
> If $a > 0$ and $a \neq 1$, then the function, defined by
>
> $\boxed{y = f(x) = a^x}$ is an **exponential function**.

The <u>domain</u> of the function is the set of all real numbers, $D = \{x/x \in R\}$, and the <u>range</u> is the set of all real numbers greater than zero, $R = \{y/y > 0\}$.

- In defining exponential functions notice that $a \neq 1$.
  - ▶ If $a = 1$, then $y = 1^x = 1$ for all values of x.
    - ◀ This is a **constant function**.

- Also notice that $a > 0$, or in other words, a is positive.
  - ▶ If a were negative, the function would not exist for all values of x.
    - ◀ For example, $y = (-1)^{1/2} = \sqrt{-1} = i$ is not defined within the set of real numbers.

Here are some examples of exponential functions: $\boxed{y = 2^x}$ $\boxed{y = 3^x}$ $\boxed{y = (0.203)^x}$

6.1.i. Graphing

The graph of an exponential function, in the form $\boxed{y = a^x}$, will either be an increasing function or a decreasing function, depending on the value of a.

- If $0 < a < 1$, the exponential function $y = a^x$ will be a decreasing function. The shape of the graph will be:

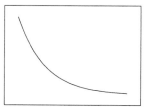

- If $a > 1$, the exponential function $y = a^x$, will be an increasing function. The shape of the graph will be:

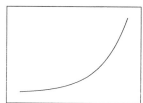

To graph an exponential function, construct a table of values.

Example **6.1.a.** *Graph* $y = f(x) = \left(\frac{3}{2}\right)^x$.

- **Solution:** Since $3/2 > 1$, the graph is increasing.

x	−2	−1	0	1	2	3
y	0.44	0.67	1	1.5	2.25	3.38

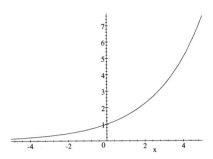

Example **6.1.b.** *Graph* $y = f(x) = \left(\frac{2}{3}\right)^x$.

- **Solution:** Since $2/3 < 1$, the graph is decreasing.

x	-2	-1	0	1	2	3
y	2.25	1.5	1	0.67	0.44	0.3

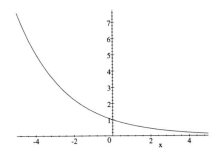

Example **6.2.** *Graph* $y = e^x$.

 • **Solution**: Remember that e is an irrational number. $e \approx 2.718281828\ldots$

x	-2	-1	0	1	2
y	0.14	0.37	1	2.72	7.39

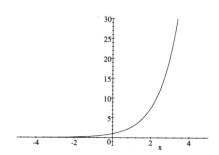

Characteristics of Exponential Function in the form:

$$f(x) = y = a^x$$

- Exponential functions are positive for all values of x.
- For all values of a, $f(0) = 1$. The y-intercept is $(0, 1)$.
  ▶ This is true since $f(0) = a^0 = 1$.
- There are no zeros for the function. In other words, the graph will not cross or touch the x-axis.
  ▶ The horizontal asymptote is $y = 0$
- For every y value there corresponds one and only one x value. In other words, the function is one-to-one.
  ▶ This implies that $f(x)$ has an inverse function.
  The next section will discuss the inverse function for the exponential function.

Exercise **6.1.** *Graph the following functions*

(a) $y = 4^x$.

Solution:

(b) $y = (\frac{1}{4})^x$.

Solution:

(c) $f(x) = \left(\frac{5}{2}\right)^x.$

Solution:

(d) $f(x) = \left(\frac{1}{3}\right)^x.$

Solution:

6.1.ii.   Other forms of exponential functions

Exponential functions may **not always appear in the form** $y = a^x$

For example, consider the function $y = 2^{-3x^2}$. To graph this function, construct a table as before.

x	-0.7	-0.6	-0.5	0	0.5	0.6	0.7
y	0.36	0.47	0.59	1	0.59	0.47	0.36

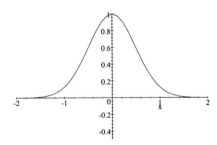

The graph increases on the interval $(-\infty, 0)$ and decreases on the interval $(0, \infty)$. Its shape is similar to the normal probability curve.

Graphs may be shifted, reflected, and stretched.

Example **6.3**. *Graph* $y = 3^x - 2$.

• Solution:

x	-2	-1	0	1	2
y	-1.9	-1.7	-1	1	7

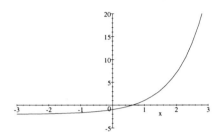

The graph has shifted down. The y-intercept is $(0,-1)$ and the x-intercept is $(0.63, 0)$.

Example **6.4**. *Graph* $y = 3^{x+1}$.

• Solution:

x	-2	-1	0	1	2
y	0.33	1	3	9	27

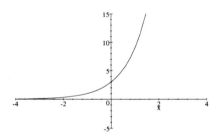

The graph has shifted to the left. The y-intercept is $(0,3)$.

Exercise **6.2**. *Graph the following functions*

(a) $y = 3^{2x}$.

Solution:

(b) $f(x) = 3^{2x-1}$.

Solution:

._218_.

(c) $f(x) = \left(\frac{1}{2}\right)^{x-3}$.

Solution:

(d) $f(x) = \left(\frac{4}{5}\right)^{x}$.

Solution:

6.1.iii.   Exponential equations

Definition

• **Exponential equations**:

Equations with variables that are exponents, are called **exponential equations**.

Some of these equations may be solved using the property given below.

• Property

If $a \neq 1$ and $a^m = a^n$, then $m = n$.

▶   For example if $2^5 = 2^x$, then $x = 5$.

Example **6.5.** *Solve for* x: $\left(\frac{1}{2}\right)^{x} = \frac{1}{4}$.

• **Solution:**

$$\left(\frac{1}{2}\right)^{x} = \frac{1}{4}$$
$$\left(\frac{1}{2}\right)^{x} = \left(\frac{1}{2}\right)^{2} \quad \text{Same base}$$
$$x = 2$$

Example **6.6.** *Solve for* x: $4^x = 32$.

• **Solution:**

$$4^x = 32$$
$$(2^2)^x = 2^5$$
$$2^{2x} = 2^5$$
$$2x = 5$$
$$x = 5/2$$

Example **6.7.** Solve for x: $\left(\frac{1}{3}\right)^x = 9$.

- Solution:

$$\left(\frac{1}{3}\right)^x = 9$$
$$(3^{-1})^x = 3^2$$
$$3^{-x} = 3^2$$
$$-x = 2$$
$$x = -2$$

Exercise **6.3. Solve the following for x.**

(a)$7^{x+6} = 7^{2x+5}$.

Solution:

(b)$27^{x-1} = 9$.

Solution:

(c)$(1/2)^x = 4$.

Solution:

(d)$10^{x+1} = 0.1$.

Solution:

(e)$e^{2x} = \frac{1}{e}$.

Solution:

(f)$2^{x+4} = 16$.

Solution:

Summary 6.1.

Exponential Function

- $y = a^x$ where $a \neq 1$, and $a > 0$
  - ▶ Domain $D = \{x/x > 0, x \neq 1\}$
  - ▶ Range $R = \{y/y > 0\}$

To graph an exponential function in the form $\boxed{y = a^x}$

- Determine if the function is increasing or decreasing.
  - ▶ $0 < a < 1$, the function is decreasing.
  - ▶ $a > 1$ the function is increasing.
- Construct a table of values.
  - ▶ **Note:** $(0, 1)$ will always be a point on the graph.

Exponential Equations

- For $a > 0$ and $a \neq 1$ if $a^m = a^n$, then $m = n$.

End of Section 6.1. Problems

Problem 6.1. *Graph the following*:

(a) $y = \left(\frac{1}{2}\right)^x + 1$.   (b) $f(x) = (7)^{2x}$.   (c) $f(x) = \left(\frac{4}{4}\right)^x + 6$.   (d) $y = (e)^x - 4$.

(e) $f(x) = (e)^{x-4}$.   (f) $y = \left(\frac{3}{2}\right)^{-x}$.   (g) $y = (4)^{x+1}$.   (h) $f(x) = (e)^{-x} + 1$.

Problem 6.2. *Graph the following*:

(a) $f(x) = (3)^{x-2} + 1$.   (b) $y = (10)^{-x}$.   (c) $f(x) = -5^x$.   (d) $f(x) = -\frac{1}{4}^x$.

(e) $f(x) = 1 \; 07^x$.   (f) $f(x) = 3^{|x|}$.   (g) $f(x) = 2^{x^2}$.   (h) $f(x) = 5 \; 2^x$.

Problem 6.3. *Solve for* x :

(a) $\left(\frac{1}{2}\right)^{x-1} = 8$.   (b) $\left(\frac{3}{2}\right)^x = \frac{4}{9}$.   (c) $4^{x^2+2x} = 64$.   (d) $7^{1-x} = \frac{1}{49}$.

(e) $e^x = \frac{1}{e^3}$.   (f) $\left(\frac{1}{4}\right)^{2x} = \frac{1}{2}$ (g) $9^x = 3$.   (h) $10^{1-2x} = 1$.

Problem 6.4. *Solve for* x :

(a) $6^{x^2} = 36$.   (b) $4^{3x+5} = 16$.   (c) $e^{4x} = \frac{1}{e^2}$.   (d) $(1/2)^x = 16$.

(e) $(4/9)^{x+1} = 2/3$.   (f) $27^{x+1} = 3$.   (g) $e^{5x+4} = e^5$.   (h) $147^{x^2-x} = 1$.

6.2. Logarithmic functions

> Definition
>
> **Logarithmic function**:
>
> A **logarithmic function** is a function in the form
>
> $$y = \log_a x$$
>
> with $a > 0$, $a \neq 1$ and $x > 0$.

This function is read "*y is equal to the logarithm of x to the base a.*"

- The domain of the function is the set of all positive real numbers, $D = \{x/x > 0\}$.
- The range is the set of all real numbers, $R = \{y/y > 0\}$.

To understand this function, it is important to know the following:

$Base^{Exponent}$	= Value	$\Rightarrow \log_{Base} Value$	= Exponent
3^2	= 9	$\Rightarrow \log_3 9$	= 2
10^{-4}	= $\frac{1}{10,000}$	$\Rightarrow \log_{10}(\frac{1}{10,000})$	= -4

- **Note:** When the **base is ten**, the logarithm is written without the base number, $\log(\frac{1}{10,000}) = -4$. Thus, when a base is not shown explicitly, it should be taken that base ten is being used.
  - ▶ This is a called a **common logarithm**.

$Base^{Exponent}$	= Value	$\Rightarrow \log_{Base} Value$	= Exponent
e^3	= 20.086	$\Rightarrow \log_e 20.086$	= 3

- **Note:** When the **base is** e, the logarithm is written without the base number, and ln is used in place of log. For example, $\ln 20.086 = 3$; the base is understood to be e.
  - ▶ This is called a **natural logarithm**.
    - ◀ In other words $\log_e = \ln$.

The exponential function can be written in logarithmic form as illustrated below:

y	= a^x		$\Rightarrow \log_a y$	= x
	Exponential Function		Logarithmic Form	

The logarithmic function can be written in exponential form as illustrated below:

y	$= \log_a x$		$\Rightarrow a^y$	$= x$
	Logarithmic Form	Exponential Function		

As discussed in the last section, the exponential function $y = a^x$ is one to one. That is for every y value there corresponds one and only one x value. Therefore, it can be shown that the **exponential** function and the **logarithmic** functions are **inverses** of each other.

Example **6.8**. *Write the following in logarithmic form* $6^3 = 216$.

- **Solution:** $\log_6 216 = 3$.

Example **6.9**. *Write the following in logarithmic form* $y = 5^x$.

- **Solution:** $\log_5 y = x$.

Exercise **6.4**. *Write the following in logarithmic form*:

(a) $7^3 = 343$.

Solution:

(b) $243 = 3^x$.

Solution:

(c) $y = 2^{x+1}$.

Solution:

Example **6.10**. *Write the following in exponential form* $\log_5 625 = 4$.

- **Solution:** $5^4 = 625$.

Example **6.11**. *Write the following in exponential form* $\log_4 x = y$.

- **Solution:** $y^4 = x$.

Exercise **6.5**. *Write the following in exponential form*:

(a) $\log_3 81 = 4$.

Solution:

(b) $\log z = m$.

Solution:

(c) $\ln 4 = 1.386$.

Solution:

6.2.i. Graphing

■ The graph of a logarithmic function in the form of $\boxed{y = \log_a x}$ will be either be an increasing function or a decreasing function depending on the base, a.

▶ If $0 < a < 1$, the function will be decreasing,

▶ and if $a > 1$, the function will be increasing.

■ To graph a logarithmic function:

▶ Write the logarithmic function in exponential form.

▶ Determine if the function is increasing or decreasing.

► Construct a table of values by substituting values for y and solving for x.

Example **6.12.a.** *Graph* $y = \log_3 x$.

- **Solution:**

$$y = \log_3 x$$
$$\Rightarrow 3^y = x$$

Since the base $a = 3 > 0$, the function is increasing.

x	0.11	0.33	1	3	9
y	-2	-1	0	1	2

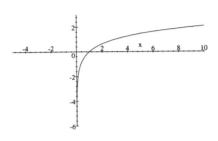

Example **6.12.b.** *Graph* $y = \log_{1/3} x$.

- **Solution:**

$$y = \log_{1/3} x$$
$$\Rightarrow (1/3)^y = x$$

Since the base is $0 < 1/3 < 1$, the function is decreasing.

x	9	3	1	0.33	0.11
y	-2	-1	0	1	2

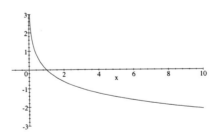

Characteristics of the Logarithmic Function in the form:

$$\boxed{f(x) = \log_a x :}$$

- The logarithmic function is positive for all values of y. The vertical asymptote is, $x = 0$.
- For all values of a, $f(1) = 0$. The x intercept (zero of the function) is $(1, 0)$.
- For every y value there corresponds one and only one x value. In other words, the function is one-to-one.
  - ▶ This implies that the logarithmic function has an inverse, namely, the exponential function.

Example **6.13**. *Graph* $y = 2^x$ *and* $y = \log_2 x$ *on the same graph.*

- **Solution**:

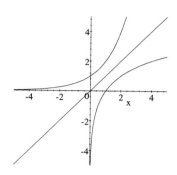

- The **graphs are mirror images** of each other with respect to the line $y = x$ (dashed line in graph), indicating that the two functions are **inverses** of each other.

    Logarithmic functions may not always appear in the form $\boxed{y = \log_a x}$. Shifting, stretching, and reflecting may also occur.

Example **6.14**. *Graph* $y = \log_2(x + 3)$.

- **Solution**:

$$y = \log_2(x + 3)$$
$$2^y = x + 3$$

x	-2.8	-2.5	-2	-2.8	1
y	-2	-1	0	-2	2

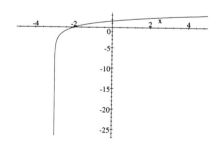

Example 6.15. Graph $y = \log x - 4$.

- **Solution:**

$$y = \log x - 4$$

Note: This is not the same as $y = \log(x - 4)$, but means $(\log x) - 4$

$$10^{y+4} = x$$

x	1000	100	10	1	0.1
y	-1	-2	-3	-4	-5

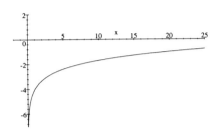

Example 6.16. Graph $y = \log_{1/4}(x - 2)$.

- **Solution:**

$$y = \log_{1/4}(x - 2)$$

$$(1/4)^y = x - 2$$

x	18	6	3	2.25	2.063	2.016
y	-2	-1	0	1	2	3

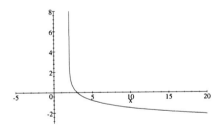

Exercise **6.6.** *Graph each of the following:*

(a) $y = \log x$.

Solution:

(b) $y = \ln x + 3$.

Solution:

(c) $y = \ln x$.

Solution:

(d) $y = \ln(x + 3)$.

Solution:

(e) $y = \log_2 x - 5$.

Solution:

(f) $y = \log_3(x + 4)$.

Solution:

6.2.ii. Logarithmic Equations

In order to solve logarithmic equations it is important to know the properties of logarithms.

Properties

Logarithms

Let a be a positive real number, $a > 0$, and $a \neq 1$. Let m and n
be any positive real numbers and r any real number,

then for **logarithms**:

$\log_a mn = \log_a m + \log_a n$

$\log_a \frac{m}{n} = \log_a m - \log_a n$

$\log_a m^r = r \log_a m$

$\log_a a = 1$

$\log_a 1 = 0$

Properties

Natural logarithms:

$\ln mn = \ln m + \ln n$

$\ln \frac{m}{n} = \ln m - \ln n$

$\ln m^r = r \ln m$

$\ln e = 1$

$\ln 1 = 0$

- There are **two basic** types of logarithmic equations:
  - ▶ **those that have "log" in every term**
  - ▶ **and those that do NOT.**

First, consider equations that can be written as: $\boxed{\log_a x_1 = \log_a x_2}$

If $\log_a x_1 = \log_a x_2$, then $x_1 = x_2$.

Example 6.17. Solve $\log_3(x+4) = \log_3 8$ for x.

- **Solution:**

$$\log_3(x+4) = \log_3 8$$
$$x + 4 = 8$$
$$x = 4$$

$$\text{Check} \quad \log_3(x+4) = \log_3 8$$
$$\log_3(4+4) = \log_3 8 \quad \text{Substituting } x = 4$$
$$\log_3 8 = \log_3 8$$

Example **6.18.** *Solve* $\ln x^2 = \ln(x+2)$ *for* x.

  • **Solution**:

$$\ln x^2 = \ln(x+2)$$
$$x^2 = x+2$$
$$x^2 - x - 2 = 0$$
$$(x-2)(x+1) = 0$$
$$x - 2 = 0, \quad x + 1 = 0$$
$$x = 2, \quad x = -1$$

Check:
$$\ln x^2 = \ln(x+2)$$
$$\ln 2^2 = \ln(2+2) \quad \text{Substituting } x = 2$$
$$\ln 4 = \ln 4$$
$$\ln(-1)^2 = \ln(-1+2) \quad \text{Substituting } x = -1.$$
$$\ln 1 = \ln 1$$

Thus $x = 2$ or $x = -1$.

Example **6.19.** *Solve* $\log_{\frac{1}{4}}(2x+3) = \log_{\frac{1}{4}} x$ *for* x.

  • **Solution**:

$$\log_{\frac{1}{4}}(2x+3) = \log_{\frac{1}{4}} x$$
$$2x + 3 = x$$
$$x + 3 = 0$$
$$x = -3. \text{ However...}$$

Check:

$$\log_{\frac{1}{4}}(2x+3) = \log_{\frac{1}{4}} x$$
$$\log_{\frac{1}{4}}[2(-3)+3] = \log_{\frac{1}{4}}(-3) \quad \text{Substituting } x = -3$$
$$\log_{\frac{1}{4}}(-6+3) = \log_{\frac{1}{4}}(-3)$$
$$\log_{\frac{1}{4}}(-3) = \log_{\frac{1}{4}}(-3). \text{ But ...}$$

Remember: $\log_a x = y \Rightarrow a^y = x$ with $x > 0$ and $a > 0$ and $a \neq 1$. That is, a positive number raised to any power is positive. Thus, x must be positive and $x \neq -3$.

Now, consider logarithmic functions where each term does NOT have logarithms.

  • In this case, all terms that have logarithms are placed on one side of the equation and simplified using the properties of logarithms. The equation is then expressed in exponential form and solved.

Example **6.20.** *Solve for* $\log_{\frac{1}{2}} x = 3$ *for* x.

  • **Solution**:

$$\log_{\frac{1}{2}} x = 3$$
$$(1/2)^3 = x \quad \text{Exponential form}$$
$$1/8 = x$$

Example **6.21.** *Solve* $\log(x+2) = 2$ *or x.*

- **Solution:**

$$\log(x+2) = 2 \quad \text{Base 10 is understood}$$
$$10^2 = x + 2 \quad \text{Exponential form}$$
$$100 = x + 2$$
$$98 = x$$

Example **6.22.** *Solve* $\log_x 16 = 2$ *for x.*

- **Solution:**

$$\log_x 16 = 2$$
$$x^2 = 16$$
$$x = \sqrt{16}$$
$$x = 4, \; x - 4$$

Since x in the original equation is in the base and the base must be positive, $x = 4$, but $x \neq -4$.

Exercise **6.7.** *Solve the following for x:*

(a) $\log x + \log(x-3) = 1$.

Solution:

(b) $\log_2(x+5) = 3$.

Solution:

(c) $\log_{\frac{1}{2}} x = \log_{\frac{1}{2}}(6 - x^2)$.

Solution:

(d) $\ln(4x+5) = \ln(2x+7)$.

Solution:

(e) $\log \; x^2 = \log \; (x+2)$.

Solution:

(f) $\log_4(x-3) - 2 = 0$.

Solution:

6.2.iii. Exponential Equations

Logarithms are often used to **solve exponential equations** when the same base is not easily obtained. In this case, we take the common or natural logarithm of both sides of the equation.

Example **6.23.** *Solve* $10^x = 42$. *for* x.

• **Solution**:

$$10^x = 42$$
Since the base is ten, take the common logarithm of both sides of the equation.

$$\log 10^x = \log 42$$
Use the properties of logarithms to simplify.

$$x \log 10 = \log 42$$

$$x(1) = \log 42$$

$$x = \log 42$$

$$x \approx 1.6232$$

Example **6.24.** *Solve* $7^{x+2} = 13$ *for* x.

• **Solution**:

$$7^{x+2} = 13$$

$$\log 7^{x+2} = \log 13$$

$$(x+2)\log 7 = \log 13$$

$$x + 2 = \frac{\log 13}{\log 7} \quad \textbf{Be careful}: \frac{\log 13}{\log 7} \neq \log\left(\frac{13}{7}\right)$$

$$x + 2 \neq 1.3181$$

$$x \neq -0.6819$$

Example **6.25.** *Solve* $e^{2x} = 5$ *for* x.

• **Solution**:

$$e^{2x} = 5$$

$$\ln e^{2x} = \ln 5 \quad \text{Since the base is e, take the}$$

$$2x \ln e = \ln 5 \quad \text{natural logarithm of both sides}$$

$$2x(1) = \ln 5 \quad \text{of the equation.}$$

$$2x = \ln 5$$

$$2x = \ln 5 \approx 1.60943$$

$$x = 0.80471$$

Exercise **6.8.** *Solve the following for* x:

(a) $2^{x+4} = 7$.

Solution:

(b) $5^{x-2} = 9$.

Solution:

(c) $10^x = 19$.

Solution:

(d) $e^{3x+1} = 15$.

Solution:

(e) $(\frac{1}{2})^{x-4} = 17$.

Solution:

(f) $e^{0.04x} = 100$.

Solution:

Summary 6.2.

Logarithmic Functions

- $y = \log_a x$ with $a > 0$, $a \neq 1$ is a logarithmic function.
- Domain: $D = \{x/x > 0\}$
- Range: $R = \{y/y > 0\}$
- **Common logarithms** have base ten: $\log_{10} x = \log x$
- **Natural logarithms** have base e: $\log_e x = \ln x$
- **To graph a logarithmic function in the form** $y = \log_a x$
  - ▶ Write the logarithmic function in exponential form.
    $$y = \log_a x \implies a^y = x$$
  - ▶ Determine if the function is increasing or decreasing.
    - ◀ $0 < a < 1$, the function is decreasing.
    - ◀ $a > 1$, the function is increasing.
- Construct a table of values.
  - ▶ Note: $(1, 0)$ will always be a point on the graph.

End of Section 6.2. Problems

Problem 6.5. *Graph each of the following*:

(a) $y = \log_{\frac{1}{3}} x - 5$. (b) $y = \log x + 1$. (c) $y = \ln(x+4)$. (d) $y = \log_6(x-3)$.

(e) $y = \ln x - 4$. (f) $y = \log_{\frac{1}{2}}(x-5)$. (g) $y = \ln x - \frac{1}{2}$. (h) $y = \log_5 x - 7$.

Problem 6.6. *Graph each of the following*:

(a) $y = \ln(x^2)$. (b) $y = \log_{\frac{1}{4}} x$. (c) $y = \ln(x-1)+1$. (d) $y = \log(x^2)$.

(e) $y = \ln e^x$. (f) $y = \log_4(x+2)$.

Problem 6.7. *Solve the following for* x:

(a) $\log_x 27 = 3$.

(b) $\log_x 16 = 4$.

(c) $\log_{\frac{1}{4}}(4x-1) = \log_{\frac{1}{4}}(2x+3)$.

(d) $\log x - \log(x+1) = \log 4$.

(e) $\ln x + \ln(x-5) = \ln 6$.

(f) $\ln(x+4) + \ln(x+3) = \frac{1}{2}\ln 144$.

(g) $\log_{12}(x+7) = 2$.

(h) $\log x - \log(x+4) = 2$.

Problem 6.8. *Solve the following for* x:

(a) $0 = 80 - 17\log(x+1)$. (b) $0 = 90 - 15\log(x+1)$. (c) $5 + 2\ln x = 6$. (d) $7\log 4x = 9$.

(e) $2\ln 3x = 5$. (f) $\ln 5x = 3$. (g) $\log 4x + 1 = 12$. (h) $\ln(3x) - \ln(x-2) = \ln(x^2)$.

Problem 6.9. *Solve the following for* x:

(a) $3^{2x+3} = 6$. (b) $4^{7x-1} = 125$. (c) $e^{x+1} = 7$. (d) $e^{5x-91} = 29.4$.

(e) $e^{x^2} = 4$. (f) $10^{x+2} = 95$. (g) $10^{x-4} = 17$. (h) $e^{x-1} = 7$.

Problem 6.10. *Solve the following for* x:

(a) $e^{1.04x} = 1000$. (b) $e^{-0.004x} = 10.95$. (c) $4e^x = 92$. (d) $1.75e^{-x} = 0.5$.

(e) $5^{\frac{x}{2}} = 0.2$. (f) $4^{-3x} = 0.15$.

6.3. Applications

There are many applications of exponential and logarithmic functions. Their use in sciences and in business is shown in the following examples.

EXPONENTIAL FUNCTIONS

Example **6.26**. *In 1978 a certain lake contained 801 fish of a particular species. In 1990 the lake contained 1720 fish of this species. Assuming exponential growth, answer each of the following:*

(a) *Write the equation for this population of fish.*

(b) *How long will it take for there to be 2804 fish in the lake?*

(c) *How many of these fish will be there in 1999?*

- **Solution:**

(a) $P(t) = P_0 e^{kt}$ is the model given for the equation, where P_0 represents the number at the beginning, t represents time and k is a constant. The first step is to determine k.

$$P_0 = 801$$

$$t = 1990 - 1978 = 12$$

At the end of those 12 years, there are 1720 fish. Thus,

$$1720 = 801 e^{k(12)}$$

$$\frac{1720}{801} = e^{k(12)}$$

$$2.14732 = e^{k(12)}$$

$$\ln 2.14732 = \ln(e^{k(12)}) \qquad \text{Remember } \ln e = 1.$$

$$0.764219 = k(12)$$

$$\frac{0.764219}{12} = k$$

$$k \approx 0.06$$

(b) $t = ?$ when there are 2804 fish. Thus,

$$2804 = 801 e^{0.06t}$$

$$\frac{2804}{801} = e^{0.06t}$$

$$3.501 = e^{0.06t}$$

$$\ln(3.501) = \ln(e^{0.06t}) \qquad \text{Remember : } \ln e = 1$$

$$1.25294 = 0.06t$$

$$\frac{1.25294}{0.06} = t$$

$$t \approx 20.9 \text{ years}$$

(c) In 1999, $t = 1999 - 1978 = 21$, so

$$P(21) = 801 e^{0.06(21)}$$

$$= 801 e^{1.26}$$

$$= 801(3.5254)$$

$$\approx 2824 \text{ fish}$$

Example **6.27.** *A doctor injects 10 mg. (milligrams) of a drug into a patient. After two hours, there was 4.7 mg of the drug present. Assuming exponential decay, answer each of the following:*

(a) Write the equation describing the behavior of the drug.

(b) How much of the drug in mg. is present in the patient's bloodstream 2.8 hours after the injection?

(c) When will there be 0.8 mg. of the drug left in the patient's bloodstream?

- **Solution:**

  (a) $P(t) = P_0 e^{kt}$, $P_0 = 10$, $t = 2$ hours. At the end of two hours, there is 4.7 mg of the drug present. Thus,

  $$4.7 = 10e^{k(2)}$$

  $$\frac{4.7}{10} = e^{k(2)}$$

  $$0.47 = e^{k(2)}$$

  $$\ln(.47) = \ln(e^{k(2)})$$

  $$-0.75502 = k(2)$$

  $$\frac{-0.75502}{2} = k$$

  $$k \approx -0.378$$

  Therefore, the model is $\quad P(t) = 10e^{-0.378t}$

  (b) $\quad t = 2.8$, thus,

  $$P(2.8) = 10e^{-0.378(2.8)}$$

  $$= 10e^{-1.0584}$$

  $$= 10(0.34701)$$

  $$= 3.47\,\text{mg}$$

- (c) $t = ?$, when 0.8 mg. is in the bloodstream.

  $$0.8 = 10e^{-0.378t}$$

  $$\frac{0.8}{10} = e^{-0.378t}$$

  $$0.08 = e^{-0.378t}$$

  $$\ln 0.08 = \ln(e^{-0.378t})$$

  $$-2.52573 = -0.378t$$

  $$\frac{-2.52573}{-0.378} = t$$

  $$t \approx 6.68 \text{ hours}$$

Example **6.28.** *Neil Rocker, a famous guitarist, bought a guitar in 1965 for $145.95. In 1979, he sold it for $17,452.00. Assuming exponential growth, answer each of the following:*

(a) Write the equation.

(b) How long will it take for the guitar to be worth $28,000?

(c) How much was the guitar worth in 1990?

- **Solution**:

  (a) $P(t) = P_0 e^{kt}$, $P_0 = 145.95$, $t = 1979 - 1965 = 14$. Thus,

  $$17452 = 145.95 e^{k(14)}$$

  $$\frac{17442}{145.95} = e^{k(14)}$$

  $$119.58 = e^{k(14)}$$

  $$\ln(119.58) = \ln e^{k(14)}$$

  $$4.78395 = k(14)$$

  $$0.342 = k$$

- Therefore, $P(t) = 145.95 e^{0.342t}$

  (b) What is the value of t when the guitar is worth $28,000?

  $$28000 = 145.95 e^{0.342t}$$

  $$191.85 = e^{0.342t}$$

  $$\ln 191.85 = \ln e^{0.342t}$$

  $$5.2567 = 0.342t$$

  $$15.37 = t$$

- In other words $t \approx 15.4$ years

  (c) $t = 1990 - 1965 = 25$

  $$P(25) = 145.95 e^{0.342(25)}$$

  $$= 145.95 e^{8.55}$$

  $$= 145.95(5166.75)$$

  $$= \$754,087.81$$

INTEREST PROBLEMS

Simple interest is computed using the formula $I = Prt$, where I is the interest, P is the principal, r is the rate per year and t is the time in years. If the final amount (A) is needed for simple interest, add the interest and the principal. Thus,

$$A = P + I \Rightarrow A = P + Prt \Rightarrow A = P(1 + rt)$$

For compound interest the formula is

$$A = P(1 + \tfrac{r}{n})^{nt}$$

where A is the final amount, P is the principal invested, r is the rate per year, n is the number of periods, and t is the time in years. Periods can be as follows:

Annually	$n = 1$
Semiannually	$n = 2$
Quarterly	$n = 4$
Monthly	$n = 12$
Daily	$n = 365$

Example **6.29**. *Find the balance if $25,000 is invested for 3 years at 7.5% interest and the money is compounded quarterly.*

- **Solution**: $P = 25000$, $r = 7.5\% = 0.075$, $t = 3$ and $n = 4$. Substituting into the formula:

$$A = P(1 + \frac{r}{n})^{nt}$$
$$= 25000(1 + \frac{0.075}{4})^{4(3)}$$
$$= 25000(1.01875)^{12}$$
$$= 25000(1.2497164)$$
$$= \$31,242.91$$

Example **6.30**. *Find the balance if $10,000 is invested for 6 months at 8% compounded monthly.*

- **Solution**: $P = 10000$, $r = 8\% = 0.08$, $t = 1/2 = 0.5$ and $n = 12$:

$$A = 10000(1 + \frac{0.08}{12})^{12(0.5)}$$
$$= 10000(1.0967)^6$$
$$= 10000(1.0406726)$$
$$= \$10,406.73$$

- When interest is compounded continuously, the formula is $A = Pe^{rt}$, where A is the ending balance, P is the principal, r is the interest rate, and t is the time in years.

Example **6.31**. *Find the ending balance when $6,012 is invested at 6.5% compounded continuously for 6 years.*

- **Solution**: $P = 6012$, $r = 6.5\% = 0.065$, and $t = 6$. Substituting into the formula:

$$A = 6012e^{0.065(6)} = 6012e^{0.39}$$
$$= 6012(1.47698)$$
$$= \$8879.61$$

Example **6.32**. *Find the balance if $10,000 is invested for 6 months at 8% compounded continuously.*

- **Solution**: $A = 10000$, $r = 8\% = 0.08$ and $t = 1/2 = 0.5$:

$$A = 10000e^{0.08(0.5)} = 10000e^{0.04}$$
$$= 10000(1.0408107)$$
$$= \$10,408.11$$

LOGARITHMIC FUNCTIONS

Example 6.33. *A certain population of beavers in a forest in the Rocky Mountains appears to be growing according to the formula* $P(t) = 12,504 + 1800\ln(t+2)$, *where* t *is time in years from now.*

(a) *How many beavers are there initially* $(t = 0)$?

(b) *How long will it take for there to be 15,000 beavers?*

- **Solution:**
  (a)

$$P(t) = 12,504 + 1800\ln(t+2)$$
$$P(0) = 12,504 + 1800\ln(0+2)$$
$$= 12,504 + 1800\ln 2$$
$$= 12,504 + 1800(.69315)$$
$$= 12,504 + 1,247.7$$
$$= 13,752 \text{ beavers}$$

(b) $t = ?$, when there are 15,000 beavers?

$$P(t) = 12,504 + 1800\ln(t+2)$$
$$15,000 = 12,504 + 1800\ln(t+2)$$
$$15,000 - 12,504 = 1800\ln(t+2)$$
$$\frac{2496}{1800} = \ln(t+2)$$
$$1.38667 = \ln(t+2) \qquad \text{By the definition of logarithm with base } e.$$
$$e^{1.38667} = t + 2$$
$$4.001 = t + 2$$
$$2.001 = t \approx 2 \text{ years}$$

Exercise 6.9. *With a principal of $10,000 at an annual interest rate of 8.25%, find the balance after 10 years if compounded:*

(a) *monthly.*

Solution:

(b) *continuously.*

Solution:

Exercise **6.10.** *Do the problem above with a principal of $2000. Is there as much of a difference between compounded monthly and compounded continuously? If you invested $2135 for 5 years at 10.2%, what would your balance be if the money were compounded*:

(a) *monthly.*

Solution:

(b) *quarterly.*

Solution:

(c) *yearly.*

Solution:

Exercise **6.11.** *Upon the birth of a child, grandparents make a deposit of $20,000 in a trust fund that pays 7% interest compounded continously. What is the balance in this account on the child's 21st birthday? What amount must be deposited today in a money market account so that the investment will grow to $30,000 in 18 years at 7.5% compounded continuously?*

Solution:

Exercise **6.12.** *A certain population of dolphins grows according to the model* $P(t) = 250e^{0.029t}$, *where t is the time in years and $t = o$ corresponds to 1987. Use this model to estimate this dolphin population in*:

(a) 1992.

Solution:

(b) 2000.

Solution:

Exercise **6.13.** *After t years, the value of your car that cost $12,500 is given by* $V(t) = 12,500(\frac{3}{4})^t$. *Find the value of your car*:

(a) 1 *year after purchase.*

Solution:

(b) 3 *years after purchase.*

Solution:

(c) 10 *years after purchase.*

Solution:

Chapter 7.

Systems of Equations and Inequalities

Introduction

In this chapter we discuss the problem of solving equations and inequalities involving two or more variables. Although there are several methods for solving such systems, we shall only discuss the **two** main approaches: the **method of substitution** and the **method of elimination**. We shall also use a graphical (geometric) approach to further understand the essence of solutions to systems of equations. Although some of these methods date back to the ancient times, not only are they still used today, they form the basis for the more modern methods which are applicable only to certain classes of equations. We shall explore some of these modern methods (matrices and determinants) in the next chapter.

We begin with a definition.

| Definition |

System

A collection of two or more equations each of which contains one or more variable is referred to as a system of equations.

The following are examples of systems of equations:

$$(1) \quad \begin{cases} x + y = 1 \\ 2x - y = 5 \end{cases} \qquad \text{(This system has two variables.)}$$

| Definition |

Naming systems

A system with m equations and n unknowns is referred to as an $m \times n$ system.

Thus, system (1) is a 2×2 system.

This system, and system (3) below, are classified as **linear**, since the **degree** of each term is not greater than **one**. System (2) below is a **nonlinear** system because one of the terms (x^2) has degree more than one.

$$(2) \quad \begin{cases} x^2 + y = 2 \\ x - y = 4 \end{cases}$$

$$(3) \quad \begin{cases} x + y - 3z = 1 \\ 2x - y + 5z = 4 \end{cases} \qquad \text{(This system contains three variables.)}$$

| Definition |

Solution

A set of values, one for each variable, is said to be a **solution** of a given system if the set satisfies **each** equation in the system.

For the system (1) $\begin{cases} x + y = 1 \\ 2x - y = 5 \end{cases}$

the set of values $\{x = 2, \quad y = -1\}$ is a <u>solution</u> since

$$(2) + (-1) = 1$$
$$\text{and} \quad 2(2) - (-1) = 5$$

- |Note| The set $\{x = -1, \quad y = 2\}$ satisfies the first equation in (1): $\quad x + y = 1$, since $(-1) + (2) = 1$

However, the same set does not satisfy the second equation in (1): $\quad 2x - y = 5 \quad$ because $2(-1) - (5) = -7 \neq 5$.

Hence, the set is **not** a solution for the system (1).

Similarly, the set of values $\{x = 2, \quad y = -2\}$ is a solution for system (2)

$$\begin{cases} x^2 + y = 2 \\ x - y = 4 \end{cases}$$

since

$$(2)^2 + (-2) = 2$$
$$\text{and} \quad (2) - (-2) = 4$$

Furthermore, another solution for system (2) is $\{x = -3, \quad y = -7\}$. (*Can you verify this?*).

7.1. Linear Systems

| *Definition* |

Linear System

A system of equations is said to be linear if each equation in the system is linear. That is, each equation is of degree **one**. Thus, each term in the system is of degree one or zero.

For example, the expression $3x + 5xy - 7\sqrt{z} + 2x^2$ has four terms. The first term, $3x$, has **degree one**; each of the second and last terms, $5xy$ and $2x^2$, has **degree two**; and the third term has **no degree**, since $\sqrt{z} \ (= z^{\frac{1}{2}})$ cannot be part of a polynomial. The expression is therefore <u>nonlinear</u> since it contains nonlinear terms.

In contrast, each of the following equations is **linear**:

$$3x - \sqrt{2}\,y = 7$$
$$\frac{2}{5}x + 6y = 3$$

Linear systems of equations are further classified according to the number of solutions they have. In order to understand the reasoning behind such classification, we will examine linear equations in only two variables.

7.1.i. A Graphical Approach

Consider the linear system in the variables x and y : $\begin{cases} x+y = 1 \\ 2x-y = 5 \end{cases}$

First observe that each equation is a **straight line**. We have seen that the set of values $\{x = 2, \quad y = -1\}$ is a solution for the system. In other words, the values satisfy both equations. Thus, the point $P(2,-1)$ in the x-y plane is on each of the straight lines. Consequently, this is the **point of intersection** of the two lines. We graph both lines on the same pair of axes:

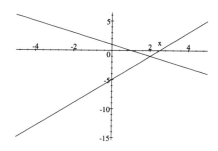

Observe that the lines intersect at $(2,-1)$. This provides us with the **geometric interpretation of a solution** for a linear system of two equations in two variables: the **point of intersection** of the two lines.

• | **Note:** | it is also the interpretation of a solution of a system **even if** the graphs are **not straight** lines.

So far, we have made two important observations for linear system of two equations in two variables, namely:

■   the graph of each equation is a straight line in the plane;

■   a solution is a point of intersection of the straight lines.

These observations lead us to **three** distinct possibilities for the number of solutions.

7.1.ii. Three classes of linear system in two variables

| **Case 1.** | **Two non-parallel lines**.

If two straight lines are not parallel, they necessarily **intersect at exactly one point**. Hence, a system of two non-parallel straight lines has a _unique_ (one only) solution. We classify such a system as **consistent** and **independent**. We identify this class of systems by checking to see that the straight lines have different slopes.

Lets consider the following linear system:

$$\begin{cases} x+y = 1 \\ 2x-y = 5 \end{cases}$$

We put each equation in the form $y = mx + b$, where m is the slope and b is the y-intercept.

$$\begin{cases} y = -x + 1 \\ y = 2x - 5 \end{cases}$$

Clearly, the first and second lines have slopes $m_1 = -1$, and $m_2 = 2$, respectively. Thus the system is **consistent** and **independent**. As we have already seen, the unique solution (point of intersection) is

$P(2,-1)$

Exercise 7.1. *Find approximate solutions to each of the following systems using the graphical method*:

(a) $\begin{cases} x - 3y = 6 \\ 2x + 3y = 3 \end{cases}$

(b) $\begin{cases} 7x - 5y = 10 \\ 3x - 2y = 6 \end{cases}$

Solution:

Solution:

Case 2. **Two parallel lines.**

In this case, the two lines **do not intersect** since they are parallel. Consequently, a system of parallel straight lines **has no solution** because the lines have no point(s) of intersection. Such systems are referred to as **inconsistent**, and can be identified by their characteristics of having the **same slope** but **different** *y*-**intercepts** (the features of distinct parallel lines).

For example, consider the system:

$$\begin{cases} 2x + y = 2 \\ 2x + y = -4 \end{cases}$$

We can rearrange the equations in the appropriate form:

$$\begin{cases} y = -2x + 2 \\ y = -2x + -4 \end{cases}$$

Clearly, the slope of each line is -2, but the *y*-intercepts are 2 and -4, respectively. Thus the lines are parallel but distinct and the system is **inconsistent**. The graphs are:

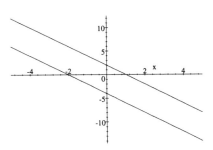

As the graphs show, the two lines do not intersect. Equivalently, the system is **inconsistent** and has **no solution**.

Exercise 7.2. *Find approximate solutions to the following system using the graphical method:*

$2x + y = -2$
$4x + 2y = 5$

Solution:

Case 3. **Two coincident lines.**

A third situation occurs when the two lines are **coincident**. That is, they are essentially the same line. Since the lines intersect at infinitely many places, the system has infinitely many solutions. Such a system is said to be **consistent** (it has solutions) and **dependent** (the straight lines are not distinct). We can **identify** a dependent system from the lines having the **same slope** and the **same** -intercept.

For example, the system

$$\begin{cases} 2x - y = 3 \\ -4x + 2y = -6 \end{cases}$$

Rearranging each equation in the slope-intercept form, we obtain

$$\begin{cases} y = 2x - 3 \\ y = \frac{4}{2}x - \frac{6}{2} \end{cases}$$

It is clear at this point that the second equation simplifies to give the first one. Their common graph is:

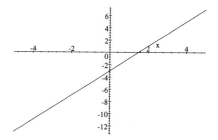

Observation

- In **Case 3** every point on the first line is also a point on the second line. Since solutions are points common to both lines, this means that every point on each line is a solution. (We shall see later how to represent such multiple solutions as we examine algebraic methods for solving linear systems.)

- The graphical method of solution, though useful, may help us to find *approximate solutions only* since it is not always feasible to interpret points of intersection on a graph with precision.

Exercise *7.3. Find approximate solution to the following system using the graphical method*:

$$y = 2x - 3$$
$$-5y = -10x + 15$$

Solution:

7.1.iii. The Method of Substitution

This is one of the algebraic methods that can be employed to solve a system of equations. It is perhaps the earliest known method for solving systems. The main idea is to solve for one of the variables in any one of the equations (i.e. make the variable the subject), and to substitute this expression in the remaining equation(s). We then solve the resulting equation(s). A final step is to do a "back substitution" for the first variable which was made the subject. An example makes this process clearer.

Example 7.1. *Solve the system* $\begin{cases} 2x - y = 5 ....(\mathbf{1}) \\ 3x + 2y = 4 ....(\mathbf{2}) \end{cases}$

- **Solution**: We can readily solve for y in $(\mathbf{1})$:

$$y = 2x - 5$$

Having made y the subject using $(\mathbf{1})$, we must substitute this expression for y in $(\mathbf{2})$:

$$3x + 2(2x - 5) = 4 ....(\mathbf{3})$$

Observe that the resulting equation $(\mathbf{3})$ is linear and involves x only. We solve it:

$$3x + 4x - 10 = 4$$
$$7x = 14$$
$$x = 2$$

We now perform a "back substitution" on the original subject:

$$y = 2x - 5 = 2(2) - 5 = 4 - 5 = -1$$

Thus, the solution is

$$\boxed{(x, y) = (2, -1).}$$

We may **check our solution**: If $x = 2$ and $y = -1$,

$$\begin{cases} 2x - y = 2(2) - (-1) = 5 \quad ....(1) \\ 3x + 2y = 3(2) + 2(-1) = 4 \quad ....(2) \end{cases}$$

The set of values satisfies **each** equation and so is a solution for the system. Speaking geometrically, the point $P(2, -1)$ is on both lines. Since the lines are non-parallel (*why?*), this is the only point of intersection. That is, the system is **consistent** and **independent**.

| **Note**: | after substitution, the resulting equation, $x = 2$, is a *conditional equation*. This is the indicator that we have a *unique solution*. (Recall from Chapter 2 that a linear **conditional equation** in **one variable** has exactly **one solution**).

Since the geometric interpretation of a solution is a point of intersection, it is customary to represent solutions as ordered pairs. This representation is extended to situations with more than two variables.

Example **7.2.** *Solve the system*

$$\begin{cases} x - 2y = 5 \dots\dots\dots\dots\dots(\mathbf{1}) \\ 2x - 4y = 3 \dots\dots\dots\dots\dots(\mathbf{2}) \end{cases}$$

- **Solution**: Observe, first, that it is most convenient to solve for x in $(\mathbf{1})$:

$$x = 2y + 5.$$

Substituting in $(\mathbf{2})$, we get

$$2(2y + 5) - 4y = 3 \dots\dots\dots\dots\dots(\mathbf{3})$$

We attempt to solve the resulting equation:

$$4y + 10 - 4y = 3$$
$$10 = 3$$

Equivalently, $7 = 0$ $\rightarrow\leftarrow$

Since 7 and 0 are not equal, we have a **contradiction**! ($\rightarrow\leftarrow$) The natural conclusion here is that the system is *inconsistent* and has *no solution*. This is always the case when the resulting equation is a *contradiction*. We may verify (independently) that the two lines do not intersect by checking to see that they are distinct and parallel.

Exercise **7.4.** *Solve each of the following systems by the method of substitution:*

(a) $\begin{cases} x - 3y = 6 \\ 2x + 3y = 3 \end{cases}$ (b) $\begin{cases} x - 3y = 6 \\ 2x - 6y = 1 \end{cases}$

Solution: **Solution**:

Example **7.3.** *Solve the system*

$$\begin{cases} -4x + 2y = -6 \ \dots\dots\dots\dots\dots(\mathbf{1}) \\ 2x - y = 3 \ \dots\dots\dots\dots\dots(\mathbf{2}) \end{cases}$$

- **Solution**: Can you see that we can easily solve for y in $(\mathbf{2})$?

$$y = 2x - 3$$

Substituting for y in $(\mathbf{1})$, we obtain

$$-4x + 2(2x - 3) = -6 \dots\dots\dots\dots\dots(\mathbf{3})$$

This is the resulting equation. We solve it:

$$-4x + 4x - 6 = -6$$
$$-6 = -6$$
or $0 = 0$ $\checkmark$

As you can see, this is a *trivial truth* as the left side equals the right side unconditionally. The conclusion here is that the two lines are *coincident* and the system has infinitely many solutions. The resulting equation is an **identity**. Verify that the slopes of the two lines are equal and so are their y-intercepts. Thus, the system is *consistent* but *dependent*.

Recall that in this case CASE 3 every point on one line is also a point on the second line. From equation (3), we know that both equations can be represented as $y = 2x - 3$. Consequently, a point on the lines has coordinates $(t, 2t - 3)$, where t is any real number.

■ **Set notation**:

▶ The solutions for the system form a set:

$$S = \{(t, 2t - 3)/t \in \mathbf{R}\}.$$

In other words, each solution to the system is an ordered pair of real numbers such that the second is 3 less than twice the first number. Examples of such pairs are:

$(0, -3)$ because if $t = 0$, then $2t - 3 = -3$;

$(-1, -5)$ since $t = -1, \Rightarrow 2t - 3 = -5$

Other solutions are: $(1, -1)$, $\left(\frac{1}{2}, -2\right)$, and $(2, 1)$. We can generate many more solutions by arbitrarily choosing values for $x = t$ and by calculating $y = 2t - 3$.

Exercise *7.5. Solve each of the following systems by the method of substitution:*

(a) $\begin{cases} -3x + y = 5 \\ 2x - 3y = -8 \end{cases}$

(b) $\begin{cases} \frac{1}{2}x - \frac{2}{3}y = 1 \\ -3x + 4y = -6 \end{cases}$

Solution:

Solution:

7.1.iv. The Method of Elimination .

This algebraic method is similar to the method of substitution. Before examining the method, we lay down the ground rules for manipulating systems of equations. These are allowable operations on the systems.

They are:

- Reordering the rows (by interchanging rows).
- Multiplying (dividing) a row by a non-zero constant.
- Multiplying one row by a non-zero constant and adding to another row.

These operations **do not alter** a system. The method consists of using a combination of these operations to *eliminate* a variable from the system. We give an example to show how this is done.

Example *7.4. Solve, using the method of elimination, the linear system*

$$\begin{cases} 2x - y = 5 ....R_1 \\ 3x + 2y = 4 ....R_2 \end{cases}$$

- **Solution**: In this case we observe that y has coefficients of opposite signs in the two equations. We multiply R_1 by 2. The system becomes

$$\begin{cases} 4x - 2y = 10 ....R_1 \\ 3x + 2y = 4 ....R_2 \end{cases}$$

Notice that we can now eliminate y by adding R_1 and R_2

$$R_1 + R_2 : 7x = 14 \dots\dots\dots\dots\dots(a)$$

We solve the resulting equation (a) for x

$$x = 2$$

and we substitute the value of x in **either** of the original equations in the system to solve for y:

$$2(2) - y = 5 \Rightarrow y = -1$$

or $\quad 3(2) + 2y = 4 \Rightarrow y = -1$

Notice that we needed to substitute the value of x in only one of the equations as either one would yield the same result. The solution is $\boxed{(x,y) = (2,-1)}$.

Note: After eliminating one variable, we have a **conditional** equation leading to a unique solution.

Example 7.5. *Solve, using the method of elimination, the linear system*

$$\begin{cases} 2x + 3y = -1 \dots\dots\dots\dots\dots R_1 \\ 3x + 5y = 1 \dots\dots\dots\dots\dots R_2 \end{cases}$$

- **Solution**:

Multiply R_1 by 3, and multiply R_2 by -2,

$$\begin{cases} 3R_1 : 6x + 9y = -3 \\ -2R_2 : -6x - 10y = -2 \end{cases}$$

Observe that the x-column will be *eliminated* when we add the two lines. This is the aim of choosing the constants 3 and -2 to multiply R_1 and R_2, respectively. Adding, we get a conditional equation:

$$(3R_1) + (-2R_2) \ :$$
$$0 - y = -5$$
$$y = 5$$

We substitute the value of y in **either** of the equations in the original system to find x:

$$2x + 3(5) = -1$$
$$2x = -15 - 1$$
$$2x = -16$$
$$x = -8$$

The unique solution is $\boxed{(x,y) = (-8,5)}$.

Exercise 7.6. *Use the method of elimination to solve the system*:

(a) $\begin{cases} \frac{x}{2} - \frac{y}{3} = 1 \\ x + y = 2 \end{cases}$

(b) $\begin{cases} 2x + 3y = 6 \\ 3x + 2y = -6 \end{cases}$

Solution:

Solution:

Example 7.6. *Solve the system*

$$\begin{cases} 2x - 3y = -5 & \dots\dots\dots\dots\dots R_1 \\ -4x + 6y = 5 & \dots\dots\dots\dots\dots R_2 \end{cases}$$

- **Solution**: Multiply R_1 by 2, the system becomes:

$$\begin{cases} 4x - 6y = -10 \\ -4x + 6y = 5 \end{cases}$$

We are ready to add:

$$2R_1 + R_2 : 0 + 0 = -5 \qquad \rightarrow\leftarrow$$

Notice the *contradiction*($\rightarrow\leftarrow$). We conclude that the system is inconsistent and has no solution.

  Note: As in the method of substitution, if the resulting equation is an identity (a trivial truth), the system is consistent but dependent.

Example 7.7. *Solve the system*

$$\begin{cases} 4x - 2y = 1 & \dots\dots\dots\dots\dots R_1 \\ -2x + y = -\frac{1}{2} & \dots\dots\dots\dots\dots R_2 \end{cases}$$

- **Solution**: Multiply R_2 by 2. The resulting system is:

$$\begin{cases} 4x - 2y = 1 \\ -4x + 2y = -1 \end{cases}$$

Adding the two equations:

$$R_1 + 2R_2 :$$
$$0 + 0 = 0$$

The resulting equation is an identity. Thus, we have a solution set. For the solution set, we solve for one of the variables in, say, (R_2):

$$y = 2x - \frac{1}{2} = \frac{4x - 1}{2}$$

Thus $S = \left\{ \left(t, \frac{4t-1}{2} \right) / t \in \mathbf{R} \right\}$

Exercise 7.7. *Use the method of elimination to solve the system*:

(a) $\begin{cases} \frac{1}{2}x - \frac{2}{3}y = 1 \\ -3x + 4y = 5 \end{cases}$

(b) $\begin{cases} 2x - 2y = -1 \\ -x + y = \frac{1}{2} \end{cases}$

Solution:

Solution:

7.1.v. Linear Equations with Three Variables

We conclude this section by solving a linear system of three equations in three variables.

Example **7.8.** *Solve, using the method of elimination, the system*

$$\begin{cases} 2x - y + 3z = -3 ....R_1 \\ x + 2y + 5z = 0 ....R_2 \\ 3x - 2y - 3z = 2 ....R_3 \end{cases}$$

- **Solution**:

Step 1. Using R_1, we solve for the variable y,

$$y = 2x + 3z + 3 ....(a)$$

Step 2. We substitute the expression in **both** R_2 and R,

$$\begin{cases} x + 2(2x + 3z + 3) + 5z = 0 ....R_2 \\ 3x - 2(2x + 3z + 3) - 3z = 2 ....R_3 \end{cases}$$

Observe that we have reduced the problem from a 3×3 to a 2×2 system. Next, we collect terms in each equation:

$$\begin{cases} 5x + 11z = -6 ....R_3 \\ -x - 9z = 8 ....R_4 \end{cases}$$

We now solve the 2×2 system.

Step 3. From R_4, we solve for x :

$$x = -9z - 8 ....(b)$$

Substituting into R_3, we have

$$5(-9z - 8) + 11z = -6$$

Solving for z,

$$-45z - 40 + 11z = -6$$
$$-34z = 34$$
$$z = -1$$

Step 4. We substitute the value of z into (b) to solve for x.

$$x = -9z - 8 ....(b)$$
$$= -9(-1) - 8$$
$$= 1$$

Similarly, we substitute the values of x and z into (a) to solve for y.

$$y = 2x + 3z + 3 ....(a)$$
$$= 2(1) + 3(-1) + 3$$
$$= 2$$

The solution to the system is $\boxed{(x, y, z) = (1, 2, -1)}$.

Example 7.9. *Solve the same system using the method of elimination:*

$$\begin{cases} 2x - y + 3z = -3 \dots\dots\dots\dots\dots R_1 \\ x + 2y + 5z = 0 \dots\dots\dots\dots\dots R_2 \\ 3x - 2y - 3z = 2 \dots\dots\dots\dots\dots R_3 \end{cases}$$

- **Solution:**

  Step 1. We may interchange rows as follows:

  $$(1) \quad \begin{cases} x + 2y + 5z = 0 \dots\dots\dots\dots\dots R_1 \\ 2x - y + 3z = -3 \dots\dots\dots\dots\dots R_2 \\ 3x - 2y - 3z = 2 \dots\dots\dots\dots\dots R_3 \end{cases}$$

  The rationale for this interchange is to have the row with 1 as the coefficient of x (or any other variable) as the first row. Generally, this tends to ease the required thought process as well as the arithmetic. In this example, we have targeted x for elimination in R_2 and R_3. If we choose to eliminate x in R_2 using R_1, we *must* use the same R_1 to eliminate x in R_3.

  Step 2. In order to eliminate x in R_2, we multiply R_1 by (-2) and add to R_2 and *this sum replaces R_2*. That is, we add the two rows:

  $$-2x - 4y - 10z = 0 \dots\dots\dots\dots\dots R_1$$
  $$2x - y + 3z = -3 \dots\dots\dots\dots\dots R_2$$

  to obtain

  $$-5y - 7y = -3 \dots\dots\dots\dots\dots R_2$$

  In addition, we describe our operations using the short-hand
  $$-2R_1 + R_2$$

  Similarly, the operation
  $$-3R_1 + R_3$$

  eliminates x in R_3 and the sum replaces R_3. Thus, our new system is:

  $$(2) \quad \begin{cases} x + 2y + 5z = 0 \dots\dots\dots\dots\dots R_1 \rightarrow R_1 \\ -5y - 7z = -3 \dots\dots\dots\dots\dots -2R_1 + R_2 \rightarrow R_2 \\ -8y - 18z = 2 \dots\dots\dots\dots\dots -3R_1 + R_3 \rightarrow R_3 \end{cases}$$

  Of course, we now eliminate either y or z from the last two rows in (2).

  Step 3. We treat the subsystem comprising of the last two rows as a new problem. We first multiply R_3 by $-\frac{1}{2}$ so as to remove the common factor 2. The subsystem is equivalent to:

  $$-5y - 7z = -3 \dots\dots\dots\dots\dots R_2$$
  $$4y + 9z = -1 \dots\dots\dots\dots\dots R_3$$

  The operation $4R_2 + 5R_3$ will eliminate y.

  $$-20y - 28z = -12 \dots\dots\dots\dots\dots 4R_2 \rightarrow R_2$$
  $$20y + 45z = -5 \dots\dots\dots\dots\dots 5R_3 \rightarrow R_3$$

  Adding, we get
  $$17z = -17$$

  This replaces either of the rows R_2, R_3. We can now solve for z. Indeed our system has become:

  $$x + 2y + 5z = 0 \dots\dots\dots\dots\dots R_1$$
  $$-5y - 7z = -3 \dots\dots\dots\dots\dots R_2$$
  $$z = -1 \dots\dots\dots\dots\dots R_3$$

  Observe that we have retained R_1 in its original form.

  Step 4. We are now ready to solve for the remaining variables in an orderly fashion. We substitute the value of z in R_2.

$$-5y - 7(-1) = -3$$
$$5y = 10$$
$$y = 2$$

Similarly, we substitute for y and z in R_1.

As expected, the solution of the system is $\boxed{(x, y, z) = (1, 2, -1).}$

Exercise 7.8. *Solve the following system*:

(a) $\begin{cases} x - y + 2z = 1 \\ -x + 2y + z = 0 \\ x + 3y + z = 5 \end{cases}$

(b) $\begin{cases} 3x + 4y + 2z = 1 \\ 2x + 3y + z = 1 \\ 6x + y + 5z = 1 \end{cases}$

Solution:

Solution:

Summary 7.1.

■ A linear system is one in which each equation is *linear* with respect to all the variables. Each term in each equation has degree *one at most*.

■ For the linear system involving two variables, the graph of each equation is a straight line in the plane. The geometric interpretation of a solution: the point(s) of *intersection of the straight lines*.

▶ There are **three** types of linear systems:

◀ A system is *consistent* and *independent* if it has a *unique solution*. The straight lines have different slopes. The resulting equation (after substitution/elimination) is *conditional*.

◀ A system is *inconsistent* if it has *no solution*. The straight lines are parallel but distinct. They have the same slope but different y-intercepts. Here, the resulting equation is a *contradiction*.

◀ A system is *consistent* but *dependent* if it has *infinitely many solutions*. The straight lines are coincident. They have the same slope as well as y-intercept. The resulting equation is an *identity*.

■ We may solve a linear system using the *graphical method* (approximate solutions only); the method of *elimination*, or the method of *substitution*.

End of Section 7.1. Problems

Problem 7.1. Find approximate solutions to each system by graphing:

(a) $\begin{cases} x + y = 1 \\ x + 3y = 5 \end{cases}$ (b) $\begin{cases} x - 2y = 1 \\ 2x - 3y = 3 \end{cases}$ (c) $\begin{cases} \frac{2x}{3} - \frac{3y}{4} = 7 \\ \frac{5x}{6} + \frac{y}{2} = 3 \end{cases}$

(d) $\begin{cases} x + 3y = 3 \\ 2x + 6y = 9 \end{cases}$ (e) $\begin{cases} 2x + 3y = 4 \\ 6x - y = 2 \end{cases}$ (f) $\begin{cases} x + y = 1 \\ 2x - y = 2 \end{cases}$

Problem 7.2. Solve the following systems by substitution:

(a) $\begin{cases} x - 2y = 0 \\ 2x - y = 5 \end{cases}$ (b) $\begin{cases} 3x + y = 1 \\ 3x + 2y = 5 \end{cases}$ (c) $\begin{cases} 4x + 3y = -4 \\ 2x + y = -1 \end{cases}$

(d) $\begin{cases} -2x + 3y = 1 \\ 3x - 2y = 6 \end{cases}$ (e) $\begin{cases} x - 3y = 2 \\ 3x - 9y = 5 \end{cases}$ (f) $\begin{cases} \frac{3x}{4} - \frac{y}{2} = 1 \\ \frac{x}{4} - \frac{3y}{8} = -2 \end{cases}$

Problem 7.3. Solve each system by elimination:

(a) $\begin{cases} x - 2y = -3 \\ 3x + 5y = 8 \end{cases}$ (b) $\begin{cases} 2x + y = 4 \\ 3x - 2y = 3 \end{cases}$ (c) $\begin{cases} 2x - y = 3 \\ x + 2y = 9 \end{cases}$

(d) $\begin{cases} \frac{x}{2} + \frac{y}{3} = \frac{3}{4} \\ \frac{x}{3} - \frac{y}{2} = -\frac{1}{6} \end{cases}$ (e) $\begin{cases} 3x - 4y = 6 \\ 3x - 4y = 12 \end{cases}$ (f) $\begin{cases} 3x + 2y = 4 \\ x + 3y = 6 \end{cases}$

Problem 7.4. Solve the following systems by any method:

(a) $\begin{cases} 3x + 4y = 9 \\ 2x - 2y = 3 \end{cases}$ (b) $\begin{cases} x + 3y = 11 \\ 2x + 5y = 17 \end{cases}$ (c) $\begin{cases} 5x - 3y = 2 \\ 2x + 2y = 7 \end{cases}$

(d) $\begin{cases} x + y + 2z = 11 \\ 2x - y - z = 0 \\ 3x + y + 4z = 25 \end{cases}$ (e) $\begin{cases} x + y - z = 1 \\ x + y + z = 0 \\ x - y + z = 3 \end{cases}$ (f) $\begin{cases} x + y + z = 1 \\ 3x + 2y - 2z = -1 \\ 2x + y - 3z = 5 \end{cases}$

7.2. Applications

In this section, we discuss a few of the numerous types of applications of linear systems. Since these are word problems, we suggest that you follow a definite strategy.

■ One such strategy is to:
  ▶ read the whole problem;
  ▶ determine what is to be found and name these as variables;
  ▶ clearly label what each variable represents in the problem;
  ▶ write down any formulas (if any) connecting the variables.
  ▶ read the problem again; this time, write down equations connecting the variables;
  ▶ solve the system that you set up.

7.2.i. Number Problems

Example **7.10**. *Three times a number is six more than twice a second number. If three times the second number is six more than four times the first number, find the two numbers.*

● **Solution**: Let

$$x = \text{first \#}$$
$$y = \text{second \#}.$$

Then,
$$\begin{cases} 3x = 6 + 2y \\ 3y = 6 + 4x \end{cases}$$

Notice that we have chosen our variables to be the quantities that we are looking for. In this way when we solve for x and y, we are done. Rearranging the two equations, we have the system

$$\begin{cases} 3x - 2y = 6 \dots\dots\dots\dots\dots\dots.R_1 \\ -4x + 3y = 6 \dots\dots\dots\dots\dots.R_2 \end{cases}$$

We now solve the system by elimination:

$$3R_1 + 2R_2 \; : \; 3x - 8x = 18 + 12$$
$$x = 30.$$

Substituting into R_2,

$$-4(30) + 3y = 6$$
$$3y = 126$$
$$y = 42$$

Thus, the first number is 30, and the second number is 42.

Example **7.11**. *A two-digit number is three more than six times the sum of its digits. Two times the tens digit is one less than three times the units digit. Find the number.*

● Solution. Let

$$x = \text{units digit}$$
$$y = \text{tens digit}.$$

Note that in the 52, five is the tens digit and two is the units digit. In fact, the number is $5(10) + 2(1)$. Thus, in the example, the number is $10y + x$. We now interpret the first two sentences symbolically:

$$10y + x = 3 + 6(x + y)$$
$$2y = 3x - 1.$$

Rearranging these in standard form, we obtain

$$\begin{cases} -5x + 4y = 3 \ldots\ldots\ldots R_1 \\ 3x - 2y = 1 \ldots\ldots\ldots R_2 \end{cases}$$

The operation $2R_2 + R_1$ eliminates the variable y:

$$x = 5.$$

We plug the value into, say, R_2:

$$3(5) - 2y = 1$$
$$\text{i.e.} \quad 2y = 14$$
$$y = 7.$$

Thus, the number is 75.

Exercise **7.9**. *Two times a number is one less than three times a second number, while four times the first number is seven more than five times the second number. Find the two numbers.*

Solution:

Exercise **7.10**. *The sum of the digits of a two-digit number is one more than twice the tens digit. If the number is one less than ten times the units digit, find the number.*

Solution:

7.2.ii. Motion Problems

Example **7.12.** *A car travels* 20 *miles per hour faster than a truck. If the two vehicles are traveling in the same direction and the car travels* 240 *miles in the sa.ne time that the truck travels* 180 *miles, find the speeds of the two vehicles.*

Solution: Here, we must remember that constant rate motion is governed by the formula

$$\boxed{d = rt}$$

where d, r, t, are distance, rate (speed), time, respectively. Let

$$x = \text{rate of the car in mph.}$$
$$y = \text{rate of the truck in mph.}$$

Then $\quad x = y + 20$

Note that the time taken for the car to travel 240 miles is

$$\frac{d}{r} = \frac{240}{x}$$

Similarly, the time taken for the truck to travel 200 miles is

$$\frac{d}{r} = \frac{180}{y}$$

We are given that the two"times" are the same, hence the second equation.

$$\frac{240}{x} = \frac{180}{y}$$

We simplify the equation : $240y = 180x$

$$: 3x = 4y.$$

Thus, we have

$$\begin{cases} x = y + 20 \\ 3x = 4y \end{cases}$$

We may solve the system by substitution.

$$3(y + 20) = 4y$$
$$3y + 60 = 4y$$
$$y = 60$$
$$x = y + 20$$
$$= 80$$

Thus $\boxed{\text{The speed of the car is } 80 \text{ mph, and the speed of the truck is } 60 \text{ mph.}}$

Exercise **7.11.** *Sylvia started a* 320*-mile journey in bad weather and drove at a certain speed for three hours before meeting good weather. She then increased her speed by twenty miles per hour and made the rest of the journey in two hours. Find the two speeds at which Sylvia drove.*

Solution:

7.2.iii. Mixtures and Percentages

Example **7.13.** *George invested some of his money at 8% and the rest at 6%. His total income from the investments was $2400. If he had interchanged his investments, his total income would have been $2150. How much did George invest at each rate?*

- **Solution**: First, we recall that "income" here refers to "interest" and that

$$I = PRT,$$

where I, P, R, T, are the interest, principal, rate, and time, respectively. We must assume one unit of time since all discussions are relative to the same time. Thus, we let $I = PR$, and

$$x = \text{amount @ 8\%}$$
$$y = \text{amount @ 6\%}.$$

Since $I = PR$, we have

$$\begin{cases} .08x + .06y = 2400 \\ .06x + .08y = 2150 \end{cases}$$

We multiply each equation by 100 to clear the fractions:

$$\begin{cases} 8x + 6y = 240,000 \\ 6x + 8y = 215,000 \end{cases}$$

The system is equivalent to

$$\begin{cases} 4x + 3y = 120,000 \dots\dots\dots\dots\dots\dots R_1 \\ 3x + 4y = 107,500 \dots\dots\dots\dots\dots\dots R_2 \end{cases}$$

Using the method of elimination,

$$-3R_1 + 4R_2 \ : \ -9y + 16y = -360,000 + 430,000$$
$$7y = 70,000$$
$$y = 10,000$$

Substituting for y in R_1,

$$4x + 3(10,000) = 120,000$$
$$4x = 90,000$$
$$x = 22,500$$

That is:

> $22,500$ was invested at 8%, $10,000$ was invested at 6%.

Example **7.14.** *A nurse has two alcohol solutions: one at 6% concentration, the other at 20% concentration. How much of each solution must she mix to produce 49 ml. of a 12% alcohol solution?*

- **Solution**: Let

$$x = \text{amount of the 6\% solution in the mixture (in ml.)}$$
$$y = \text{amount of the 20\% solution in the mixture (in ml.).}$$

Then,

quantity of pure alcohol in x ml. @ 6% is $.06x$;

quantity of pure alcohol in y ml. @ 20% is $.2y$;

quantity of pure alcohol in 49 ml. @ 12% is $.12(49)$.

In mixture problems, we equate like quantities. Thus the system of equations is

$$\begin{cases} x + y = 49 \\ .06x + .2y = .12(49) \end{cases}$$

We may multiply the second equation by 100 to clear the fractions:

$$\begin{cases} x + y = 49 ....R_1 \\ 6x + 20y = 588 ....R_2 \end{cases}$$

Better still, we may divide R_2 by 2

$$\begin{cases} x + y = 49 ....R_1 \\ 3x + 10y = 294 ....R_2 \end{cases}$$

We now eliminate one variable:

$$-3R_1 + R_2 \; : \; -3y + 10y = -3(49) + 294$$
$$7y = 147$$
$$y = 21$$

Solving for x,

$$x + 21 = 49$$
$$x = 28$$

That is: the nurse must mix:

28ml. of the 6% solution with 21ml. of the 20% solution to get 49ml. of solution @ 12% concentration.

Note The above are examples of applications of linear systems of equations. Other areas of application include geometry, force, value, and work problems. Some of these will be found in the exercises.

Exercise 7.12. *A certain amount of a 70% alcohol solution is to be added to another quantity of a 15% alcohol solution to produce a 40 ml. alcohol solution at 37% concentration. How many ml. of each solution must be used to produce the mixture?*

Solution:

Summary 7.2.

- In order to apply systems of linear equations to solve problems,
  - ▶ read the problem completely;
  - ▶ choose your variables to be the quantities you wish to find;
  - ▶ use any formulas connecting the variables;
  - ▶ write down equations based on given information;
  - ▶ solve the system.
- For number problems involving digits, note that a two digit number has value

$$10(\text{tens digit}) + 1(\text{units digit}).$$

For instance, $37 = 10(3) + 1(7)$.

- For constant rate motion problems, the guiding formula is

$$\text{distance} = \text{rate} \times \text{time}.$$

- The formula $I = PRT$ connects the interest I, the principal P, the rate R, and the time T.

- For mixture problems, remember to equate like quantities.

End of Section 7.2. Problems

Problem 7.5. *The sum of two numbers is 46. If three times the first number is 7 less than two times the second number, find the two numbers.*

Problem 7.6. *The sum of one-third of a number and one-half of a second number is 42. If the first number is 16 more than one-third of the second number, find the two numbers.*

Problem 7.7. *A two-digit number is three times the sum of the digits. If the digits are interchanged, the new number is 2 more than ten times the units digit of the original number. Find the original number.*

Problem 7.8. *The difference of the speeds of two cars is 10 m.p.h.. The cars are 800 miles apart and are traveling towards each other. If they meet in 5 hours, find the speeds of the cars.*

Problem 7.9. *Flying with the wind, a plane can cover a distance of 2310 miles in 6 hours. Traveling against the wind, the plane covers only 1,530 miles in the same time. Find the speed of the plane instill air and the speed of the wind.*

Problem 7.10. *An office manager bought 4 boxes of Pilot pens and 2 boxes of Bic pens for $18. If she bought 2 boxes of Pilot pens and 5 boxes of Bic pens, she would have paid $17. Find the cost of a box of each type of pen.*

Problem 7.11. *A bank teller receives a deposit of 41 bills totaling $530. If the bills are $10 and $20 bills, how many of each denomination did she receive?*

Problem 7.12. *Jenny's piggy-bank has $2.90 in quarters and nickels. If the nickels were quarters and the quarters were nickels, the value of the coins would be $3.70. How many coins of each denomination are in the piggy-bank?*

Problem 7.13. *Four years ago, John's age was two-thirds of his fathers' age. Eighteen years ago, the father was twice as old as John. What are their ages now?*

Problem 7.14. *A charter train charges travellers a fixed processing cost plus a mileage fee. It costs $48 to travel for 200 miles and $61 to a 300- mile destination. Find the fixed charge and the charge per mile.*

Problem 7.15. *A pharmacist has two iodine solutions of 30% and 70% concentration. How much of each solution should be mixed to produce 50 ml. of a 54% iodine solution?*

Problem 7.16. *A retailer mixes coffee beans which sell for $1.20 per pound with coffee beans selling for $1.80 per pound to produce a blend that sells for $1.60 per pound. Determine the quantity of each type of coffee needed to produce 36 pounds of the blend.*

Problem 7.17. *How much of a 17% alcohol solution must be added to 40ml. of a 30% alcohol solution to produce a 25% solution?*

Problem 7.18. *An investor put $12,000 in two accounts. The first of these pays 8% annual interest while the other has a return of 10% annually. If her total annual income is $1040, how much did she invest at each rate?*

Problem 7.19. *The length of a rectangular plot is 25 meters more than its width. If the perimeter of the plot is 330 meters, what are the dimensions of the plot?*

7.3. Nonlinear Systems

A system is said to be **nonlinear** if at least one equation in the system is nonlinear. Thus, ᴛnere is **at least one term** in the system whose **degree** is **neither one nor zero**. Here are a few examples of nonlinear system:

$$\begin{cases} 2x - y^2 = 5 \\ 4x + 3y = 7 \end{cases}$$

$$\begin{cases} \sqrt{x} + y = -3 \\ 3x - 2y = 1 \end{cases}$$

$$\begin{cases} xy + y^2 = 6 \\ x^2 - y^2 = 3 \end{cases}$$

Can you identify the *nonlinear* terms in each system?

7.3.i. Methods of Solution

Although the graph of at least one equation in a nonlinear system is *not a straight line* (why?), the geometric interpretation of a solution remains the same: **It is a point of intersection of the graphs**.

A combination of substitution **and/or** elimination may be used to analytically solve such a system. The **method of elimination** is readily applicable if the two equations have **like terms**.

In **absence of like terms**, the **method of substitution** should be employed.
We further examine these situations by looking at some examples.

Example **7.15**. *Solve the nonlinear system*

$$\begin{cases} x^2 + y = 2 \dots\dots\dots\dots\dots R_1 \\ x - y = 4 \dots\dots\dots\dots\dots R_2 \end{cases}$$

Solution: First, observe the nonlinear term (x^2) in R_1. We can rearrange the system as follows

$$\begin{cases} y = -x^2 + 2 \dots\dots\dots\dots(1) \\ y = x - 4 \dots\dots\dots\dots(2) \end{cases}$$

We graph the two equations on the same pair of axes:

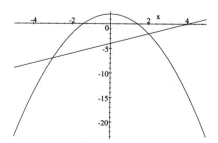

Clearly, the straight line intersects the parabola in two places. (*Can you estimate the solutions at this point?*) Thus, the system has two solutions. Let us find them analytically. Since the equations have like terms in the variable y, we may **use** the **method of elimination** although the method of substitution can also be used in this case. The operation $R_1 + R_2$ eliminates the variable y :

$$R_1 + R_2 : x^2 + x = 6 \ldots\ldots\ldots\ldots\ldots\ldots(a)$$

This resulting equation is quadratic and (in this case) can be solved by factorization:

$$x^2 + x = 6$$
$$x^2 + x - 6 = 0$$
$$(x + 3)(x - 2) = 0$$
$$x = -3, \text{ or } x = 2.$$

We now find values of y corresponding to each value of x. Here, we may use either of the equations in the original system. From R_2,

$$y = x - 4$$

Thus,

$$x = -3 \Rightarrow y = x - 4 = (-3) - 4 = -7$$
$$x = 2 \Rightarrow y = x - 4 = (2) - 4 = -2$$

Hence, the two solutions are $\boxed{(-3, -7) \text{ and } (2, -2).}$ Our algebraic solutions agree with the graphs.

Exercise 7.13. *Solve the system* $\begin{cases} 2x^2 + y^2 = 8 \\ x + y = 2 \end{cases}$

Solution:

Example 7.16. *Solve the system*

$$\begin{cases} x^2 + y^2 = 4 \ldots\ldots\ldots\ldots\ldots\ldots R_1 \\ x - y = 2 \ldots\ldots\ldots\ldots\ldots\ldots R_2 \end{cases}$$

• **Solution**: Note that the first equation is a circle while the second is a straight line. Thus, there may be no solution, one solution, or exactly 2 solutions depending on whether the straight line misses, touches, or crosses the circle. Unlike the preceding example, R_1 and R_2 **do not** have like terms. Thus we necessarily use the **method of substitution** (since the method of elimination will pose problems). We solve for, say, y in R_2.

$$y = x - 2 \ldots\ldots\ldots\ldots\ldots\ldots(a)$$

Substituting into R_1, we have

$$x^2 + (x - 2)^2 = 4$$
$$x^2 + x^2 - 4x + 4 = 4$$
$$2x^2 - 4x = 0$$
$$2x(x - 2) = 0$$
$$\text{and} \quad x = 0 \text{ or } 2$$

We now find corresponding values of y using (a)

$$x = 0 \Rightarrow y = x - 2 = (0) - 2 = -2$$
$$x = 2 \Rightarrow y = x - 2 = (2) - 2 = 0$$

As a result, the solutions are $\boxed{(0, -2) \text{ and } (2, 0).}$

Note! Having found the values of x, if we used these in R_1 to find corresponding values of y, something interesting happens:

$$x = 0, \qquad x^2 + y^2 = 4$$
$$\Rightarrow (0)^2 + y^2 = 4$$
$$\Rightarrow y = \pm 2.$$

Similarly,

$$x = 2, \qquad x^2 + y^2 = 4$$
$$\Rightarrow (2)^2 + y^2 = 4$$
$$\Rightarrow y^2 = 0$$
$$\Rightarrow y = 0.$$

That is: the points $(0,2)$, $(0,-2)$ and $(2,0)$ are solutions. However, a quick check of the points in R_2 reveals that $(0,-2)$ and $(2,0)$ are indeed solutions but $(0,2)$ is *not a solution* since

$$x = 0, y = 2$$
$$\Rightarrow x - y = (0) - (2) = -2 \neq 2$$

Thus, $(0,2)$ is an **extraneous solution**. This indicates that when *nonlinear equations involve squares (or square roots), we should check our solutions*. We now graph the system to see the points of intersection geometrically:

$$x^2 + y^2 = 4 ....R_1$$
$$x - y = 2 ....R_2$$

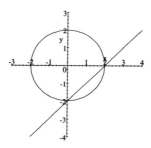

The graph confirms our analytical solutions.

We conclude this section by reflecting on the possibilities for second degree systems of two equations in two variables.

- ■ If the equations are **dependent**, we have infinitely many solutions. These may be readily identified since one equation is a (constant) multiple of the other equation.
- ■ In absence of dependency, the number of solutions ranges from *none* to *four*. We use parabolas and circles to show the possibilities:

$$\begin{cases} x^2 + y^2 = 1 \\ x^2 + y = -1 \end{cases}$$

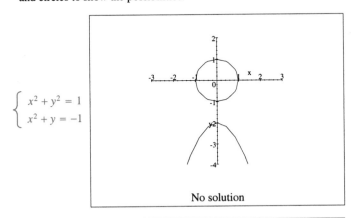

No solution

$$\begin{cases} x^2 + y^2 = 1 \\ x^2 + y = -1 \end{cases}$$

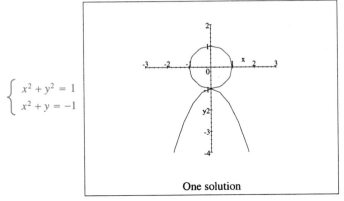

One solution

$$\begin{cases} x^2 + y^2 = 1 \\ x^2 + y = 0. \end{cases}$$

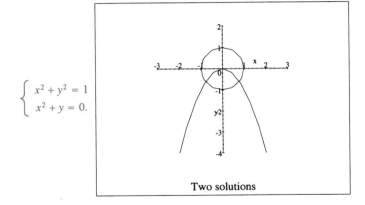

Two solutions

$$\begin{cases} x^2 + y^2 = 1 \\ 2x^2 + y = 1 \end{cases}$$

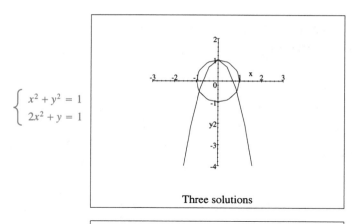

Three solutions

$$\begin{cases} x^2 + y^2 = 1 \\ 4x^2 + y = 2 \end{cases}$$

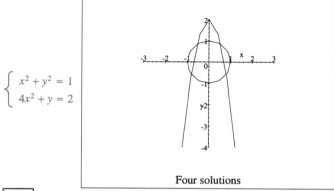

Four solutions

Note The above analysis indicates that we need diligence in dealing with nonlinear systems. Further development of this topic, while exciting, is beyond our present scope.

Summary 7.3.

- For a nonlinear 2×2 system :
    - ▶ There is at least one nonlinear term in an equation;
    - ▶ The solution(s) are points of intersection of the two curves;
    - ▶ We use the method of elimination if the two equations have like terms.
    - ▶ If there are no like terms, the method of substitution is preferred.
    - ▶ Remember to check your solutions for extraneous solutions.

End of Section 7.3. Problems

Problem 7.20. *Solve each of the following systems.*

(a) $\begin{cases} x^2 + y^2 = 5 \\ 2x^2 + y = 0 \end{cases}$
(b) $\begin{cases} x^2 + 2y^2 = 8 \\ 2x^2 - y^2 = 6 \end{cases}$

Problem 7.21. *Solve each of the following systems:*

(a) $\begin{cases} x^2 - y = 4 \\ x + y = 2 \end{cases}$
(b) $\begin{cases} x^2 - y^2 = 9 \\ x + y = 3 \end{cases}$
(c) $\begin{cases} 2x^2 + y^2 = 6 \\ -x^2 + y^2 = 3 \end{cases}$

(d) $\begin{cases} 4x^2 + 9y^2 = 72 \\ 4x - 3y^2 = 0 \end{cases}$
(e) $\begin{cases} x - y = 1 \\ xy = 1 \end{cases}$
(f) $\begin{cases} x^2 - y = 0 \\ x^2 - 2x + y = 6 \end{cases}$

Problem 7.22. *Solve each of the following systems:*

(a) $\begin{cases} x^2 + y^2 = 9 \\ x^2 + y = 3 \end{cases}$
(b) $\begin{cases} x^2 + y^2 = 4 \\ x^2 + y = -1 \end{cases}$
(c) $\begin{cases} x^2 + y^2 = 1 \\ x^2 + y = 0 \end{cases}$

(d) $\begin{cases} x^2 + y^2 = 1 \\ 2x^2 + y = 1 \end{cases}$
(e) $\begin{cases} x^2 + y^2 = 1 \\ 4x^2 + y = 2 \end{cases}$ 3
(f) $\begin{cases} x^2 - 3y^2 = 4 \\ x^2 + y^2 = 4 \end{cases}$

Problem 7.23. *Solve each of the following systems:*

(a) $\begin{cases} x = y^2 \\ x^2 = y \end{cases}$
(b) $\begin{cases} 2x^2 + y^2 = 8 \\ x^2 - y^2 = 4 \end{cases}$
(c) $\begin{cases} x^2 + y^2 = 4 \\ x^2 + 2x + y^2 = 0 \end{cases}$

(d) $\begin{cases} x^2 - y^2 = 4 \\ 2x^2 + 3y^2 = 6 \end{cases}$
(e) $\begin{cases} x^3 - y^3 = 26 \\ x - y = 2 \end{cases}$
(f) $\begin{cases} x + y = 2 \\ x^3 + y^3 = 56 \end{cases}$

7.4. Systems of Linear Inequalities (Two Variables)

Consider the inequality $2x + y > 4$. The set of points $P(x, y)$ in the plane that satisfy the inequality all lie above the line $2x + y = 4$.

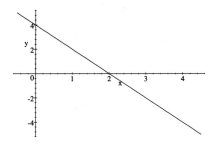

Thus, the line is a boundary of the set of points in question. Similarly, for the inequality

$$2x + y < 4$$

all points lying below the line, $2x + y = 4$, satisfy the inequality.

- In order to correctly locate the set that satisfies a linear inequality, we:
  - ▶ Replace the inequality sign with equality. The graph of the equation is a straight line. This is the *boundary* of the set that satisfies the inequality.
  - ▶ Solve the inequality for the variable y.
    - ◀ If the inequality reads $y > (*)$, then the desired set lies **above the boundary** (line).
    - ◀ If, however, the inequality reads $y < (*)$, then the desired set lies **below the boundary** (line).
  - ▶ The form $y > (*)$ and $y < (*)$ will be referred to as the **standard form**.
- The desired set *includes the boundary* if the inequality is *non-strict*. That is: the inequality is of the form $\{\geq\}$ or $\{\leq\}$. In this event, we draw the line as a "continuous" graph.
  - ▶ If the inequality is *strict*, then the desired set excludes the boundary and we only "dot" the line to show exclusion.
- It is customary to pick a point in the desired set and to test the **original inequality**. If one point in the region satisfies the inequality, then all points in that region satisfy the inequality. Remember that the boundary (line) divides the plane into two regions.

7.4.i. A method

Consider a system of linear inequalities:

$$\begin{cases} 2x + y > 4 \\ x - y < 0 \end{cases}$$

Put each inequality in **standard form**:

$$\begin{cases} y > -2x + 3 ....L_1 \\ y > x ....L_2 \end{cases}$$

Graph the two lines $y = -2x + 3$ and $y = x$.

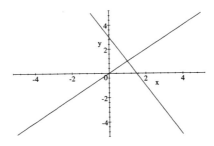

For L_1, we may use the test point $P(4,0)$:
$$2x + y = 2(4) + (0) > 3$$
Thus, the point satisfies the inequality.

Since P lies above the line, all points above the line satisfy the inequality. Similarly we may use the point $Q(2,0)$ to test and see that all points above the second line satisfy the inequality. Notice that P satisfies both inequalities but Q does not satisfy the L_1. Hence, P is in the solution set while Q is not a solution.

In general, a solution for the system is any point that satisfies **both** inequalities. Thus the solution set is the set of all points in the plane that lie above the two lines. Clearly, there are infinitely many solutions.

Example 7.17. Graph the region that satisfies the inequalities

$$\begin{cases} x + y > 2 \\ 2x + 3y \le 12 \\ 2x \ge 1 \end{cases}$$

Solution: First we put each inequality in standard form:

$$\begin{cases} y > -x + 2 \dots\dots\dots\dots \left(\text{above the line, strict}\right) \\ y \le -\frac{2}{3}x + 4 \dots\dots\dots \left(\text{below the line, non-strict}\right) \\ x \ge \frac{1}{2} \dots\dots\dots\dots\dots(\text{the right half, non-strict}). \end{cases}$$

Then we graph each line and shade to include the line in each non-strict case but we exclude the line if the inequality is strict (indicated by dotting the line):

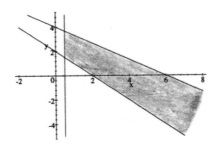

The solution (region) is shaded.

Exercise **7.14**. *In each case, graph the region that satisfies the system*:

(a) $\begin{cases} 3x + y \leq 6 \\ x + y < 3 \\ x \geq 0 \\ y \geq 0. \end{cases}$

(b) $\begin{cases} 3x + 2y \leq 6 \\ 3x + y \geq -3 \\ x < 2 \\ y \geq 1 \end{cases}$

Solution:

Solution:

7.4.ii. Linear Programming

We shall apply some of what we have learned about systems of linear inequalities to understand the basic problem of **linear programming**.

Linear programing is a mathematical tool for decision making and is often used in business/industry for a variety of needs such as creating telephone lines and airline scheduling, in agriculture, and the military. In practice, a linear programming problem may involve hundreds of variables and inequalities, but we shall limit our examples to two variables and a few inequalities for simplicity. And we shall use a graphical approach to solve the problems.

> **Definition**
>
> In a linear programming problem, the **objective function** is the function to be optimized (i.e. minimize or maximize). *Conditions* on the variables in terms of which the objective function is defined are called **constraints**.
>
> The conditions which are not specified in the problem but must be included for the problem to make sense are referred to as *natural constraints*.
>
> A linear programming problem in the variables x and y consists of two distinct parts:
>
> 1. optimizing a *linear* objective function of the form $\boxed{z = ax + by}$,
>
>   where a and b are real numbers that are not both zero; the optimization is subject.
>
> 2. constraints which can be expressed as linear inequalities in x and/or y.

Example **7.18**. *A manufacturer produces two models of lawn mowers, M and K. Model M earns $50 profit while model K earns a profit of $70. Each week, the manufacturer can produce at most 300 units of model M and 200 of model K. However, due to limitations, the manufacturer can produce at most a total of 400 lawn mowers in one week. How many mowers of each model should the manufacturer produce to make the maximum profit?*

• **Solution**: Let

$$x = \text{\# of model M mowers produced in one week}$$
$$y = \text{\# of model K mowers produced in one week}$$
$$P = \text{total profit earned.}$$

Then, from the given information:

▶ the profit for the week in question is
$$F(x,y) = 50x + 70y$$

▶ the following limitations apply:
$$(*) \quad \begin{cases} x + y \le 400 \\ x \le 300 \\ y \le 200 \end{cases}$$

$\boxed{\text{Note}}$ Although we are not specifically reminded in the problem, x and y cannot be negative since these are numbers of mowers. Thus, we add the limitations:
$$(**) \quad \begin{cases} x \ge 0 \\ y \ge 0 \end{cases}$$

- The problem is to find numbers x and y that **maximize** the function F subject to the **constraints** $(*)$ and $(**)$. We now name the pertinent quantities formally.

- **A graphical approach**

$$\text{Maximize} \quad F(x,y) = 50x + 70y$$

$$\text{subject to} \quad \begin{cases} x + y \le 400 \\ x \le 300 \\ y \le 200 \\ x \ge 0, \quad y \ge 0. \end{cases}$$

We seek a region in the plane that satisfies each of the constraints.

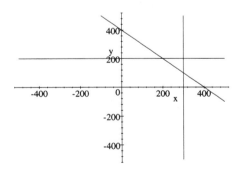

- We refer to this region as the **feasible region**. We are looking for a point $P(a,b)$ in the feasible region for which the function F has its maximum value. We note that the feasible region is **convex** (that is: any two points in the region can be joined by a straight line which lies completely in the region). We state (without proof) a key result that is necessary to complete our problem.

$\boxed{\text{Maximum-Minimum Principle}}$ In a convex region R, a linear function of the form $z = ax + by$ takes its **maximum** and **minimum** values on the boundary of R. In fact, the extreme values occur at the vertices of R.

We now complete the solution to our problem. The **Maximum-Minimum Principle** dictates that we should evaluate the function F at all vertices of the feasible region. The maximum (and minimum) of F is the largest of the values of F at the vertices. Thus we proceed to find the vertices which are points of intersection of the lines that constitute the boundary of the feasible region. For,

$$\begin{cases} x + y = 400 \\ x = 300 \\ y = 200 \\ x = 0, \quad y = 0. \end{cases}$$

the points of intersection are

$$A(0,0), \ B(300,0), \ C(300,100), \ D(200,200), \ E(0,200).$$

Evaluating F at these points, we have

Points	$F = 50x + 70y$
$A(0,0)$	0
$B(300,0)$	$15,000$
$C(300,100)$	$22,000$
$D(200,200)$	$24,000$
$E(0,200)$	$14,000$

Clearly, the maximum of F is $\$24,000$ and this occurs at $D(200,200)$. That is, the manufacturer should produce:

> 200 of model M and 200 of model K to make the maximum possible profit of $\$24,000$.

Example 7.19. *Suppose that a man and his wife have a small business producing two kinds of sweaters: Type I and Type II each of which is in high demand. They make a profit of $\$12$ on each Type I sweater and $\$15$ on each Type II sweater. Two machines A and B are used (jointly) to make each type. In a working day, the man operates machine A for no more than 12 hours while his wife operates machine B for no longer than 10 hours and each person works with one machine only. The respective amounts of time spent on the machines for each type of sweater are given in the table below:*

Machine	Type I	Type II
A	45 mins	45 mins
B	20 mins	1 hr

Determine the number of sweaters of each type to be produced per day for the maximum possible profit. How much daily profit can this couple make from the business?

Solution: Let

$$x = \text{number of Type I sweaters}$$
$$y = \text{number of Type II sweaters.}$$

Then, the daily total profit is given by

$$z = 12x + 15y.$$

We now convert the times from minutes to hours so that the constraints are

$$\frac{3}{4}x + \frac{3}{4}y \leq 12$$
$$\frac{1}{3}x + y \leq 10.$$

Note that the first constraint expresses the condition that the man works machine A for $\frac{3}{4}$ hour on each sweater for at most 12 hours in one day. Similarly, the second condition is the given constraint that machine B is operated on each Type I sweater for $\frac{1}{3}$ hour and on each Type II sweater for 1 hour for an aggregate of at most 10 hours daily. In addition, since x and y are number of sweaters, we add the natural constraints:

$$x \geq 0, \quad y \geq 0.$$

Thus the linear programming problem is to

$$\text{maximize} \quad z = 12x + 15y$$

$$\text{subject to} \quad \begin{cases} \frac{3}{4}x + \frac{3}{4}y \leq 12 \\ \frac{1}{3}x + y \leq 10 \\ x \geq 0, \quad y \geq 0. \end{cases}$$

In order to find the feasible region, first, we rewrite the constraints in standard form having cleared all fractions:

$$\begin{cases} x + y \leq 16 \\ x + 3y \leq 30 \\ x \geq 0, \quad y \geq 0. \end{cases}$$

The graph of the region follows:

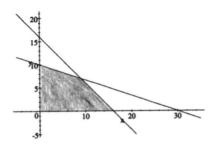

The feasible region (shaded) lies in the first quadrant below both lines. Next, we find the vertices of the feasible region by solving the following equations in pairs:

$$x + y = 16$$
$$x + 3y = 30$$
$$x = 0, \quad y = 0.$$

The vertices are readily found to be $P(0,0), Q(16,0), R(9,7), S(0,10)$. We now compute the value of the objective function, z, at each vertex.

Vertex	$z = 12x + 15y$
$P(0,0)$	$0 + 0 = 0$
$Q(16,0)$	$12 \cdot 16 + 0 = 192$
$R(9,7)$	$12 \cdot 9 + 15 \cdot 7 = 213$
$S(0,10)$	$0 + 15 \cdot 10 = 150$

As the table shows, the daily maximum profit is \$213 and is attained if the couple produces 9 sweaters of Type I and 7 sweaters of Type II.

Exercise 7.15. *Find the maximum and minimum of the function* $z = 27x + 32y$ *in the region that satisfies the system*

$$3x + 2y \leq 12$$
$$x + 2y \leq 6$$
$$3x + 4y \geq 12.$$

> Solution:

Exercise 7.16. *Solve the linear programming problem*

$$\text{maximize} \quad z = 120x + 105y$$

$$\text{subject to} \quad \begin{cases} x + 2y \leq 12 \\ 3x + 2y \leq 24 \\ x \geq 0, \quad y \geq 0. \end{cases}$$

> Solution:

Summary 7.4.

- In order to solve a linear programming problem graphically we must:
  - ▶ graph the feasible region (that satisfies all the constraints);
  - ▶ find the coordinates of the vertices of the region;
  - ▶ evaluate the objective function at each of the vertices; and
  - ▶ identify the vertex that produces the maximum/ minimum of the function.
  We are now ready for an example that is less transparent.

End of Section 7.4. Problems

Problem 7.24. In each case, graph the region that satisfies the given system:

(a) $\begin{cases} 3x - 2y < 6 \\ x + y \geq 8 \end{cases}$ (b) $\begin{cases} x - y \leq 4 \\ 2x + y \geq 6 \end{cases}$ (c) $\begin{cases} 3x - y \geq -9 \\ 3x + y \leq -3 \end{cases}$

(d) $\begin{cases} 2x + y \geq -2 \\ y - 2x \leq -4 \end{cases}$ (e) $\begin{cases} 3x - 2y \leq 6 \\ 2x - 3y < -6 \end{cases}$ (f) $\begin{cases} 3x + 2y \geq 12 \\ 2x - y \leq 6 \end{cases}$

Problem 7.25. In each case, graph the region that satisfies the given system:

(a) $\begin{cases} x - 3y \geq -6 \\ x + y \leq 7 \\ x > 1 \end{cases}$
(b) $\begin{cases} x + y > 4 \\ 2x + y \geq 6 \\ x \geq 0, \quad y \geq 0 \end{cases}$
(c) $\begin{cases} 2x + 3y \leq 6 \\ 2x + y \geq 2 \\ x \geq 0, \quad y \geq 1 \end{cases}$

(d) $\begin{cases} 2x + y \leq 8 \\ 3x - 2y \leq 12 \\ x \geq 0, \quad y \geq -2 \end{cases}$
(e) $\begin{cases} x + 2y \leq 6 \\ -x + y < -6 \\ x \geq 0, \quad y < 2 \end{cases}$
(f) $\begin{cases} 2x - y \geq -3 \\ x + y \leq 5 \\ x \geq 1, \quad y > 0 \end{cases}$

Problem 7.26. Minimize the function $z = 0.21x + 0.14y$ subject to the constraints:

$2x + y \geq 22$
$6x + 5y \geq 90$
$3x + 2y \leq 60$
$x \geq 0, \quad y \geq 0$

Problem 7.27 Maximize the function $z = 51x + 22y$ subject to the constraints:

$3x + y \leq 18$
$2x + y \leq 14$
$x \geq 0, \quad y \geq 0$

Problem 7.28. Find both the maximum and minimum of $z = 3x + 2y$ in the region:

$2x + 3y \leq 12$
$x + 3y \leq 9$
$2x - y \leq 2$
$x \geq 0, \quad y \geq 1$

Problem 7.29. A farmer has a daily supply of 180 quarts of milk and 150 quarts of cream which he uses to make ice cream and yogurt. Each quart of ice cream requires 0.4 quart of milk and 0.2 quart of cream and sells for $0.5. A quart of yogurt requires 0.2 quart of milk and 0.5 quart of cream and sells for $0.65. How many quarts of ice cream and yogurt should the farmer produce daily to realize maximum sales?

Problem 7.30. A merchant mixes Louisiana coffee and Columbian coffee to package a strong blend and a mild blend of coffee. She uses 0.5 pound of each type to obtain a pound of the mild blend which she sells for a profit of $1.50. A pound of her strong blend sells for a profit of $2.75 and is a mixture of 0.25 of the Louisiana coffee and 0.75 of the Columbian coffee. If she gets a weekly supply of 100 pounds of Louisiana coffee and 125 pounds of Columbian coffee, how many pounds of each blend must she make to maximize her weekly profit?

Problem 7.31. Dennen Hauling Company owns two trucks: a mid-size and a large. They use the trucks to haul loads of sand to construction sites. The mid-size truck has a capacity of 30 cubic meters and costs $45 per load to operate. On the other hand, the large truck with a capacity of 50 cubic meters has an operating cost of $55 per load. Each working day, the company must deliver at least 300 cubic meters using a minimum of 8 loads. How many loads must the company deliver in each truck per day to minimize the operating cost? What is the minimum cost?

Problem 7.32. In the preceding exercise, if the operating costs per load are $45 for the mid-size truck and $60 for the large truck, how many loads must the company deliver per day per truck so as to minimize their cost?

End of Chapter 7. Problems

Problem 7.33. *For each of the following systems, determine if the system is consistent and independent, inconsistent, or consistent and dependent. Find approximate solutions for the systems that are consistent and independent by the method of graphing:*

(a) $\begin{cases} x + 3y = 7 \\ 5x + 15y = 9 \end{cases}$
(b) $\begin{cases} x + y = 7 \\ 2x + 3y = 7 \end{cases}$
(c) $\begin{cases} -2x + 10y = 2 \\ 3x - 15y = -3 \end{cases}$

(d) $\begin{cases} -x + 2y = -6 \\ 2x - 4y = 7 \end{cases}$
(e) $\begin{cases} 2x + y = 5 \\ 4x - y = 7 \end{cases}$
(f) $\begin{cases} 2x - 7y = 6 \\ 4x - 14y = 12 \end{cases}$

Problem 7.34. *Solve each system by the method of substitution:*

(a) $\begin{cases} 2x + 3y = 6 \\ 3x - y = -2 \end{cases}$
(b) $\begin{cases} 2x + y = -1 \\ 4x - y = 4 \end{cases}$
(c) $\begin{cases} 4x + 5y = 3 \\ 2x + 3y = 2 \end{cases}$

(d) $\begin{cases} x - 6y = 11 \\ 2x + 5y = 12 \end{cases}$
(e) $\begin{cases} 2x - 2y = 7 \\ x - 2y = -4 \end{cases}$
(f) $\begin{cases} 3x - 2y = 4 \\ 2x + y = -2 \end{cases}$

Problem 7.35. *Solve each system by the method of elimination:*

(a) $\begin{cases} 2x - 4y = -14 \\ x + 6y = 5 \end{cases}$
(b) $\begin{cases} x - 3y = 7 \\ 2x + 3y = 8 \end{cases}$
(c) $\begin{cases} 3x + 4y = 12 \\ 2x - 3y = 12 \end{cases}$

(d) $\begin{cases} -3x + y = -13 \\ 2x - 3y = 11 \end{cases}$
(e) $\begin{cases} 3x - y = -9 \\ 7x - 2y = 20 \end{cases}$
(f) $\begin{cases} \frac{1}{3}x + \frac{1}{2}y = 1 \\ -\frac{1}{2}x + \frac{1}{4}y = \frac{3}{2} \end{cases}$

Problem 7.36. *Solve each system by any method:*

(a) $\begin{cases} \frac{2}{3}x + \frac{5}{6}y = 6 \\ \frac{1}{6}x - \frac{7}{12}y = -8 \end{cases}$
(b) $\begin{cases} \frac{1}{2}x + \frac{3}{4}y = -1 \\ \frac{3}{4}x - \frac{1}{2}y = 13 \end{cases}$
(c) $\begin{cases} \frac{2}{5}x + 2y = \frac{7}{2} \\ \frac{2}{5}x - y = -1 \end{cases}$

(d) $\begin{cases} x - 4y - 2z = -15 \\ 3x - y - z = -12 \\ 2x + 5y - 2z = -9 \end{cases}$
(e) $\begin{cases} x + 4y - 4z = 3 \\ -x - y + 6z = 4 \\ 3x + 3y + 8z = 1 \end{cases}$
(f) $\begin{cases} 2x + 3y - z = -4 \\ x - 2y + 2z = -6 \\ 3x - 2y - z = -1 \end{cases}$

Problem 7.37. *Solve each nonlinear system:*

(a) $\begin{cases} x - y = 1 \\ x^2 + y^2 = 13 \end{cases}$

(b) $\begin{cases} 2x^2 + y^2 = 22 \\ 4x^2 - 3y^2 = 24 \end{cases}$

(c) $\begin{cases} x^2 + 3y^2 = 7 \\ 4x^2 - 5y^2 = 11 \end{cases}$

(d) $\begin{cases} x^2 - 2y^2 = 6 \\ 3x^2 + y^2 = 11 \end{cases}$

(e) $\begin{cases} x + y = 4 \\ x^2 + y = -8 \end{cases}$ 4

(f) $\begin{cases} x^2 + 4y^2 = 7 \\ 2x^2 + 7y^2 = 10 \end{cases}$

Problem 7.38. *In each case, find the region that satisfies the system of inequalities:*

(a) $\begin{cases} 2x + 3y \le 2 \\ x - y \ge 1 \end{cases}$.

(b) $\begin{cases} x - 2y \ge 4 \\ 2x - y \le 6 \end{cases}$

(c) $\begin{cases} 2x - 3y \ge 6 \\ 3x + y < 3 \end{cases}$

(d) $\begin{cases} 2x + y < 4 \\ 2x - 5y \le -8 \\ x \ge 0, \quad y \ge 1 \end{cases}$

(e) $\begin{cases} 2x + y \le 10 \\ -x + 3y \le 12 \\ x \ge 0, \quad y \ge 0 \end{cases}$

(f) $\begin{cases} 3x + 4y > 12 \\ 3x + 2y < 6 \\ x \ge 0, \quad y \ge 0 \end{cases}$

Problem 7.39. *Determine the maximum of the function* $z = 3x + 4y$ *in the region*

$x + 2y \le 400$

$3x + 2y \le 480$

$x \ge 0, \quad y \ge 0.$

Problem 7.40. *Determine the maximum of the function* $z = 12x + 9y$ *subject to the constraints*

$x + y \le 50,000$

$0 \le x \le 20,000$

$y \ge x, \quad y \ge 15,000.$

Problem 7.41. *Determine the minimum of the function* $z = 3x + 4y$ *subject to the constraints*

$2x + 3y \ge 12$

$x + y \le 16$

$x \ge 0, \quad y \ge 0.$

Problem 7.42. *Determine the minimum of the function* $z = 4x + 5y$ *in the region*

$x + y \ge 3$

$x + y \le 9$

$3x + 3y \le 6$

$x \ge 0, \quad y \ge 0.$

Problem 7.43. *In each case, determine the maximum and minimum values (where they exist) of the given function subject to:*

$2x - 3y \le 0$

$x - 4y \ge -20$ (a)$z = x + 2y$. (b)$z = 5x - 7y$.

$4x - y \ge 20$

Problem 7.44. *In each case, determine the maximum and minimum values (where feasible) of the given function subject to:*

$3x + y \ge 300$

$x + y \ge 200$

$x + 3y \ge 360$ (a)$z = 2x + y$. (b)$z = 7x - 5y$.

$5x + 6y \le 1,800.$

The following are application problems:

Problem 7.45. *The sum of a two-digit number and its tens digit is* 40. *Five times the tens digit is one more than twice the units digit. Find the number.*

Problem 7.46. *A nurse wants to mix a* 30% *alkaline solution and a* 10% *alkaline solution to obtain a mixture at* 22% *concentration. How many mls. of each solution must she use to produce* 300 *mls. of the desired mixture?*

Problem 7.47. *Flying with the wind, an airplane makes a journey of* 3500 *miles in* 5 *hours. However, on the return journey (against the wind), the trip takes* 2 *hours longer. Find the speed of the airplane in still air and the speed of the wind.*

Problem 7.48. *Nena has* 95 *coins in pennies, nickels and dimes amounting to* $5.75. *The number of pennies added to the number of dimes is five more than twice the number of nickels. How many coins of each type does Nena have?*

Problem 7.49. *A manufacturer produces two types of locks, a premium lock and a dead bolt. Each premium lock takes* 30 *minutes to smith,* 30 *minutes to refinish and assemble and sells for a profit of* $10. *His dead bolt takes* 1 *hour to smith,* 30 *minutes to refinish and assemble and sells for a profit of* $14. *During each week, the manufacturer has a maximum of* 200 *man-hours for smithing and at most* 160 *man-hours in refinishing and assembly. How many locks of each type should the manufacturer produce per week to maximize his profit?.*

Problem 7.50. *Two types of chicken feed, Type I and Type II each contain three desired nutrients* N_1, N_2, N_3. *Suppose that the number of mgs. in each gram of feed is given in the table:*

Type	N_1	N_2	N_3
I	40	15	5
II	40	5	15

A breeder wishes to mix the two types of feed to obtain a mixture that contains at least 90 *grams of* N_1, *at least* 15 *grams of* N_2, *and at least* 18 *grams of* N_3. *If Type I costs* $0.4 *per gram and Type II costs* $0.3 *per gram, how many grams of each feed should the breeder use to produce a mixture that meets his requirements with minimum cost?*

Chapter 8.

Matrices and Determinants

8.1. Matrices and Systems of Linear Equations

Recall the system of linear equations

$$\begin{aligned} x + 2y &= 5 \\ 2x - 3y &= 3 \end{aligned}$$

from the previous chapter; we used the Method of Elimination to find that; $x = 3$, $y = 1$ is the solution to this system.

We notice that the names of the **variables** x and y are really **unimportant**; the system

$$\begin{aligned} w + 2z &= 5 \\ 2w - 3z &= 3 \end{aligned}$$

would have the solution $w = 3, z = 1$.

We would compute this solution using exactly the same steps we used with the variables x and y.

■ The **coefficients** on the variables and the **constants** on the right side of the equations determine the solution to the system.

In other words, to find the solution to a system of linear equations we need only keep track of the **coefficients and constants**. Here we will use a rectangular array of numbers called a **matrix**, to keep track.

For example, we could rewrite the system

$$\begin{aligned} x + 2y &= 5 \\ 2x - 3y &= 3 \end{aligned} \quad \text{as the matrix} \quad \begin{bmatrix} 1 & 2 & 5 \\ 2 & -3 & 3 \end{bmatrix}.$$

▶ This matrix is called the **augmented matrix** of the system.

Example **8.1**. *Find the coefficient matrix for the system*

$$\begin{aligned} x + 2y &= 5 \\ 2x - 3y &= 3 \end{aligned}$$

• **Solution**:

$$\begin{bmatrix} 1 & 2 \\ 2 & -3 \end{bmatrix}.$$

•

> | Definition |
>
> We say that a matrix with m rows and n columns has **size** $m \times n$, which we read "m by n."

For example, the matrix $\begin{bmatrix} 1 & 2 & 5 \\ 2 & -3 & 3 \end{bmatrix}$ is 2×3;

And the matrix $\begin{bmatrix} 1 & 2 & 4 & -1 \\ 2 & 3 & 0 & -2 \\ -1 & 0 & 1 & 3 \end{bmatrix}$ is 3×4.

Elements

The entries in the matrix are called

The (i,j)-**element** of a matrix is the entry located in row i and column j of the matrix.

For example, consider the matrix $\begin{bmatrix} 1 & -1 & 0 \\ 2 & 4 & -2 \end{bmatrix}$. The $(1,2)$-element (row1, column2) of the matrix is -1, while the $(2,3)$-element is -2.

Main diagonal:

The **main diagonal** of a matrix is the (i,i)-elements of the matrix; that is, $(1,1),(2,2),(3,3),...(i,i)$.

For example, the main diagonal of the following matrix is underlined:

$$\begin{bmatrix} \underline{1} & 2 & -1 \\ -1 & \underline{3} & 0 \\ 0 & 7 & \underline{-2} \end{bmatrix}$$ that is, $(1,1),(2,2),(3,3),...(i,i)$

- In the previous section, we used the following operations to solve systems of linear equations:
  - ► Changing the order of the equations.
  - ► Multiplying an equation by any non-zero constant.
  - ► Multiplying an equation by a constant, add the result to a second equation, then replace the second equation with the result of the addition.
- These operations correspond to operations on the rows of the matrix, which we call **ELEMENTARY ROW OPERATIONS**.
  - ► Interchanging the rows of the matrix.
  - ► Multiplying a row of the matrix by any non-zero constant.
  - ► Multiplying a row of the matrix by a constant, add the result to a second row of the matrix, then replacing that second row with the result of the addition.

We will use the following notation for elementary row operations:

OPERATION	NOTATION
Interchanging the rows of the matrix.	$Ri \Leftrightarrow Rj$
Multiplying a row of the matrix by any non-zero constant.	$c \cdot Ri \Rightarrow Ri$
Multiplying a row of the matrix by a constant, add the result to a second row of the matrix, then replacing that second row with the result of the addition.	$c \cdot Ri + Rj \Rightarrow Rj$

In the next example, we show how these operations are applied to matrices, while also performing the corresponding operations on the corresponding system of linear equations.

For example,

$$\begin{bmatrix} 1 & -3 & 2 & -5 \\ 2 & 5 & -5 & 3 \\ -1 & 2 & -1 & 4 \end{bmatrix} \equiv \begin{array}{rcl} x - 3y + 2z &=& -5 \\ 2x + 5y - 5z &=& 3 \\ -x + 2y - z &=& 4 \end{array}$$

$-2 \cdot R1 + R2 \Rightarrow R2$, thus:

$$\begin{bmatrix} 1 & -3 & 2 & -5 \\ 0 & 11 & -9 & 13 \\ -1 & 2 & -1 & 4 \end{bmatrix} \equiv \begin{array}{rcl} x - 3y + 2z &=& -5 \\ 11y - 9z &=& 13 \\ -x + 2y - z &=& 4 \end{array}$$

$1 \cdot R1 + R3 \Rightarrow R3$, thus:

$$\begin{bmatrix} 1 & -3 & 2 & -5 \\ 0 & 11 & -9 & 13 \\ 0 & -1 & 1 & -1 \end{bmatrix} \equiv \begin{array}{rcl} x - 3y + 2z &=& -5 \\ 11y - 9z &=& 13 \\ -y + z &=& -1 \end{array}$$

$R2 \Leftrightarrow R3$, thus:

$$\begin{bmatrix} 1 & -3 & 2 & -5 \\ 0 & -1 & 1 & -1 \\ 0 & 11 & -9 & 13 \end{bmatrix} \equiv \begin{array}{rcl} x - 3y + 2z &=& -5 \\ -y + z &=& -1 \\ 11y - 9z &=& 13 \end{array}$$

$-1 \cdot R2 \Rightarrow R2$, thus:

$$\begin{bmatrix} 1 & -3 & 2 & -5 \\ 0 & 1 & -1 & 1 \\ 0 & 11 & -9 & 13 \end{bmatrix} \equiv \begin{array}{rcl} x - 3y + 2z &=& -5 \\ y - z &=& 1 \\ 11y - 9z &=& 13 \end{array}$$

$-11 \cdot R2 + R3 \Rightarrow R3$, thus:

$$\begin{bmatrix} 1 & -3 & 2 & -5 \\ 0 & 1 & -1 & 1 \\ 0 & 0 & 2 & 2 \end{bmatrix} \equiv \begin{array}{rcl} x - 3y + 2z &=& -5 \\ y - z &=& 1 \\ 2z &=& 2 \end{array}$$

$\frac{1}{2} \cdot R3 \Rightarrow R3$, thus:

$$\begin{bmatrix} 1 & -3 & 2 & -5 \\ 0 & 1 & -1 & 1 \\ 0 & 0 & 1 & 1 \end{bmatrix} \equiv \begin{array}{rcl} x - 3y + 2z &=& -5 \\ y - z &=& 1 \\ z &=& 1 \end{array}$$

At this point the system is much more easily solved. We can see immediately that $z = 1$, and can plug that value into the second equation to find that $y = 2$. Plugging the values we have found for y and z into the first equation yields $x = -1$.

A matrix in the form of the last matrix above is said to be in **row echelon form**; matrices in row echelon form correspond to systems which are easily solved.

Row echelon form:

A matrix is in **row echelon form** if it has the following three properties:

I All rows consisting entirely of zeros are at the bottom of the matrix.

II The first non-zero element in each row is in a column to the right of the first non-zero element in the row above it.

III All entries in a column below the first non-zero element in a row are zero.

For example, the matrix

$$\begin{bmatrix} 2 & -1 & 3 & 4 \\ 0 & 1 & -2 & 7 \\ 0 & 0 & 1 & -5 \end{bmatrix}$$

<u>is</u> in row echelon form.

The matrix

$$\begin{bmatrix} 2 & -1 & 3 & 4 \\ 0 & 1 & -2 & 7 \\ 0 & 2 & 1 & -5 \end{bmatrix}$$

is <u>not</u> in row echelon form, since there is a non-zero element below the first non-zero element in the second row.

The word **echelon** means **staircase-like** formation, and we can see that the first non-zero entries form a staircase which moves down and to the right through the matrix.

■ Matrices in row echelon form yield systems that can be solved fairly easily.

We may continue performing row operations on the matrix to make the system even <u>easier</u> to solve.

Consider the example above where the matrix

$$\begin{bmatrix} 1 & -3 & 2 & -5 \\ 2 & 5 & -5 & 3 \\ -1 & 2 & -1 & 4 \end{bmatrix} \text{ was converted to } \begin{bmatrix} 1 & -3 & 2 & -5 \\ 0 & 1 & -1 & 1 \\ 0 & 0 & 1 & 1 \end{bmatrix}$$

We continue this example.

$$\begin{bmatrix} 1 & -3 & 2 & -5 \\ 0 & 1 & -1 & 1 \\ 0 & 0 & 1 & 1 \end{bmatrix} \equiv \begin{array}{rcl} x - 3y + 2z &=& -5 \\ y - z &=& 1 \\ z &=& 1 \end{array}$$

$1 \cdot R3 + R2 \Rightarrow R2$, thus:

$$\begin{bmatrix} 1 & -3 & 2 & -5 \\ 0 & 1 & 0 & 2 \\ 0 & 0 & 1 & 1 \end{bmatrix} \equiv \begin{array}{rcl} x - 3y + 2z &=& -5 \\ y &=& 2 \\ z &=& 1 \end{array}$$

$-2 \cdot R3 + R1 \Rightarrow R1$, thus:

$$\begin{bmatrix} 1 & -3 & 0 & -7 \\ 0 & 1 & 0 & 2 \\ 0 & 0 & 1 & 1 \end{bmatrix} \equiv \begin{array}{rcl} x - 3y &=& -7 \\ y &=& 2 \\ z &=& 1 \end{array}$$

$3 \cdot R2 + R1 \Rightarrow R1$, thus:

$$\begin{bmatrix} 1 & 0 & 0 & -1 \\ 0 & 1 & 0 & 2 \\ 0 & 0 & 1 & 1 \end{bmatrix} \equiv \begin{array}{rcr} x & & = -1 \\ & y & = 2 \\ & z = & 1 \end{array}$$

This final system has a solution which may be read directly from the matrix: $x = -1$, $y = 2$, and $z = 1$.

Row operations have done the work of plugging in the values and solving for the variables which we did earlier.

■   A matrix in this final form is said to be in **reduced row echelon form**.

> | Definition |
>
> **Reduced row echelon form**:
>
> A matrix is in **reduced row echelon form** if it is in row echelon
>
> form and also has the following two properties:
>
> The first non-zero element in each row is a 1.
>
> The first non-zero element in each row is the only non-zero element in its column.

8.1.i.   Gaussian elimination

Our goal, here, is to take the augmented matrix of a system of linear equations and to use row operations to produce a matrix in row echelon or reduced row echelon form. We will thus need a system of rules, or algorithm, for producing the row and reduced row echelon forms of a matrix.

This algorithm is often called **Gaussian elimination**, or **Gauss-Jordan reduction**. We will explain this algorithm while performing it on a particular matrix.

Example **8.2**. *Apply row operations to transform the following matrix into row echelon form, then reduced row echelon form.*

$$\begin{bmatrix} 0 & 1 & -5 & 3 \\ 4 & 3 & 1 & 5 \\ -2 & -1 & 3 & -7 \end{bmatrix}$$

- **Solution**:
- **STEP 1**: Start with the first column on the left. Interchange rows if necessary to place a non-zero entry in the top row.

  The left-most non-zero column in this case is the <u>first</u> column; since there is currently a 0 at the top of this column, we interchange rows 1 and 3 to place -2 at the top of this column. ($R1 \Leftrightarrow R2$) The following matrix is the result.

$$\begin{bmatrix} -2 & -1 & 3 & -7 \\ 4 & 3 & 1 & 5 \\ 0 & 1 & -5 & 3 \end{bmatrix}$$

- **STEP 2**: Use row operations to produce zeros in all positions below this entry.

  To eliminate the 4 which lies below the -2, we use the row operation $2 \cdot R1 + R2 \Rightarrow R2$. Which produces:

$$\begin{bmatrix} -2 & -1 & 3 & -7 \\ 0 & 1 & 7 & -9 \\ 0 & 1 & -5 & 3 \end{bmatrix}$$

- **STEP 3**: Ignore the row which you have just used to eliminate the other entries and any rows above it ($R1$); ignore the column from which you have just eliminated and any columns to its left (column1).

  Repeat steps 1 and 2 to the matrix that remains until there are no more rows remaining.

  The matrix we are now considering is listed in boldface below:

$$\begin{bmatrix} -2 & -1 & 3 & -7 \\ 0 & 1 & 7 & -9 \\ 0 & 1 & -5 & 3 \end{bmatrix}$$

- Performing Step 1, we begin with the second column. As it already has a non-zero entry (a 1) at its top, we do not need to interchange rows. Step 2 tells us to create zeros below this 1; we only need the operation $-1 \cdot R2 + R3 \Rightarrow R3$ to do this. We get the matrix

$$\begin{bmatrix} -2 & -1 & 3 & -7 \\ 0 & 1 & 7 & -9 \\ 0 & 0 & -12 & 12 \end{bmatrix}$$

  Since there are now no more rows to eliminate beneath, our process stops. The matrix is now in row echelon form.

  To produce the reduced row echelon form, we do Step 4.

- **STEP 4**: Beginning at the bottom row of the matrix, use row operations to produce zeros above the first non-zero element in each row. Work up the column, then to the next column on its left. Also multiply rows by constants if necessary to ensure that the first non-zero element in each row is a 1.

  We begin by multiplying row 3, by $-\frac{1}{12}$

$$(-\frac{1}{12} \cdot R3 \Rightarrow R3)$$

  which produces:

$$\begin{bmatrix} -2 & -1 & 3 & -7 \\ 0 & 1 & 7 & -9 \\ 0 & 0 & 1 & -1 \end{bmatrix}$$

  We now eliminate the entries above the 1 in the third row with the operations

$$-7 \cdot R3 + R2 \Rightarrow R2$$
$$and -3 \cdot R3 + R1 \Rightarrow R1$$

  These operations produce:

$$\begin{bmatrix} -2 & -1 & 0 & -4 \\ 0 & 1 & 0 & -2 \\ 0 & 0 & 1 & -1 \end{bmatrix}$$

- Moving leftward, we see that the first non-zero element in the second row is already 1, so we may continue and eliminate the -1 above it with the operation

$$1 \cdot R2 + R1 \Rightarrow R1 :$$
$$\begin{bmatrix} -2 & 0 & 0 & -6 \\ 0 & 1 & 0 & -2 \\ 0 & 0 & 1 & -1 \end{bmatrix}$$

- We finish by multiplying the first row by $-\frac{1}{2}$,

$$(-\frac{1}{2} \cdot R1 \Rightarrow R1),$$

  thus making a non-zero non-zero entry a 1.

$$\begin{bmatrix} 1 & 0 & 0 & 3 \\ 0 & 1 & 0 & -2 \\ 0 & 0 & 1 & -1 \end{bmatrix}$$

■ The matrix is now in reduced row echelon form.

We can now use this Gaussian elimination algorithm to find solutions to systems of linear equations.

Example 8.3. *Find the solution, if it exists, of the system*

$$\begin{aligned} 2x &- y = 5 \\ 3x &- 7y = 4 \end{aligned}$$

• **Solution**: The augmented matrix for this system is

$$\begin{bmatrix} 2 & -1 & 5 \\ 3 & -7 & 4 \end{bmatrix}$$

To simplify our calculations we begin by multiplying <u>row 1</u> by $\frac{1}{2}$

$$(\tfrac{1}{2} \cdot R1 \Rightarrow R1),$$

$$\begin{bmatrix} 1 & -\frac{1}{2} & \frac{5}{2} \\ 3 & -7 & 4 \end{bmatrix}$$

As there is already a non-zero element in the upper left corner, we eliminate beneath it with the operation

$$-3 \cdot R1 + R2 \Rightarrow R2$$

This produces the matrix

$$\begin{bmatrix} 1 & -\frac{1}{2} & \frac{5}{2} \\ 0 & -\frac{11}{2} & -\frac{7}{2} \end{bmatrix}$$

Again to simplify our work, we multiply row 2 by $-\frac{2}{11}$

$$(-\tfrac{2}{11} \cdot R2 \Rightarrow R2)$$

$$\begin{bmatrix} 1 & -\frac{1}{2} & \frac{5}{2} \\ 0 & 1 & \frac{7}{11} \end{bmatrix}$$

We now can eliminate above the 1 in row 2 with the operation

$$\tfrac{1}{2} \cdot R2 + R1 \Rightarrow R1,$$

which gives us the matrix

$$\begin{bmatrix} 1 & 0 & \frac{31}{11} \\ 0 & 1 & \frac{7}{11} \end{bmatrix} \text{ i.e. } \begin{bmatrix} x+y = ? \\ x+y = ? \end{bmatrix}$$

This matrix is in reduced row echelon form; the solution to the system is $x = \frac{31}{11}, y = \frac{7}{11}$.

Exercise 8.1. *Find the solution, if it exists, of the system*

$$
\begin{aligned}
x & & & + & z & = & -2 \\
2x & - & 2y & + & z & = & 0 \\
x & + & y & + & 2z & = & 3
\end{aligned}
$$

Solution:

Example 8.4. *Find the solution, if it exists, of the system*

$$
\begin{aligned}
x & + & y & - & 2z & = & 1 \\
2x & + & 2y & - & 3z & = & -1 \\
3x & + & 3y & + & z & = & 2
\end{aligned}
$$

- **Solution:** The augmented matrix for this system is

$$
\begin{bmatrix}
1 & 1 & -2 & 1 \\
2 & 2 & -3 & -1 \\
3 & 3 & 1 & 2
\end{bmatrix}
$$

Since we have a 1 in the upper left corner, we may begin eliminating beneath it with the operations

$$-2 \cdot R1 + R2 \Rightarrow R2$$
$$and -3 \cdot R1 + R3 \Rightarrow R3$$

We now have the matrix

$$
\begin{bmatrix}
1 & 1 & -2 & 1 \\
0 & 0 & 1 & -3 \\
0 & 0 & 7 & -1
\end{bmatrix}
$$

Ignoring the first row and first column, we find that there are no more non-zero terms remaining in the second column. We thus proceed to the third column, where we eliminate below the 1 with

$$-7 \cdot R2 + R3 \Rightarrow R3 :$$

$$
\begin{bmatrix}
1 & 1 & -2 & 1 \\
0 & 0 & 1 & -3 \\
0 & 0 & 0 & 20
\end{bmatrix}
$$

As an equation, the third row now says that

$$0x + 0y + 0z = 20,$$
$$\text{or } 0 = 20$$

Since this statement is <u>never true</u>, our system <u>fails</u> to have a solution. In other words, the system is <u>inconsistent</u>.

Exercise 8.2. *Find the solution, if it exists, of the system*

$$
\begin{aligned}
x + y - 2z &= 0 \\
2x + 2y - 3z &= 1 \\
3x + 3y + z &= 7
\end{aligned}
$$

Solution:

Example 8.5. *Find the solution, if it exists, of the system*

$$
\begin{aligned}
x + y + 2z + 5w &= 2 \\
2x + 3y + z + w &= 5 \\
2y - z + 3w &= 13 \\
5y + z + w &= 19
\end{aligned}
$$

Solution: The augmented matrix for this system is

$$
\begin{bmatrix}
1 & 1 & 2 & 5 & 2 \\
2 & 3 & 1 & 1 & 5 \\
0 & 2 & -1 & 3 & 13 \\
0 & 5 & 1 & 1 & 19
\end{bmatrix}
$$

We begin Gaussian elimination with the operation

$$-2 \cdot R1 + R2 \Rightarrow R2 :$$

$$
\begin{bmatrix}
1 & 1 & 2 & 5 & 2 \\
0 & 1 & -3 & -9 & 1 \\
0 & 2 & -1 & 3 & 13 \\
0 & 5 & 1 & 1 & 19
\end{bmatrix}
$$

We now ignore the first row and first column, and eliminate beneath the first 1 in the second row with

$$-2 \cdot R2 + R3 \Rightarrow R3$$
$$and -5 \cdot R2 + R4 \Rightarrow R4 :$$

$$
\begin{bmatrix}
1 & 1 & 2 & 5 & 2 \\
0 & 1 & -3 & -9 & 1 \\
0 & 0 & 5 & 21 & 11 \\
0 & 0 & 16 & 46 & 14
\end{bmatrix}
$$

We now ignore the second row and second column; we change the 5 in the third row to a 1 with the operation

$$\tfrac{1}{5}R3 \Rightarrow R3 \;:$$

$$\begin{bmatrix} 1 & 1 & 2 & 5 & 2 \\ 0 & 1 & -3 & -9 & 1 \\ 0 & 0 & 1 & \frac{21}{5} & \frac{11}{5} \\ 0 & 0 & 16 & 46 & 14 \end{bmatrix}$$

and eliminate beneath it with the operation

$$-16 \bullet R3 + R4 \Rightarrow R4 \;:$$

$$\begin{bmatrix} 1 & 1 & 2 & 5 & 2 \\ 0 & 1 & -3 & -9 & 1 \\ 0 & 0 & 1 & \frac{21}{5} & \frac{11}{5} \\ 0 & 0 & 0 & -\frac{106}{5} & -\frac{106}{5} \end{bmatrix}$$

We clean up the fourth row with

$$-\frac{5}{106} \bullet R4 \Rightarrow R4 \;:$$

$$\begin{bmatrix} 1 & 1 & 2 & 5 & 2 \\ 0 & 1 & -3 & -9 & 1 \\ 0 & 0 & 1 & \frac{21}{5} & \frac{11}{5} \\ 0 & 0 & 0 & 1 & 1 \end{bmatrix}$$

and proceed to eliminate above the fourth row with

$$-\frac{21}{5} \bullet R4 + R3 \Rightarrow R3,$$
$$9 \bullet R4 + R2 \Rightarrow R2,$$
$$and -5 \bullet R4 + R1 \Rightarrow R1 \;:$$

$$\begin{bmatrix} 1 & 1 & 2 & 0 & -3 \\ 0 & 1 & -3 & 0 & 10 \\ 0 & 0 & 1 & 0 & -2 \\ 0 & 0 & 0 & 1 & 1 \end{bmatrix}$$

To eliminate above the 1 in the third row,

$$3 \bullet R3 + R2 \Rightarrow R2$$
$$and -2 \bullet R3 + R1 \Rightarrow R1$$

$$\begin{bmatrix} 1 & 1 & 0 & 0 & 1 \\ 0 & 1 & 0 & 0 & 4 \\ 0 & 0 & 1 & 0 & -2 \\ 0 & 0 & 0 & 1 & 1 \end{bmatrix}$$

and the operation $-1 \bullet R2 + R1 \Rightarrow R1$ will finish the process:

$$\begin{bmatrix} 1 & 0 & 0 & 0 & -3 \\ 0 & 1 & 0 & 0 & 4 \\ 0 & 0 & 1 & 0 & -2 \\ 0 & 0 & 0 & 1 & 1 \end{bmatrix} \quad \text{i.e.} \quad \begin{bmatrix} x+y+z+w = ? \\ x+y+z+w = ? \\ x+y+z+w = ? \\ x+y+z+w = ? \end{bmatrix}$$

Thus this system has the solution $x = -3$, $y = 4$, $z = -2$, $w = 1$.

Exercise 8.3. *Solve the following system of linear equations by converting the augmented matrix to reduced row echelon form.*

$$
\begin{aligned}
x &- y &+ 2z &+ &w &= 8 \\
3x &+ 2y &- z & & &= -3 \\
-x &+ y & &- &3w &= 4 \\
2x &- 3y &+ z &- &w &= 11
\end{aligned}
$$

Solution:

Summary 8.1.

■ Matrices are used to store information from a system of linear equations. We may solve the system by performing elementary row operations on the augmented matrix of the system.

■ Matrices in row echelon form and reduced row echelon form are associated with systems that are easy to solve; the Gaussian elimination algorithm is used to convert a matrix into row echelon or reduced row echelon form.

End of Section 8.1. *Problems*

Problem 8.1. *Solve the following systems of linear equations by converting the augmented matrix to reduced row echelon form.*

(a)

$$
\begin{aligned}
x &- 2y &= 3 \\
3x &- 5y &= 11
\end{aligned}
$$

(b)

$$
\begin{aligned}
x &+ 2y &- z &= 6 \\
3x &+ 8y &+ 9z &= 10 \\
2x &- y &+ 2z &= -2
\end{aligned}
$$

(c)

$$
\begin{aligned}
x &+ 2y & &= 4 \\
-x &+ 3y &+ 3z &= -2 \\
 &y &+ z &= 0
\end{aligned}
$$

Problem 8.2. *Solve the following system of linear equations by converting the augmented matrix to reduced row echelon form.*

(a)

$$
\begin{aligned}
2x &- y &= 5 \\
3x &+ 4y &= 13
\end{aligned}
$$

(b)

$$
\begin{aligned}
x &- y &+ z &= 7 \\
2x &+ y &- 3z &= 1 \\
-x &- 3y &+ 2z &= 3
\end{aligned}
$$

(c)

$$3x + 2y - 4z = -11$$
$$-5x - 2y + z = -7$$
$$2x + y - z = 0$$

(d)

$$x + 3y - z + w = 1$$
$$2x - y + 2z + 3w = 3$$
$$y + z - 4w = 13$$
$$-3x + 2y + 4z - 2w = -5$$

Problem 8.3.

(a)

$$2x - y = 5$$
$$3x + 4y = 13$$

(b)

$$x + 2y = -3$$
$$5x - 3y = 11$$

(c)

$$3x + y = -2$$
$$-7x - 2y = 6$$

(d)

$$x - y + z = 7$$
$$2x + y - 3z = 1$$
$$-x - 3y + 2z = 3$$

(e)

$$x + 2y + z = -3$$
$$2x - 3y + z = 10$$
$$7x + 4y + 3z = 7$$

(f)

$$-x + 2y - z = 0$$
$$3x + 4y + 2z = 1$$
$$-4x + 6y + z = -5$$

(g)

$$x + 3y - 4z + w = 5$$
$$-2x + y - w = -12$$
$$3x - 2y + z - 2w = 4$$
$$x + y - 5z - 3w = -6$$

(h)

$$2x + 3y + z - w = -4$$
$$4x - 5y + 2z + w = 7$$
$$-2x + y - z - w = -6$$
$$6x + y + 7z - 4w = -5$$

(i)

$$x + y + z + w = 3$$
$$-2x - y + 3z - w = -2$$
$$4x + 5y - 2z + 2w = 14$$
$$x - y + 3z - 5w = -21$$

8.2. Matrix Operations

In the previous section, we have used matrices as a shorthand way to solve systems of linear equations. However, we can also treat matrices as mathematical objects in their own right. In this section we will explore how to add, subtract, and multiply matrices and how these operations relate to each other. We will find that our theoretical explorations will allow us to develop other ways to solve systems of linear equations and to find other applications for matrices.

Notation

We begin by developing some notation.

- We may represent a matrix in **three** ways.
    - ▶ A matrix may be represented by an uppercase letter such as A, B, C, etc.
    - ▶ If we write the (i,j)-element of the matrix as a_{ij} (or b_{ij}, c_{ij}, etc.), then the entire matrix may be represented by the notation $[a_{ij}]$ (or $[b_{ij}]$, or $[c_{ij}]$, etc.)
    - ▶ A matrix may be represented by a complete listing of its elements:

$$\begin{bmatrix} a_{11} & a_{12} & \cdots & a_{1n} \\ a_{21} & a_{22} & \cdots & a_{2n} \\ \vdots & \vdots & \ddots & \vdots \\ a_{m1} & a_{m2} & \cdots & a_{mn} \end{bmatrix}$$

- There is the notion of **equality of matrices**: two matrices are **equal** if all of their corresponding elements are equal. $\boxed{\text{Note:}}$ That for all of the corresponding elements to be equal, the two matrices must be the same size.

$\boxed{Definition}$

Equality of matrices:

Two $m \times n$ matrices $A = [a_{ij}]$ and $B = [b_{ij}]$ are **equal** if $a_{ij} = b_{ij}$ for $1 \le i \le m$ and $1 \le j \le n$.

For example,

$$\begin{bmatrix} a_{11} & a_{12} \\ a_{21} & a_{22} \end{bmatrix} = \begin{bmatrix} 1 & 2 \\ 0 & -1 \end{bmatrix} \text{ if } a_{11} = 1, a_{12} = 2, a_{21} = 0, \text{ and } a_{22} = -1.$$

- We **add** two matrices of the same size by adding their corresponding elements.

$\boxed{Definition}$

Adding two matrices:

If $A = [a_{ij}]$ and $B = [b_{ij}]$ are two $m \times n$ matrices, then their **sum** $A + B$ is the $m \times n$ matrix $A + B = [a_{ij} + b_{ij}]$.

For example,

$$\begin{bmatrix} 3 & -1 \\ 1 & 4 \end{bmatrix} + \begin{bmatrix} 1 & 2 \\ 0 & -1 \end{bmatrix} = \begin{bmatrix} 3+1 & -1+2 \\ 1+0 & 4+(-1) \end{bmatrix} + \begin{bmatrix} 4 & 1 \\ 1 & 3 \end{bmatrix}$$

- If two matrices have different sizes, we cannot add them.
  For example, The matrix

$$\begin{bmatrix} 3 & -1 & -5 \\ 1 & 4 & 2 \end{bmatrix} + \begin{bmatrix} 1 & 2 \\ 0 & -1 \end{bmatrix}$$

does **not** exist.

> **Definition**
>
> **Zero matrix** :
>
> The **zero matrix** is the $m \times n$ matrix of all zeros;
> we will denote this matrix by O.

For example, $\begin{bmatrix} 0 & 0 & 0 \\ 0 & 0 & 0 \end{bmatrix}$ is the 2×3 zero matrix.

and

$$\begin{bmatrix} 1 & 3 & -7 \\ -2 & 4 & -1 \end{bmatrix} + \begin{bmatrix} 0 & 0 & 0 \\ 0 & 0 & 0 \end{bmatrix}$$
$$= \begin{bmatrix} 1 & 3 & -7 \\ -2 & 4 & -1 \end{bmatrix}$$

We will define two types of multiplication involving matrices.

Matrix multiplication I

We may multiply a matrix by a real number c. This type of multiplication is called **scalar multiplication**; we multiply each element in the matrix by c.

> **Definition**
>
> **Scalar multiplication;:**
>
> If $A = [a_{ij}]$ is an $m \times n$ matrix and c is any real number,
> then the **scalar multiple** cA is the $m \times n$ matrix $cA = [ca_{ij}]$.

For example,

$$3 \cdot \begin{bmatrix} 1 & 3 & -7 \\ -2 & 4 & -1 \end{bmatrix} = \begin{bmatrix} 3 \cdot 1 & 3 \cdot 3 & 3 \cdot -7 \\ 3 \cdot -2 & 3 \cdot 4 & 3 \cdot -1 \end{bmatrix}$$
$$= \begin{bmatrix} 3 & 9 & -21 \\ -6 & 12 & -3 \end{bmatrix}$$

- **Scalar multiplication** and addition of matrices have many properties similar to multiplication and addition of real numbers.

  ■ **Properties of Matrix Addition and Scalar Multiplication**

  If A, B, and C are $m \times n$ matrices, O is the $m \times n$ zero matrix, and c and d are real numbers, then:

Property 1	$A + B = B + A$
Property 2	$A + (B + C) = (A + B) + C$
Property 3	$A + O = A$
Property 4	$(cd)A = c(dA)$
Property 5	$1 \cdot A = A$
Property 6	$c(A + B) = cA + cB$
Property 7	$(c + d)A = cA + dA$

For example, let $A = \begin{bmatrix} 1 & -1 \\ 2 & 3 \end{bmatrix}$, $B = \begin{bmatrix} 3 & 7 \\ -1 & 2 \end{bmatrix}$, and $C = \begin{bmatrix} -3 & 1 \\ 4 & -1 \end{bmatrix}$.

Then

$$A + B = \begin{bmatrix} 4 & 6 \\ 1 & 5 \end{bmatrix}$$

$$and\ B + C = \begin{bmatrix} 0 & 8 \\ 3 & 1 \end{bmatrix}$$

$$so\ A + (B + C) = \begin{bmatrix} 1 & 7 \\ 5 & 4 \end{bmatrix}$$

$$and\ (A + B) + C = \begin{bmatrix} 1 & 7 \\ 5 & 4 \end{bmatrix}$$

Note: $A + (B + C) = (A + B) + C$, as stated in **Property 2** above.

Also, let A and B be as above, and let $c = 2$.

Then $A + B = \begin{bmatrix} 4 & 6 \\ 1 & 5 \end{bmatrix}$.

Thus

$$c(A + B) = \begin{bmatrix} 8 & 12 \\ 2 & 10 \end{bmatrix}$$

$$and\ cA + cB = \begin{bmatrix} 2 & -2 \\ 4 & 6 \end{bmatrix} + \begin{bmatrix} 6 & 14 \\ -2 & 4 \end{bmatrix}$$

$$= \begin{bmatrix} 8 & 12 \\ 2 & 10 \end{bmatrix}$$

so $c(A + B) = cA + cB$, as **Property 6** would predict.

Again, let A be as above, $c = 2$ and $d = 3$.

Then $(c + d)A = \begin{bmatrix} 5 & -5 \\ 10 & 15 \end{bmatrix}$, while

$$cA + dA = \begin{bmatrix} 2 & -2 \\ 4 & 6 \end{bmatrix} + \begin{bmatrix} 3 & -3 \\ 6 & 9 \end{bmatrix}$$

$$= \begin{bmatrix} 5 & -5 \\ 10 & 15 \end{bmatrix}$$

Thus $(c + d)A = cA + dA$, as stated in **Property 7**.

Matrix multiplication II

The second type of multiplication we will define is called **matrix multiplication**; here we will multiply a matrix by another matrix. The definition is much more involved than one might expect, but it will be useful in various applications.

> **Definition**
>
> **Matrix multiplication**:
>
> If $A = [a_{ij}]$ in an $m \times n$ matrix and $B = [b_{ij}]$ is an $n \times p$ matrix,
>
> then the **matrix product** $C = AB$ is the $m \times p$ matrix,
>
> with $c_{ij} = a_{i1}b_{1j} + a_{i2}b_{2j} + a_{i3}b_{3j} + \cdots a_{in}b_{nj}$.

Example **8.6**. Let $A = \begin{bmatrix} 1 & -1 \\ 2 & 3 \end{bmatrix}$ and $B = \begin{bmatrix} 3 & 7 \\ -1 & 2 \end{bmatrix}$. *Using the above definition, calculate the product* $C = AB$.

Solution: By the definition,

$$c_{11} = a_{11}b_{11} + a_{12}b_{21} = (1)(3) + (-1)(-1) = 4$$
$$c_{12} = a_{11}b_{12} + a_{12}b_{22} = (1)(7) + (-1)(2) = 5$$
$$c_{21} = a_{21}b_{11} + a_{22}b_{21} = (2)(3) + (3)(-1) = 3$$
$$c_{22} = a_{21}b_{12} + a_{22}b_{22} = (2)(7) + (3)(2) = 20,$$

so $C = \begin{bmatrix} 4 & 5 \\ 3 & 20 \end{bmatrix}$.

In practice, what we are doing is moving along the i-th row of the matrix A and pairing those elements off with the elements going down the j-th column of the matrix B.

For example, to calculate c_{12} above we move along the first row of A and the second column of B, as shown below.

$$\begin{bmatrix} 1 & -1 \\ 2 & 3 \end{bmatrix} \bullet \begin{bmatrix} 3 & 7 \\ -1 & 2 \end{bmatrix} = \begin{bmatrix} 4 & 5 \\ 3 & 20 \end{bmatrix}$$

We multiply these corresponding elements together, then add all of these products to get the (i,j)-element of AB.

> **Note:** for the matrix product to be defined, the number of columns in the first matrix must equal the number of rows in the second matrix. That is, if A has n columns, in order for AB to be defined B must have n rows.

For example,

$$\begin{bmatrix} 1 & -1 \\ 2 & 3 \end{bmatrix} \bullet \begin{bmatrix} 3 & 7 & -2 \\ -1 & 2 & 5 \end{bmatrix} = \begin{bmatrix} 4 & 5 & -7 \\ 3 & 20 & 11 \end{bmatrix}$$

while

$$\begin{bmatrix} 3 & 7 & -2 \\ -1 & 2 & 5 \end{bmatrix} \bullet \begin{bmatrix} 1 & -1 \\ 2 & 3 \end{bmatrix}$$

is undefined.

Notice that the order in which we multiply matrices matters; this is unlike multiplication of numbers. Also recall from the definition that if A is $m \times n$ and B is $n \times p$, then the product AB is defined and AB is $m \times p$.

For example,

$$\begin{bmatrix} 6 & -9 & -2 \\ -6 & -7 & 1 \end{bmatrix} \cdot \begin{bmatrix} -1 & 2 & 4 \\ -8 & 7 & -3 \\ 6 & 0 & 3 \end{bmatrix} = \begin{bmatrix} 54 & -51 & 45 \\ 68 & -61 & 0 \end{bmatrix}$$

Note: a 2×3 matrix multiplied by a 3×3 matrix is a 2×3 matrix.

Also,

$$\begin{bmatrix} 1 & 5 & -2 \\ 3 & -2 & 1 \end{bmatrix} \cdot \begin{bmatrix} -3 \\ 1 \\ -1 \end{bmatrix} = \begin{bmatrix} 4 \\ -12 \end{bmatrix}$$

Note: a 2×3 matrix multiplied by a 3×1 matrix is a 2×1 matrix.

And again,

$$\begin{bmatrix} 1 & 2 \\ 2 & 4 \end{bmatrix} \cdot \begin{bmatrix} 6 & 4 \\ -3 & -2 \end{bmatrix} = \begin{bmatrix} 0 & 0 \\ 0 & 0 \end{bmatrix}$$

and

$$\begin{bmatrix} 6 & 4 \\ -3 & -2 \end{bmatrix} \cdot \begin{bmatrix} 1 & 2 \\ 2 & 4 \end{bmatrix} = \begin{bmatrix} 14 & 28 \\ -7 & -14 \end{bmatrix}$$

In this case, both products are defined but are unequal. So even if the products AB and BA both make sense, they still may not be equal. Also Note: the product of two non-zero matrices may be the zero matrix, as in the first equation above.

- Apparently matrix multiplication fails to have many of the properties we associate with multiplication of real numbers. Since matrix multiplication fails to have these properties, we should be careful when using this operation.

  For example,

$$\begin{bmatrix} 1 & -1 \\ 2 & 3 \end{bmatrix} \cdot \begin{bmatrix} 1 & 0 \\ 0 & 1 \end{bmatrix} = \begin{bmatrix} 1 & -1 \\ 2 & 3 \end{bmatrix}$$

- Multiplying by the matrix $\begin{bmatrix} 1 & 0 \\ 0 & 1 \end{bmatrix}$ has not changed the value of the first matrix. We call the second matrix the **identity matrix of order 2**.

Definition

Identity matrix of order:

The **identity matrix of order** n is the $n \times n$ matrix that has 1's on its main diagonal and 0's elsewhere.
It is denoted by the symbol I_n, or by I if n is understood.

By this definition,

$$I_n = \begin{bmatrix} 1 & 0 & 0 & \cdots & 0 \\ 0 & 1 & 0 & \cdots & 0 \\ 0 & 0 & 1 & \cdots & 0 \\ \vdots & \vdots & \vdots & \ddots & \vdots \\ 0 & 0 & 0 & \cdots & 1 \end{bmatrix}$$

Even though AB fails to equal BA in most cases, matrix multiplication still possesses some useful properties.

Properties of Matrix Multiplication

If A, B, and C are appropriately sized matrices, c is a real number, and I is an appropriately sized identity matrix, then:

Property	Definition
1	$A(BC) = (AB)C$
2	$A(B + C) = AB + AC$
3	$(A + B)C = AC + BC$
4	$c(AB) = (cA)B = A(cB)$
5	$IA = A$
6	$AI = A$

Now that we have defined and studied matrix multiplication a bit, we may ask why this operation has such a peculiar definition. In the following example, we show that we may represent a system of linear equations by using matrix multiplication.

Consider the system of linear equations

$$
\begin{aligned}
x - 3y + 2z &= -5 \\
2x + 5y - 5z &= 3 \\
-x + 2y - z &= 4
\end{aligned}
$$

which we used in an earlier section. If we let

$$
A = \begin{bmatrix} 1 & -3 & 2 \\ 2 & 5 & -5 \\ -1 & 2 & -1 \end{bmatrix}
$$

be the coefficient matrix for this system, and let

$$
X = \begin{bmatrix} x \\ y \\ z \end{bmatrix}
$$

then the product AX is the 3×1 matrix

$$
\begin{bmatrix} x - 3y + 2z \\ 2x + 5y - 5z \\ -x + 2y - z \end{bmatrix}
$$

If we in addition let

$$
B = \begin{bmatrix} -5 \\ 3 \\ 4 \end{bmatrix}
$$

be the matrix containing the constants on the right hand side of each equation, then we find that we may rewrite the system of equations in the form $AX = B$.

- This is true in general: if A is the coefficient matrix for a system of m linear equations in n unknowns, X is the $n \times 1$ matrix containing the unknowns, and B is the $n \times 1$ matrix containing the constants from the right hand sides of the equations, then the system may be written in the form $AX = B$. We shall see how to use this notation to solve the system in the next section.

Exercise **8.4**. Find $A + B$, $A - B$, and $A + 2B$ if

$$A = \begin{bmatrix} 2 & -1 \\ 3 & 7 \end{bmatrix} \quad \text{and} \quad B = \begin{bmatrix} -3 & 4 \\ 5 & -2 \end{bmatrix}$$

Solution:

Exercise **8.5**. Find (*if possible*) AB and BA if

$$A = \begin{bmatrix} -1 & 3 \\ 2 & 4 \end{bmatrix} \quad \text{and} \quad B = \begin{bmatrix} 0 & 2 \\ -3 & 1 \end{bmatrix}$$

Solution:

Exercise **8.6**. Find (*if possible*) AB and BA if

$$A = \begin{bmatrix} 1 & 2 & -1 \\ 3 & -1 & 0 \end{bmatrix} \quad \text{and} \quad B = \begin{bmatrix} -4 & 1 \\ 5 & 2 \end{bmatrix}$$

Solution:

Exercise **8.7**. Write the following system in the form $AX = B$.

$$
\begin{array}{rrrrr}
3x & - & y & + & 7z & = & 4 \\
2x & + & 3y & - & 5z & = & -3 \\
x & + & y & - & z & = & 11
\end{array}
$$

Solution:

Summary 8.2.

- We may define what it means for two matrices to be equal, how to add two matrices, how to multiply a matrix by a real number, and how to multiply two matrices together.
- These operations share many properties with addition and multiplication of real numbers, but matrix multiplication fails to have several familiar properties.
- Matrix multiplication gives us another way of converting a system of linear equations into matrix language.

End of Section 8.2.Problems

Problem 8.4. *Perform the following operation*:

(a) *Find* $A + B$, $A - B$, *and* $-3A + 2B$ *if*

$$A = \begin{bmatrix} 2 & 1 \\ -5 & -8 \end{bmatrix} \text{ and } B = \begin{bmatrix} 9 & -6 \\ -2 & 6 \end{bmatrix}$$

(b) *Find (if possible)* AB *and* BA *if*

$$A = \begin{bmatrix} -2 & 8 \\ -5 & -4 \end{bmatrix} \text{ and } B = \begin{bmatrix} 9 & 1 \\ -4 & 7 \end{bmatrix}$$

(c) *Find (if possible)* AB *and* BA *if*

$$A = \begin{bmatrix} -2 & 8 & -1 \\ -4 & 3 & -6 \end{bmatrix} \text{ and } B = \begin{bmatrix} -9 & -6 & 4 \\ -6 & 5 & -9 \\ 0 & -7 & -1 \end{bmatrix}$$

(d) *Write the system*

$$\begin{array}{rcrcrcrcr} x & - & 7y & - & z & + & w & = & 4 \\ & & 2y & - & 3z & - & 5w & = & -3 \\ 3x & + & 11y & + & z & + & 2w & = & 11 \\ -2x & - & y & & & + & 6w & = & 23 \end{array}$$

in the form $AX = B$.

Problem 8.5. *Find* $A + B$, $A - B$, *and* $-3A + 2B$ *if*

$$A = \begin{bmatrix} 2 & 1 \\ -5 & -8 \end{bmatrix} \text{ and } B = \begin{bmatrix} 9 & -6 \\ -2 & 6 \end{bmatrix}$$

Problem 8.6. *Find* $A + B$, $A - B$, *and* $-3A + 2B$ *if*

$$A = \begin{bmatrix} 7 & 3 \\ 2 & -1 \\ -5 & 6 \end{bmatrix} \text{ and } B = \begin{bmatrix} 1 & 6 \\ 7 & 2 \\ -4 & 0 \end{bmatrix}$$

Problem 8.7. *Find (if possible)* AB *and* BA *for the following*:

(a)

$$A = \begin{bmatrix} -2 & 8 \\ -5 & -4 \end{bmatrix} \text{ and } B = \begin{bmatrix} 9 & 1 \\ -4 & 7 \end{bmatrix}$$

(b)

$$A = \begin{bmatrix} 6 & -6 \\ -2 & -8 \end{bmatrix} \text{ and } B = \begin{bmatrix} 9 & 9 & 2 \\ -1 & -9 & 5 \end{bmatrix}$$

(c)

$$A = \begin{bmatrix} -2 & 8 & -1 \\ -4 & 3 & -6 \end{bmatrix} \text{ and } B = \begin{bmatrix} -9 & -6 & 4 \\ -6 & 5 & -9 \\ 0 & -7 & -1 \end{bmatrix}$$

(d)

$$A = \begin{bmatrix} -1 & -6 & 4 \\ 7 & 9 & 1 \\ -8 & -6 & 6 \end{bmatrix} \text{ and } B = \begin{bmatrix} -1 & -5 & 3 & -9 \\ -9 & -4 & -8 & 1 \\ 7 & -6 & -6 & 8 \end{bmatrix}$$

Problem 8.8. *Write the following systems of linear equations in the form* $AX = B$.

(a)

$$\begin{array}{rcrcrcrcr} x & - & 7y & - & z & + & w & = & 4 \\ & & 2y & - & 3z & - & 5w & = & -3 \\ 3x & + & 11y & + & z & + & 2w & = & 11 \\ -2x & - & y & & & + & 6w & = & 23 \end{array}$$

(b)

$$\begin{array}{rcrcrcrcr} x & + & y & + & z & + & w & = & 3 \\ -2x & - & y & + & 3z & - & w & = & -2 \\ 4x & + & 5y & - & 2z & + & 2w & = & 14 \\ x & - & y & + & 3z & - & 5w & = & -21 \end{array}$$

8.3. The Inverse of a Matrix–Systems Revisited

We noted at the end of the last section that to solve a system of linear equations we can solve the matrix equation

$$AX = B \text{ for } X.$$

This is not as simple as it would be if A, X, and B were numbers: if they were we would just divide by A and be done. We cannot do that in this case since A, X, and B are matrices, but we can do something similar.

Suppose that we could find a matrix C such that $\boxed{CA = I.}$ Then we could multiply both sides of $AX = B$ by C. We would get

$$AX = B$$
$$C(AX) = CB$$
$$(CA)X = CB$$
$$IX = CB$$
$$X = CB$$

So, if we can find such a C, we could simply compute CB to find the solution for our system.

■   We will call C the **inverse** of A.

Definition

inverse of a matrix:

Let A be an $n \times n$ matrix. If there exists a matrix C such that $AC = CA = I_n$, then C is the **inverse** of A.

We denote the inverse of A with the notation $C = A^{-1}$, which is read "A inverse."

If a matrix A has an inverse, A is called **invertible**.

Note: only square matrices may have inverses by the definition.

Example **8.7**. *Show that the matrices*

$$A = \begin{bmatrix} 1 & 2 \\ 2 & 3 \end{bmatrix} \quad and \quad C = \begin{bmatrix} -3 & 2 \\ 2 & -1 \end{bmatrix}$$

are inverses.

Solution: We compute that

$$AC = \begin{bmatrix} 1 & 2 \\ 2 & 3 \end{bmatrix} \cdot \begin{bmatrix} -3 & 2 \\ 2 & -1 \end{bmatrix} = \begin{bmatrix} 1 & 0 \\ 0 & 1 \end{bmatrix} = I_2$$

and

$$CA = \begin{bmatrix} -3 & 2 \\ 2 & -1 \end{bmatrix} \cdot \begin{bmatrix} 1 & 2 \\ 2 & 3 \end{bmatrix} = \begin{bmatrix} 1 & 0 \\ 0 & 1 \end{bmatrix} = I_2,$$

so by definition A and C are inverses. We can also write $A = C^{-1}$ and $C = A^{-1}$.

We can use inverses as advertised to solve systems of linear equations.

Example **8.8.** *Use inverses to solve the system*

$$x + 2y = 1$$
$$2x + 3y = 4$$

- **Solution**: For this system the coefficient matrix A is $A = \begin{bmatrix} 1 & 2 \\ 2 & 3 \end{bmatrix}$, while the matrix of unknowns is $X = \begin{bmatrix} x \\ y \end{bmatrix}$ and the matrix of constants is $B = \begin{bmatrix} 1 \\ 4 \end{bmatrix}$. We must compute $X = A^{-1}B$; but from the previous example we know that $A^{-1} = \begin{bmatrix} -3 & 2 \\ 2 & -1 \end{bmatrix}$

so

$$X = A^{-1}B = \begin{bmatrix} -3 & 2 \\ 2 & -1 \end{bmatrix} \cdot \begin{bmatrix} 1 \\ 4 \end{bmatrix} = \begin{bmatrix} 5 \\ -2 \end{bmatrix}$$

Thus the solution to this system is $x = 5, y = -2$.

Of course, one needs to have an effective way of computing the inverse of a matrix if inverses are to be truly useful in the solving of systems of equations. We develop a method for the computing of an inverse in the following example.

Example **8.9.** *Find the inverse of the matrix*

$$A = \begin{bmatrix} 1 & 2 \\ 2 & 3 \end{bmatrix}$$

- **Solution**: Let $A^{-1} = \begin{bmatrix} a & b \\ c & d \end{bmatrix}$

To find A^{-1}, we will solve the matrix equation

$$A \cdot A^{-1} = I_2 \; for A^{-1}$$

We begin by computing $A \cdot A^{-1}$:

$$A \cdot A^{-1} = \begin{bmatrix} 1 & 2 \\ 2 & 3 \end{bmatrix} \cdot \begin{bmatrix} a & b \\ c & d \end{bmatrix}$$
$$= \begin{bmatrix} a + 2c & b + 2d \\ 2a + 3c & 2b + 3d \end{bmatrix}$$

- Since $A \cdot A^{-1} = I_2 = \begin{bmatrix} 1 & 0 \\ 0 & 1 \end{bmatrix}$ we find that we must solve the following two systems of equations:

$$\begin{array}{rcl} a & + 2c & = 1 \\ 2a & + 3c & = 0 \end{array} \quad \text{and} \quad \begin{array}{rcl} b & + 2d & = 0 \\ 2b & + 3d & = 1 \end{array}$$

These systems may be solved by performing Gaussian elimination on their augmented matrices:

For example,

$$\begin{bmatrix} 1 & 2 & 1 \\ 2 & 3 & 0 \end{bmatrix} \quad \text{and} \quad \begin{bmatrix} 1 & 0 & -3 \\ 0 & 1 & 2 \end{bmatrix}$$

- ■ Thus $a = -3, c = 2$.

$$\begin{bmatrix} 1 & 2 & 0 \\ 2 & 3 & 1 \end{bmatrix} \quad \text{and} \quad \begin{bmatrix} 1 & 0 & 2 \\ 0 & 1 & -1 \end{bmatrix}$$

■ Thus $b = 1$, $d = -1$, and (as given in the first example)

$$A^{-1} = \begin{bmatrix} -3 & 2 \\ 2 & -1 \end{bmatrix}$$

Note: The coefficient matrix in each of the systems we solved to find a, b, c, and d was just the matrix A. Since it is the same matrix in both systems, we could have saved time and effort by doing both eliminations at the same time. It would look like this:

$$\begin{bmatrix} 1 & 2 & 1 & 0 \\ 2 & 3 & 0 & 1 \end{bmatrix} \quad \text{and} \quad \begin{bmatrix} 1 & 0 & -3 & 2 \\ 0 & 1 & 2 & -1 \end{bmatrix}$$

■ In effect we have placed the matrix I_2 on the right side of A, then performed Gaussian elimination on A until it has become I_2. As we are applying the same operations to I_2, it is becoming A^{-1}. This gives us a clue as to how to find the inverse of an $n \times n$ matrix A.

■ **Algorithm for finding the inverse of a matrix A**

▶ Form a new matrix by placing the appropriate sized identity matrix I on the right side of the matrix A. The new matrix is denoted $[A|I]$; the process is called **adjoining** A and I.

▶ Perform Gaussian elimination on A while applying the row operations to the entire matrix $[A|I]$. If it is possible to reduce A to the matrix I, then $[A|I]$ will have become $[I|A^{-1}]$ in the process. If A cannot be row reduced to I, then A will not be invertible.

Exercise 8.8. *Find the inverse, if it exists, of*

$$A = \begin{bmatrix} 2 & -5 \\ 1 & 4 \end{bmatrix}$$

Solution:

Example **8.10**. *Use the above result to solve the system*

$$2x - 5y = -4$$
$$x + 4y = 7$$

- **Solution:** Since the coefficient matrix is A from the previous problem and the constant matrix B is

$$\begin{bmatrix} -4 \\ 7 \end{bmatrix}$$

$$X = A^{-1} \cdot B = \begin{bmatrix} \frac{4}{13} & \frac{5}{13} \\ -\frac{1}{13} & \frac{2}{13} \end{bmatrix} \cdot \begin{bmatrix} -4 \\ 7 \end{bmatrix} = \begin{bmatrix} \frac{19}{13} \\ \frac{18}{13} \end{bmatrix}$$

- So $x = \frac{19}{13}$, $y = \frac{18}{13}$ is the solution to this system, as you may check using Gaussian elimination on the augmented matrix for this system.

Exercise **8.9**. *Find the inverse, if it exists, of*

(a) $A = \begin{bmatrix} 1 & 2 \\ 2 & 4 \end{bmatrix}$

(b) $A = \begin{bmatrix} 1 & 4 & -3 \\ -2 & -7 & 6 \\ -3 & -7 & 10 \end{bmatrix}$

Solution:

Solution:

Exercise **8.10**. *Use the inverse which was just calculated in* **8.9.b**. *to solve the system*

$$x + 4y - 3z = -1$$
$$-2x - 7y + 6z = 1$$
$$-3x - 7y + 10z = -6$$

Solution:

The case of 2×2 matrices

There is a formula for A^{-1}, which is sometimes easier to use than calculating A^{-1} by the above method. Remember that this formula only applies to 2×2 matrices.

- Let A be a general 2×2 matrix: $A = \begin{bmatrix} a & b \\ c & d \end{bmatrix}$

We adjoin I_2 to A and row reduce until A has become I_2:

$$A = \begin{bmatrix} a & b & 1 & 0 \\ c & d & 0 & 1 \end{bmatrix} \quad \text{and} \quad \begin{bmatrix} 1 & 0 & \frac{d}{ad-bc} & \frac{-b}{ad-bc} \\ 0 & 1 & \frac{-c}{ad-bc} & \frac{a}{ad-bc} \end{bmatrix}$$

- Thus

$$A^{-1} = \begin{bmatrix} \frac{d}{ad-bc} & \frac{-b}{ad-bc} \\ \frac{-c}{ad-bc} & \frac{a}{ad-bc} \end{bmatrix}$$

$$= \frac{1}{ad-bc} \begin{bmatrix} d & -b \\ -c & a \end{bmatrix}$$

Example **8.11**. *Find the inverse, if it exists, of* $A = \begin{bmatrix} 7 & -2 \\ -2 & 4 \end{bmatrix}$

- **Solution**: By the formula,

$$A^{-1} = \frac{1}{24} \begin{bmatrix} 4 & 2 \\ 2 & 7 \end{bmatrix} = \begin{bmatrix} \frac{1}{6} & \frac{1}{12} \\ \frac{1}{12} & \frac{7}{24} \end{bmatrix}$$

Example **8.12**. *Find the inverse, if it exists, of* $A = \begin{bmatrix} 1 & 2 \\ 2 & 4 \end{bmatrix}$

- **Solution**: We already know from our above work that A does not have an inverse. Notice that our formula confirms this: for this matrix, $ad - bc = (1)(4) - (2)(2) = 0$, so our formula would have us divide by zero. Thus the formula is not valid, and A does not possess an inverse.

  - Apparently, the number $ad - bc$ gives us valuable information about whether a 2×2 matrix is invertible.
    - ▶ If $ad - bc \neq 0$, the matrix is invertible, while it fails to be invertible if $ad - bc = 0$. We shall explore this number and others like it in the remaining sections of this chapter.

Exercise 8.11. *Use the algorithm to find the inverse of, if it exists, the following. Confirm your inverse using the special 2×2 formula.*

(a)

$$\begin{bmatrix} 1 & -3 \\ -3 & 8 \end{bmatrix}$$

Solution:

(b)

$$\begin{bmatrix} 3 & -7 \\ 3 & -8 \end{bmatrix}$$

Solution:

(c)

$$\begin{bmatrix} 1 & 0 & 5 \\ 2 & 2 & 0 \\ 3 & 2 & 6 \end{bmatrix}$$

Solution:

Exercise 8.12. *Use the inverse of a matrix to solve the system*

$$\begin{aligned} x &+ 4y &- 2z &= 1 \\ 2x &- y &+ 3z &= -3 \\ &y &- z &= 2 \end{aligned}$$

Solution:

Summary 8.3.

- To solve matrix equations of the form $AX = B$ with A a square matrix, we may attempt to find the inverse of the matrix A (denoted A^{-1}); the solution to the equation is then $X = A^{-1}B$.
- Some square matrices have inverses, and are called invertible matrices; others do not.
- To find the inverse of A, we row reduce the matrix $[A|I]$, which produces $[I|A^{-1}]$ if A is invertible.
- For 2×2 matrices there is a formula for A^{-1}:
$$\blacktriangleright \quad = \tfrac{1}{ad-bc} \begin{bmatrix} d & -b \\ -c & a \end{bmatrix}$$

End of Section 8.3. Poblems

Problem 8.9.

(a) Use the algorithm of this section to find the inverse, if it exists, of

$$\begin{bmatrix} 1 & -3 \\ -2 & 5 \end{bmatrix}$$

Confirm your inverse using the special 2×2 formula.

(b) Use the algorithm of this section to find the inverse, if it exists, of

$$\begin{bmatrix} 1 & 2 \\ 2 & 7 \end{bmatrix}$$

Confirm your inverse using the special 2×2 formula.

(c) Find the inverse, if it exists, of

$$\begin{bmatrix} 1 & -1 & 3 \\ 2 & 1 & 0 \\ 3 & 1 & 2 \end{bmatrix}$$

(d) Use the inverse of a matrix to solve the system

$$
\begin{aligned}
x & & & - & 2z & = & 5 \\
3x & + & y & - & z & = & -1 \\
-4x & - & 2y & - & z & = & 3
\end{aligned}
$$

Problem 8.10. Use the algorithm of this section to find the inverse of the given matrix if it exists. Confirm your answer using the special 2×2 formula.

(a)
$$\begin{bmatrix} 1 & -3 \\ -2 & 5 \end{bmatrix}$$

(b)
$$\begin{bmatrix} 1 & 2 \\ 2 & 7 \end{bmatrix}$$

(c)
$$\begin{bmatrix} -1 & 9 \\ 3 & -27 \end{bmatrix}$$

(d)
$$\begin{bmatrix} -1 & 9 \\ 1 & 7 \end{bmatrix}$$

Problem 8.11. In Exercises 11-15, find the inverse of the given matrix if it exists.

(a)
$$\begin{bmatrix} 1 & -1 & 3 \\ 2 & 1 & 0 \\ 3 & 1 & 2 \end{bmatrix}$$

(b)
$$\begin{bmatrix} 0 & 1 & 3 \\ 5 & 5 & 4 \\ 1 & 1 & 1 \end{bmatrix}$$

(c)
$$\begin{bmatrix} 3 & 5 & 3 \\ 1 & 4 & 2 \\ 0 & -2 & -1 \end{bmatrix}$$

(d)
$$\begin{bmatrix} 1 & 0 & 2 \\ 0 & 1 & 3 \\ 3 & -2 & 0 \end{bmatrix}$$

Problem 8.12. In Exercises 16-20, use the inverse of a matrix to solve the given system of linear equations.

(a)
$$
\begin{aligned}
x & & & - & 2z & = & 5 \\
3x & + & y & - & z & = & -1 \\
-4x & - & 2y & - & z & = & 3
\end{aligned}
$$

(b)
$$
\begin{aligned}
x & + & 2y & - & 11z & = & -1 \\
x & + & 3y & - & 15z & = & 0 \\
 & - & y & + & 5z & = & 1
\end{aligned}
$$

(c)
$$
\begin{aligned}
x & + & 4y & + & 2z & = & 4 \\
 & & 2y & + & z & = & -1 \\
3x & + & 5y & + & 3z & = & 7
\end{aligned}
$$

8.4. The Determinant of a Matrix

We have seen that in the case of 2×2 matrices $A = \begin{bmatrix} a & b \\ c & d \end{bmatrix}$, the number $ad - bc$ determines whether the matrix A is invertible: if $ad - bc = 0$, the matrix is not invertible, while if $ad - bc \neq 0$ then the matrix has an inverse. We will call the quantity $ad - bc$ the **determinant** of the matrix A, but will change our notation slightly.

> $\boxed{\textit{Definition}}$
>
> **The Determinant of a Matrix**:
>
> If $A = \begin{bmatrix} a_{11} & a_{22} \\ a_{21} & a_{22} \end{bmatrix}$ is a 2×2 matrix, we define
>
> the **determinant** of A to be the quantity $a_{11}a_{22} - a_{12}a_{21}$.

■ There are several ways to denote the determinant of a matrix:

▶ $\det A$,

▶ $|A|$,

▶ $\begin{vmatrix} a_{11} & a_{12} \\ a_{21} & a_{22} \end{vmatrix}$

are all used.

For example, let $A = \begin{bmatrix} 1 & -2 \\ 5 & 3 \end{bmatrix}$

Then

$$\det A = |A| = \begin{vmatrix} 1 & -2 \\ 5 & 3 \end{vmatrix}$$
$$= (1)(3) - (5)(-2)$$
$$= 13$$

We now wonder if there are any numbers for 3×3, 4×4, or larger square matrices which would help us determine whether a matrix is invertible or not. Let us first consider the 3×3 case: let

$$A = \begin{bmatrix} a_{11} & a_{12} & a_{13} \\ a_{21} & a_{22} & a_{23} \\ a_{31} & a_{32} & a_{33} \end{bmatrix}$$

and suppose that A is invertible. We may perform a series of row operations on A and produce the following matrix:

$$\begin{bmatrix} a_{11} & a_{12} & a_{13} \\ 0 & a_{11}a_{22} - a_{12}a_{21} & a_{11}a_{23} - a_{13}a_{21} \\ 0 & 0 & a_{11} \star \end{bmatrix}$$

where

$$\star = a_{11}a_{22}a_{33} + a_{12}a_{23}a_{313} + a_{13}a_{21}a_{32}$$
$$- a_{11}a_{23}a_{32} - a_{12}a_{21}a_{33} - a_{13}a_{22}a_{31}$$

Since A is an invertible matrix, $\star$ cannot be zero.

We seem to have found a number which acts like a determinant should, although it has a rather complicated formula. In fact, we will call $\star$ the **determinant** of a 3×3 matrix. This formula is rather complex. However, it may be simplified by writing some of the terms as 2×2 determinants:

$$\star = a_{11}a_{22}a_{33} + a_{12}a_{23}a_{313} + a_{13}a_{21}a_{32}$$
$$- a_{11}a_{23}a_{32} - a_{12}a_{21}a_{33} - a_{13}a_{22}a_{31}$$
$$= (a_{11}a_{22}a_{33} - a_{11}a_{23}a_{32}) - (a_{12}a_{21}a_{33} - a_{12}a_{23}a_{31})$$
$$+ (a_{13}a_{21}a_{32} - a_{13}a_{22}a_{31})$$
$$= a_{11}(a_{22}a_{33} - a_{23}a_{32}) - a_{12}(a_{21}a_{33} - a_{23}a_{31})$$
$$+ a_{13}(a_{21}a_{32} - a_{22}a_{31})$$
$$= a_{11} \begin{vmatrix} a_{22} & a_{23} \\ a_{32} & a_{33} \end{vmatrix} - a_{12} \begin{vmatrix} a_{21} & a_{23} \\ a_{31} & a_{33} \end{vmatrix}$$
$$+ a_{13} \begin{vmatrix} a_{21} & a_{22} \\ a_{31} & a_{32} \end{vmatrix}$$

Now we introduce some notation: let A_{ij} be the $(n-1) \times (n-1)$ matrix created by removing the i-th row and j-th column from the $n \times n$ matrix A. Then for our matrix A, we have:

$$A_{11} = \begin{bmatrix} a_{22} & a_{23} \\ a_{32} & a_{33} \end{bmatrix}, A_{12} = \begin{bmatrix} a_{21} & a_{23} \\ a_{31} & a_{33} \end{bmatrix}$$

and $$A_{13} = \begin{bmatrix} a_{21} & a_{22} \\ a_{31} & a_{32} \end{bmatrix}$$

Thus we find that

$$\star = a_{11} \det A_{11} - a_{12} \det A_{12} + a_{13} \det A_{13}$$

■ This leads to a **recursive** definition for the determinant of an $n \times n$ matrix: when $n = 2$, we have already defined it. When $n = 3$, we may use the above formula to calculate the determinant using the determinants of 2×2 matrices. Likewise, when $n = 4$, we could define the determinant using determinants of 3×3 matrices.

| Definition |

Recursive definition for the determinant:

f $n \geq 2$, the **determinant** of the $n \times n$ matrix A consists of a series of terms $a_{1j} \det A_{1j}$ with plus and minus signs alternating.

That is, $\det A = a_{11} \det A_{11} - a_{12} \det A_{12} + a_{13} \det A_{13} - \cdots + (-1)^{(1+n)} a_{1n} \det A_{1n}$

Example **8.13**. *Find the determinant of the matrix*

$$A = \begin{bmatrix} 1 & 3 & -2 \\ -2 & 4 & 2 \\ 1 & -3 & -1 \end{bmatrix}$$

Solution: By the definition,

$$\det A = 1\det A_{11} - 3\det A_{12} + (-2)\det A_{13}$$

$$= 1 \begin{vmatrix} 4 & 2 \\ -3 & -1 \end{vmatrix} - 3 \begin{vmatrix} -2 & 2 \\ 1 & -1 \end{vmatrix}$$

$$+ (-2) \begin{vmatrix} -2 & 4 \\ 1 & -3 \end{vmatrix}$$

$$= 1(-4 - (-6)) - 3(2 - 2) + (-2)(6 - 4)$$

$$= 2 - 0 - 4 = -2$$

We may write the definition in even more compact notation as follows. Given an $n \times n$ matrix A, the (i,j)-**cofactor** of A is

$$C_{ij} = (-1)^{i+j} \det A_{ij}$$

The determinant of A is then

$$\det A = a_{11}C_{11} + a_{12}C_{12} + \cdots + a_{1n}C_{1n}$$

■ This is called the **cofactor expansion along the first row,**

And it is one way to calculate the determinant of A. It turns out that one may compute the determinant by using a cofactor expansion along any row or column of the matrix. That is, we may perform a **cofactor expansion along the i-th row**:

$$\det A = a_{i1}C_{i1} + a_{i2}C_{i2} + \cdots + a_{in}C_{in}$$

or a **cofactor expansion along the j-th column**:

$$\det A = a_{1j}C_{1j} + a_{2j}C_{2j} + \cdots + a_{nj}C_{nj}$$

■ Remember that the cofactor C_{ij} is just the determinant of A after its i-th row and j-th column have been removed, multiplied by either a $+1$ or -1.

An easy way to remember whether a $+1$ or -1 is to be used is to think of an $n \times n$ matrix with a checkerboard pattern of pluses and minuses:

$$\begin{bmatrix} + & - & + & \cdots \\ - & + & - & \cdots \\ + & - & + & \cdots \\ \vdots & \vdots & \vdots & \ddots \end{bmatrix}$$

If the (i,j)-element in this matrix is a $+$, then multiply $\det A_{ij}$ by 1 to get C_{ij}; if it is a $-$, then multiply by a -1.

For example, let us again consider the matrix

$$A = \begin{bmatrix} 1 & 3 & -2 \\ -2 & 4 & 2 \\ 1 & -3 & -1 \end{bmatrix}$$

If we use a cofactor expansion along the second row, we find that

$$\det A = a_{21}C_{21} + a_{22}C_{22} + a_{23}C_{23}$$

$$= (-1)a_{21}\det A_{21} + (1)a_{22}\det A_{22} + (-1)a_{23}\det A_{23}$$

$$= (-1)(-2) \begin{vmatrix} 3 & -2 \\ -3 & -1 \end{vmatrix} + (1)(4) \begin{vmatrix} 1 & -2 \\ 1 & -1 \end{vmatrix}$$

$$+ (-1)(2) \begin{vmatrix} 1 & 3 \\ 1 & -3 \end{vmatrix}$$

$$= 2(-9) + 4(1) - 2(-6)$$

$$= -18 + 4 + 12 = -2$$

while if we expand along the third column, we get

$$\det A = a_{13}C_{13} + a_{23}C_{23} + a_{33}C_{33}$$
$$= (1)a_{13}\det A_{13} + (-1)a_{23}\det A_{23} + (1)a_{33}\det A_{33}$$
$$= (1)(-2)\begin{vmatrix} -2 & 4 \\ 1 & -3 \end{vmatrix} + (-1)(2)\begin{vmatrix} 1 & 3 \\ 1 & -3 \end{vmatrix}$$
$$+ (1)(-1)\begin{vmatrix} 1 & 3 \\ -2 & 4 \end{vmatrix}$$
$$= (-2)(2) + (-2)(-6) + (-1)(10)$$
$$= -4 + 12 - 10 = -2$$

■   Since we may use any row or column for a cofactor expansion, we will want to choose that row or column which makes our computations simplest. A row or column with a lot of zeros is a good choice.

Example 8.14. *Find the determinant of*

$$A = \begin{bmatrix} 3 & 0 & 1 & 4 \\ -2 & 1 & -1 & -3 \\ 0 & 0 & -2 & 1 \\ 0 & 0 & 1 & 1 \end{bmatrix}$$

Solution: Since there are three zeros in the second column, we expand along it:

$$\det A = 0 + (1)(1)\begin{vmatrix} 3 & 1 & 4 \\ 0 & -2 & 1 \\ 0 & 1 & 1 \end{vmatrix} + 0 + 0$$

To compute this 3×3 determinant, we expand along the first column, and find that

$$\det A = (1)(1)\left[(1)(3)\begin{vmatrix} -2 & 1 \\ 1 & 1 \end{vmatrix} + 0 + 0 \right]$$
$$= (1)(1)(1)(3)(-2 - 1) = -9$$

■   Notice that we have found incidentally that the matrix in the last example does have an inverse.

Although we can use determinants to find whether a given matrix A has an inverse and thus find whether a system of the form $AX = B$ has a solution, this is never done for large systems; computing a determinant for a large matrix is a very involved computation.

For example, to compute the determinant of a 15×15 matrix, one must compute the determinant of 15 14×14 matrices. But each determinant of a 14×14 matrix requires the calculation of the determinants of 14 13×13 matrices. Thus $15 \cdot 14 = 210$ 13×13 determinants must be calculated. Continuing on this path, we will need to compute

$15 \cdot 14 \cdot 13 \cdot 12 \cdot 11 \cdot 10 \cdot 9 \cdot 8 \cdot 7 \cdot 6 \cdot 5 \cdot 4 \cdot 3 = 653,837,184,000,$ 2×2 determinants.

In any case, calculating the determinant takes more effort than Gaussian elimination, which produces more information than the determinant.

Exercise 8.13. *Find the determinant of*:

(a)

$$A = \begin{bmatrix} 7 & -3 \\ -4 & 1 \end{bmatrix}$$

Solution:

(b)

$$A = \begin{bmatrix} 2 & 1 & 0 \\ -3 & 4 & 1 \\ 1 & 5 & 2 \end{bmatrix}$$

Solution:

(c)

$$A = \begin{bmatrix} -3 & 1 & 2 \\ 1 & 0 & 1 \\ -1 & -2 & 1 \end{bmatrix}$$

Solution:

(d)

$$A = \begin{bmatrix} 1 & -2 & 0 & 1 \\ -3 & 4 & 3 & -1 \\ 0 & -1 & 0 & 2 \\ -2 & 5 & 0 & 1 \end{bmatrix}$$

Solution:

Summary 8.4.

■ The determinant $\det A$ of a square matrix is a number which tells us whether the matrix A is invertible.

■ If $\det A = 0$ then A is not invertible, while if $\det A \neq 0$ A is invertible.

■ For a 2×2 matrix $A = \begin{bmatrix} a_{11} & a_{12} \\ a_{21} & a_{22} \end{bmatrix}$ we define $\det A = a_{11}a_{22} - a_{12}a_{21}$; for larger matrices we use cofactor expansion to define $\det A$.

End of Section 8.4. Problems

Problem 8.13. *Find the determinant of*:

(a)

$$A = \begin{bmatrix} 4 & -7 \\ -3 & 6 \end{bmatrix}$$

(b)

$$A = \begin{bmatrix} 1 & 2 & -1 \\ -1 & 4 & 3 \\ 2 & 7 & -2 \end{bmatrix}$$

(c)

$$A = \begin{bmatrix} -4 & 7 & -10 \\ 3 & -3 & 4 \\ 1 & 0 & -2 \end{bmatrix}$$

(d)

$$A = \begin{bmatrix} 1 & -3 & 0 & 1 \\ 3 & -1 & 4 & 1 \\ 0 & 0 & 0 & 2 \\ -2 & 4 & 1 & 3 \end{bmatrix}$$

Problem 8.14. *Find the determinants of the following matrices.*

(a)

$$A = \begin{bmatrix} 4 & -7 \\ -3 & 6 \end{bmatrix}$$

(b)

$$A = \begin{bmatrix} 1 & -9 \\ -5 & 9 \end{bmatrix}$$

(c)

$$A = \begin{bmatrix} 1 & -5 & 0 \\ 1 & -4 & -1 \\ 4 & 2 & -5 \end{bmatrix}$$

(d)

$$A = \begin{bmatrix} -1 & -9 & 4 \\ 0 & -7 & 4 \\ 8 & -8 & 1 \end{bmatrix}$$

(e)

$$A = \begin{bmatrix} 1 & -3 & 0 & 1 \\ 3 & -1 & 4 & 1 \\ 0 & 0 & 0 & 2 \\ -2 & 4 & 1 & 3 \end{bmatrix}$$

(f)

$$A = \begin{bmatrix} 1 & -2 & -1 & 1 \\ -3 & 1 & -1 & -1 \\ 1 & 0 & -2 & 2 \\ 2 & 0 & -1 & 2 \end{bmatrix}$$

(g)

$$A = \begin{bmatrix} 1 & 2 & 5 & 1 \\ 0 & 3 & 0 & 0 \\ 0 & 4 & 2 & 1 \\ 0 & 3 & 4 & 1 \end{bmatrix}$$

(h)

$$A = \begin{bmatrix} 1 & -1 & 4 & -6 \\ -1 & 3 & -5 & 6 \\ 0 & 2 & 0 & 0 \\ 0 & 1 & -6 & 3 \end{bmatrix}$$

(i)

$$A = \begin{bmatrix} 3 & 3 & -1 & 5 \\ -1 & -2 & -6 & 4 \\ -2 & 0 & 6 & -7 \\ 0 & 0 & -3 & 0 \end{bmatrix}$$

8.5. Applications of the Determinant

Cramer's Rule

It is possible to use determinants to solve systems of linear equations. To do this we will have to apply **Cramer's Rule**.

To show how this Rule is developed, we will solve a general system of two equations in two unknowns.

Consider the system of equations

$$a_{11}x + a_{12}y = b_1$$
$$a_{21}x + a_{22}y = b_2$$

If we let

$$A = \begin{bmatrix} a_{11} & a_{12} \\ a_{21} & a_{22} \end{bmatrix} \quad \text{and} \quad X = \begin{bmatrix} x \\ y \end{bmatrix} \quad \text{and} \quad B = \begin{bmatrix} b_1 \\ b_2 \end{bmatrix}$$

then the system may be written in its $AX = B$ form, and may be solved by finding $X = A^{-1}B$. Using the formula from Section 4, we find that

$$A^{-1} = \frac{1}{a_{11}a_{22} - a_{12}a_{21}} \begin{bmatrix} a_{22} & -a_{12} \\ -a_{21} & a_{11} \end{bmatrix}$$

so,

$$X = A^{-1}B$$

$$= \frac{1}{a_{11}a_{22} - a_{12}a_{21}} \begin{bmatrix} a_{22} & -a_{12} \\ -a_{21} & a_{11} \end{bmatrix} \begin{bmatrix} b_1 \\ b_2 \end{bmatrix}$$

$$= \frac{1}{a_{11}a_{22} - a_{12}a_{21}} \begin{bmatrix} b_1 a_{22} - b_2 a_{12} \\ -b_1 a_{21} + b_2 a_{11} \end{bmatrix}$$

$$= \begin{bmatrix} \frac{b_1 a_{22} - b_2 a_{12}}{a_{11}a_{22} - a_{12}a_{21}} \\ \frac{-b_1 a_{21} + b_2 a_{11}}{a_{11}a_{22} - a_{12}a_{21}} \end{bmatrix}$$

Thus we have found that the solution, if it exists, to this system is

$$x = \frac{b_1 a_{22} - b_2 a_{12}}{a_{11}a_{22} - a_{12}a_{21}}, \quad \text{and} \quad y = \frac{-b_1 a_{21} + b_2 a_{11}}{a_{11}a_{22} - a_{12}a_{21}}$$

Now we can use determinants to simplify this notation. First we recognize that we write x and y as

$$x = \frac{\begin{vmatrix} b_1 & a_{12} \\ b_2 & a_{22} \end{vmatrix}}{\begin{vmatrix} a_{11} & a_{12} \\ a_{21} & a_{22} \end{vmatrix}}, \quad \text{and} \quad y = \frac{\begin{vmatrix} a_{11} & b_1 \\ a_{21} & b_2 \end{vmatrix}}{\begin{vmatrix} a_{11} & a_{12} \\ a_{21} & a_{22} \end{vmatrix}}$$

Notice how we can create the matrices found in the numerators of these expressions: the matrix $\begin{bmatrix} b_1 & a_{12} \\ b_2 & a_{22} \end{bmatrix}$ is formed by replacing the first column (the column containing the coefficients of x in the system) with the column of constants from the system.

Likewise the matrix $\begin{bmatrix} a_{11} & b_1 \\ a_{21} & b_2 \end{bmatrix}$ is formed by replacing the second column (the column containing the coefficients of y in the system) with the column of constants from the system.

Notation

We develop some notation: let $A_x(B)$ be the matrix resulting when we replace the column containing the coefficients of x in the system by the column B containing the constants of the system.

Also let $A_y(B)$ be the matrix resulting when we replace the column containing the coefficients of x in the system by the column B containing the constants of the system.

We can summarize our research with the following theorem, called Cramer's Rule:

Theorem

Cramer's Rule:

Let A be an invertible 2×2 matrix.

For any matrix of constants B, the solution of the system $AX = B$ is given by

$$x = \frac{\det A_x(B)}{\det A}, \quad \text{and} \quad y = \frac{\det A_y(B)}{\det A}$$

Example 8.15. Find the solution, if it exists, to the system

$$x + 2y = 5$$
$$2x - 3y = 3$$

Solution: Here the matrix A and B are

$$A = \begin{bmatrix} 1 & 2 \\ 2 & -3 \end{bmatrix} \quad \text{and} \quad B = \begin{bmatrix} 5 \\ 3 \end{bmatrix}$$

so the matrices $A_x(B)$ and $A_y(B)$ are

$$A_x(B) = \begin{bmatrix} 5 & 2 \\ 3 & -3 \end{bmatrix} \quad \text{and} \quad A_y(B) = \begin{bmatrix} 1 & 5 \\ 2 & 3 \end{bmatrix}$$

Using Cramer's Rule, we compute that

$$x = \frac{\det A_x(B)}{\det A} = \frac{-21}{-7} = 3,$$
$$y = \frac{\det A_y(B)}{\det A} = \frac{-7}{-7} = 1$$

which is exactly the result we computed in Section 1.

■   Cramer's Rule may be extended to systems of n linear equations in n unknowns, as follows.

Let u be **any** of the variables from the system, and let $A_u(B)$ be the matrix which results from replacing the column of coefficients of u with the column of constants of the system. Then we find the following:

Definition

Cramer's Rule extended:

Let A be an invertible $n \times n$ matrix.

For any matrix of constants B, the value of the variable u

in the solution of the system

$AX = B$ is given by $u = \frac{\det A_u(B)}{\det A}$

8.5.i. Using Cramer's Rule to solve a system of three equations in three unknowns.

Example 8.16. *Use Cramer's Rule to find a solution to the system of linear equations*

$$
\begin{aligned}
x &+ y + 2z = 4 \\
2x & - z = 2 \\
3x &+ y + 3z = -2
\end{aligned}
$$

- **Solution:** Here the matrix A and B are

$$
A = \begin{bmatrix} 1 & 1 & 2 \\ 2 & 0 & -1 \\ 3 & 1 & 3 \end{bmatrix} \quad \text{and} \quad B = \begin{bmatrix} 4 \\ 2 \\ -2 \end{bmatrix}
$$

so the matrices $A_x(B), A_y(B)$, and $A_y(B)$ are

$$
A_x(B) = \begin{bmatrix} 4 & 1 & 2 \\ 2 & 0 & -1 \\ -2 & 1 & 3 \end{bmatrix}
$$

$$
A_y(B) = \begin{bmatrix} 1 & 4 & 2 \\ 2 & 2 & -1 \\ 3 & -2 & 3 \end{bmatrix}
$$

$$
A_z(B) = \begin{bmatrix} 1 & 1 & 4 \\ 2 & 0 & 2 \\ 3 & 1 & -2 \end{bmatrix}
$$

Using Cramer's Rule, we compute that

$$
x = \frac{\det A_x(B)}{\det A} = \frac{4}{-4} = -1,
$$

$$
y = \frac{\det A_y(B)}{\det A} = \frac{-52}{-4} = 13,
$$

$$
z = \frac{\det A_z(B)}{\det A} = \frac{16}{-4} = -4.
$$

Areas and Volumes

The determinant of a matrix may also be used to calculate areas of regions in the plane.

To begin, consider the parallelogram P in the plane with one corner at the origin $(0,0)$, one corner at the point (a,c) –*square*, and one corner at the point (b,d) –*circle*. These choices for three of the corners forces the fourth corner to lie at the point $(a+b, c+d)$ –*cross*. See the figure below.

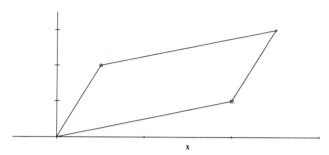

We will first consider the problem of finding the area of this parallelogram; as in the figure we assume that $a, b, c,$ and d are all positive. We add a few helpful line segments to create the following diagram:

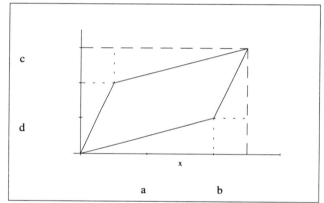

Notice that the area of the parallelogram P is equal to the area of the large rectangle minus the areas of the four triangles and two rectangles which surround P. Thus the area of the parallelogram P is:

$$\text{and } = (a+b)(c+d) - \tfrac{1}{2}ac - \tfrac{1}{2}bd - \tfrac{1}{2}ac - \tfrac{1}{2}bd - bc - bc$$

$$= (a+b)(c+d) - ac - bd - 2bc$$

$$= (ac + ad + bc + bd) - ac - bd - 2bc$$

$$= ad - bc$$

However, the quantity $ad - bc$ is just the determinant of the matrix $\begin{bmatrix} a & b \\ c & d \end{bmatrix}$. Notice that the columns of this matrix contain the coordinates of the corners (other than the origin) which we specified for the parallelogram. Thus we may suppose that the area of a given parallelogram may be written as the determinant of a particular matrix.

⌑ Note ⌑ what happens if we allow the corners of the parallelogram to lie in other quadrants than the first.

■ Following the same strategy as above, we could show that the areas of these parallelograms are always either $ad - bc$ or $-(ad - bc)$, whichever of these is positive.

We have discovered the following result.

> **Theorem**
>
> Let P be a parallelogram in the x-y plane with a corner at $(0,0)$.
> Suppose also that the corners of P adjacent to $(0,0)$ are (a,b) and (c,d).
> Then the area of the parallelogram P is $|\det A|$,
>
> where A is the 2×2 matrix $\begin{bmatrix} a & b \\ c & d \end{bmatrix}$.

Exercise *Example* **8.17.** *Find the area of the parallelogram with corners* $(0,0)$, $(1,2)$, $(3,1)$, *and* $(4,3)$.

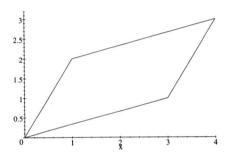

Solution: From the Figure, we see that the corners adjacent to $(0,0)$ are $(1,2)$ and $(3,1)$. Thus we find that $A = \begin{bmatrix} 1 & 3 \\ 2 & 1 \end{bmatrix}$ and the area of the parallelogram is $|\det A| = |(1)(1) - (2)(3)| = |-5| = 5$.

Exercise **8.14.** *Find the area of the parallelogram with corners* $(0,0)$, $(-2,3)$, $(5,2)$, *and* $(3,5)$.

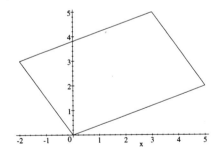

Solution:

It is possible to find the areas of triangles also, since the area of a triangle with corners $(0,0)$, (a,c), and (b,d) is one half of the area of the parallelogram having those points as three of its corners.

Exercise 8.15. *Find the area of the triangle with corners* $(0,0)$, $(-1,-2)$, *and* $(4,-1)$.

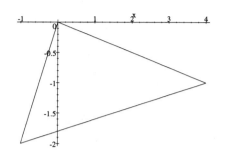

Solution:

Of course, we may ask if it is possible to find the area of **any** parallelogram or triangle using determinants, not just those shapes which happen to have a corner at $(0,0)$. This is in fact possible, if we use a technique called translation.

Example 8.18. *Find the area of the parallelogram with corners* $(-2,-1)$, $(-1,1)$, $(1,0)$, *and* $(2,2)$.

Solution: None of the corners is $(0,0)$, but we can move one of the corners to that point by adding (or subtracting) certain values from its coordinates. In this case, consider moving the point $(-2,-1)$ to the point $(0,0)$. We would add 2 to the x-coordinate and add 1 to the y-coordinate to make this happen. This is called **translating** a point. In order for the parallelogram to retain its shape, we add 2 to the x-coordinate of each of the other points, and add 1 to the y-coordinate of each of the other points. The figure below shows the original shape along with its translation; their areas are equal. Thus we have reduced the problem to one solved previously: that of finding the area of the parallelogram with corners $(0,0)$, $(1,2)$, $(3,1)$, and $(4,3)$. Thus by the above example, the area of the parallelogram is 5.

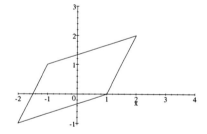 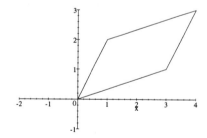

Exercise 8.16. *Find the area of the triangle with corners* $(4,-2)$, $(1,1)$, *and* $(-1,-3)$.

Solution:

If we look at three-dimensional space (volume), something similar happens. Consider a box whose sides are all parallelograms. This is called a **parallelepiped**. The cube is an example of a parallelepiped, as is a rectangular prism.

Example 8.19. *Find the volume of a cube with each side having length 2.*

- **Solution:** From geometry, we know that the volume is $2^3 = 8$.

  However, consider placing one corner of this cube at the point $(0,0,0)$ and three of its other corners at the points $(2,0,0)$, $(0,2,0)$, and $(0,0,2)$ (See the figure). Then we notice that

  $$\begin{vmatrix} 2 & 0 & 0 \\ 0 & 2 & 0 \\ 0 & 0 & 2 \end{vmatrix} = (2)\begin{vmatrix} 2 & 0 \\ 0 & 2 \end{vmatrix} = (2)(4) = 8,$$

  which is the volume of the cube.

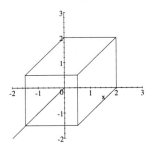

Example 8.20. *Find the volume of a rectangular prism with sides having lengths 2, 3, and 4.*

- **Solution:** From geometry, we again know that the volume is $2 \cdot 3 \cdot 4 = 24$. Now consider placing one corner of this cube at the point $(0,0,0)$ and three of its other corners at the points $(2,0,0)$, $(0,3,0)$, and $(0,0,4)$ (See the figure). Then we notice that

  $$\begin{vmatrix} 2 & 0 & 0 \\ 0 & 3 & 0 \\ 0 & 0 & 4 \end{vmatrix} = (2)\begin{vmatrix} 3 & 0 \\ 0 & 4 \end{vmatrix} = (2)(12) = 24,$$

  which is the volume of the prism.
  $(2,0,0)$, $(0,3,0)$, $(0,0,4)$

  Notice that in these cases, the volume may be expressed as a determinant; this fact is true for parallelepipeds in general.

- | Theorem |

  Let P be a parallelepiped with a corner at $(0,0,0)$.

  Suppose also that the corners of P adjacent to $(0,0,0)$ are (a,d,g), (b,e,h), and (c,f,i).

  Then the volume of the parallelepiped P is $|\det A|$,

  where A is the 3×3 matrix $\begin{bmatrix} a & b & c \\ d & e & f \\ g & h & i \end{bmatrix}$

Example 8.21. *Find the volume of the parallelepiped with one corner at* $(0,0,0)$ *and whose corners adjacent to* $(0,0,0)$ *are* $(1,2,0)$, $(0,3,1)$, *and* $(2,0,3)$.

- **Solution:**

By the formula this volume should be $|\det A|$, with $A = \begin{bmatrix} 1 & 0 & 2 \\ 2 & 3 & 0 \\ 0 & 1 & 3 \end{bmatrix}$

that is the volume is

$$|\det A| = \left| (1) \begin{vmatrix} 3 & 0 \\ 1 & 3 \end{vmatrix} \right| = (1)(9) = 9$$

We may find the volume of a triangular pyramid in a fashion analogous to the case in two dimensions, by taking one half of the volume of an appropriate parallelepiped. We may also handle an arbitrary parallelepiped or pyramid by translating it.

Example 8.22. *Find the volume of a triangular pyramid with corners* $(1,-1,1)$, $(3,-2,2)$, $(4,1,5)$, *and* $(1,-1,6)$.

- **Solution:** We begin by translating the point $(-1,1,-1)$ to the point $(0,0,0)$: we now have a pyramid with corners at $(0,0,0)$, $(2,-1,1)$, $(3,2,4)$, and $(0,0,5)$. We compute that the volume of the translated pyramid

is $\frac{1}{2}|\det A|$ with $A = \begin{bmatrix} 2 & 3 & 0 \\ -1 & 2 & 0 \\ 1 & 4 & 5 \end{bmatrix}$ thus the volume is

$$\frac{1}{2}|\det A| = \left| \frac{1}{2}(5) \begin{vmatrix} 2 & 3 \\ -1 & 2 \end{vmatrix} \right| = \left| \frac{1}{2}(5)(7) \right| = \frac{35}{2}$$

Exercise 8.17. *Use Cramer's Rule to find the solution of the system*:

(a) $\begin{aligned} 5x + 3y &= 7 \\ 2x - y &= 5 \end{aligned}$

(b) $\begin{aligned} x - y + z &= 1 \\ 2x + 2y + 3z &= 3 \\ y - 4z &= 13 \end{aligned}$

Solution:

Solution:

Exercise **8.18**. *Find the area of the parallelogram with corners* $(1,2)$, $(4,5)$, $(3,-2)$, *and* $(6,1)$.

Solution:

Exercise **8.19**. *Find the volume of a pyramid with corners* $(0,0,0)$, $(-1,2,5)$, $(2,2,2)$, *and* $(1,-3,4)$.

Solution:

Summary 8.5.

- We may use determinants to solve systems of linear equations using Cramer's Rule.
- Determinants may also be used to calculate areas of parallelograms and triangles and to calculate volumes of parallelepipeds and pyramids.

End of Section 8.5. *Problems*

Problem 8.15. *Use Cramer's Rule to find the solution of the system of linear equations:*

(a)

$$-2x + y = 4$$
$$7x - 11y = -5$$

(b)

$$3x + 2y - z = -3$$
$$-x + y = -2$$
$$2x - 3y + z = 9$$

Problem 8.16. *Find the area of the triangle with corners* $(-4,1)$, $(3,2)$, *and* $(2,-5)$.

Problem 8.17. *Find the volume of a parallelepiped with one corner at* $(0,0,0)$, *and with* $(1,3,2)$, $(-2,1,4)$, *and* $(1,-2,0)$ *as the corners adjacent to* $(0,0,0)$.

Problem 8.18. *Use Cramer's Rule to find the solution of the given system of linear equations.*

(a)

$$-2x + y = 4$$
$$7x - 11y = -5$$

(b)

$$x - 3y = 4$$
$$5x + 2y = -7$$

(c)

$$2x + y = -5$$
$$3x + 5y = 1$$

(d)

$$x - y = 7$$
$$-2x + 3y + z = -3$$
$$2y - z = 0$$

(e)

$$x + y + z = 3$$
$$-2x - y + 2z = 0$$
$$2x + 3z = 1$$

(f)

$$x - y + 5z = -1$$
$$9x + y - z = 3$$
$$5x + y - z = 4$$

Problem 8.19. *Find the area of the parallelogram with corners:*

(a)

$(0,0), (-1,2), (3,-1)$, and $(2,1)$.

(b)

$(0,0), (4,-3), (1,3)$, and $(5,0)$.

(c)

$(1,-2), (5,-3), (2,0)$, and $(6,-1)$.

Problem 8.20. *Find the volume of a parallelepiped with one corner at* $(0,0,0)$, *and with* $(1,3,2)$, $(-2,1,4)$, *and* $(1,-2,0)$ *as the corners adjacent to* $(0,0,0)$.

Problem 8.21. *Find the volume of a parallelepiped with one corner at* $(2,-1,3)$, *and with* $(-3,0,1)$, $(4,1,0)$, *and* $(1,-2,1)$ *as the corners adjacent to* $(2,-1,3)$.

Problem 8.22. *Find the volume of a pyramid with corners* $(0,0,0), (2,1,0), (-1,2,2)$, *and* $(-3,4,1)$.

Problem 8.23. *Find the volume of a pyramid with corners* $(1,2,-1), (4,0,1), (2,-1,3)$, *and* $(4,2,0)$.

End of Chapter 8. Problems

Problem 8.24. *Use the Method of Elimination to determine whether each of the following systems of equations has a solution; if the system has a solution, find it.*

(a)

$$x + 2y = 3$$
$$3x + 5y = 10$$

(b)

$$x - 3y = -1$$
$$-2x + 7y = 3$$

(c)

$$2x + 5y = 4$$
$$3x + y = -7$$

(d)

$$x + y - 2z = 0$$
$$-4y + z = 4$$
$$-4x + 9y + 5z = -13$$

(e)

$$x + 3y - z = -2$$
$$-2x - y + z = 3$$
$$3x + 4y - 2z = 1$$

(f)

$$3x - 5y + 2z = -3$$
$$x + 7y - z = 13$$
$$-2x + y + z = 2$$

Problem 8.25. *Solve the given systems of linear equations by converting the augmented matrix to reduced row echelon form.*

(a)

$$3x - 5y = 7$$
$$x + y = 5$$

(b)

$$8x + y = -1$$
$$-x + 2y = 15$$

(c)

$$x - 4y = -1$$
$$-2x + y = -5$$

(d)

$$x + 2y + z = -3$$
$$-x + y + 3z = 2$$
$$x - 4z = -7$$

(e)

$$3x + y + 2z = 10$$
$$2y - z = -15$$
$$-x + y + z = 3$$

(f)

$$3x + y - 2z = 1$$
$$y + z = -3$$
$$-6x + y + 7z = 2$$

(g)

$$x + y + z = 6$$
$$2x + 3y + 2z = 9$$
$$x - y + 5z = 8$$

(h)

$$x - 2y + z - 3w = 14$$
$$x - 3z + 2w = -3$$
$$-3x - y + z + 4w = 0$$
$$-2x + y + w = -12$$

(i)

$$x - 5y + 3z - 2w = 3$$
$$2x + z - w = -2$$
$$y - 2z + w = 1$$
$$2x + 11y - 6z + 3w = -5$$

Problem 8.26. *Find* $A + B$, $A - B$, *and* $5A - 2B$ *for the given matrices* A *and* B.

(a)
$A = \begin{bmatrix} 0 & 5 \\ -7 & 4 \end{bmatrix}$ and $B = \begin{bmatrix} 1 & 9 \\ -7 & -5 \end{bmatrix}$

(b)
$A = \begin{bmatrix} -3 & -8 & -1 \\ -8 & 9 & 1 \end{bmatrix}$ and $B = \begin{bmatrix} -2 & -7 & 6 \\ -8 & -1 & 5 \end{bmatrix}$

(c)
$A = \begin{bmatrix} -1 & -1 \\ -9 & -6 \\ -8 & 2 \end{bmatrix}$ and $B = \begin{bmatrix} 4 & 0 \\ 8 & 5 \\ -4 & 4 \end{bmatrix}$

(d)
$A = \begin{bmatrix} -6 & -3 & -5 \\ -4 & -7 & 8 \\ 4 & 3 & 0 \end{bmatrix}$ and $B = \begin{bmatrix} -1 & -8 & -4 \\ 7 & -2 & -5 \\ 0 & -1 & 0 \end{bmatrix}$

Problem 8.27. *Find (if possible)* AB *and* BA.

(a)
$A = \begin{bmatrix} 7 & -2 \\ -4 & 9 \end{bmatrix}$ and $B = \begin{bmatrix} -2 & 6 \\ -1 & 1 \end{bmatrix}$

(b)
$A = \begin{bmatrix} 0 & -6 & 0 \\ -5 & 6 & -5 \end{bmatrix}$ and $B = \begin{bmatrix} 3 & -1 \\ 0 & 8 \end{bmatrix}$

(c)
$A = \begin{bmatrix} 4 & -4 \\ 2 & -1 \\ -5 & 0 \end{bmatrix}$ and $B = \begin{bmatrix} -1 & -8 \\ -3 & 1 \end{bmatrix}$

(d)
$A = \begin{bmatrix} -1 & 6 \\ 4 & -3 \\ 2 & 3 \end{bmatrix}$ and $B = \begin{bmatrix} -2 & 8 & 0 \\ 2 & -3 & 9 \end{bmatrix}$

Problem 8.28. *Find the inverse of the given matrix if it exists.*

(a)

$$\begin{bmatrix} 1 & 2 \\ 5 & -3 \end{bmatrix}$$

(b)

$$\begin{bmatrix} 1 & -1 \\ 3 & -7 \end{bmatrix}$$

(c)

$$\begin{bmatrix} 2 & -1 \\ 3 & 4 \end{bmatrix}$$

(d)

$$\begin{bmatrix} 1 & 0 & 2 \\ 4 & 1 & 8 \\ 2 & -1 & 3 \end{bmatrix}$$

(e)

$$\begin{bmatrix} 3 & -1 & 2 \\ 1 & 4 & -1 \\ 1 & -9 & 4 \end{bmatrix}$$

(f)

$$\begin{bmatrix} 1 & 4 & 3 \\ 2 & 2 & 3 \\ 1 & 2 & 0 \end{bmatrix}$$

Problem 8.29. *Use the inverse of a matrix to solve the given system of linear equations.*

(a)

$$3x + y + 2z = 6$$
$$2x + y = -3$$
$$x - y + 3z = 0$$

(b)

$$x + 2y + 5z = 4$$
$$y + z = -2$$
$$2x - 3y + z = 6$$

(c)

$$2x + y - z = 1$$
$$5x + 2y - 3z = 0$$
$$2y + z = -3$$

(d)

$$x - 2y = -4$$
$$2x - 3y + z = 2$$
$$x + y + 5z = 8$$

Problem 8.30. *Find the determinants of the given matrices.*

(a)

$$A = \begin{bmatrix} 7 & -2 \\ -3 & -5 \end{bmatrix}$$

(b)

$$A = \begin{bmatrix} -4 & 1 \\ 4 & -8 \end{bmatrix}$$

(c)

$$A = \begin{bmatrix} 8 & 4 \\ -6 & 4 \end{bmatrix}$$

(d)

$$A = \begin{bmatrix} 9 & -3 & -7 \\ 0 & 5 & 1 \\ -1 & 2 & 0 \end{bmatrix}$$

(e)

$$A = \begin{bmatrix} 8 & 2 & 6 \\ -1 & 0 & 2 \\ 3 & 0 & 8 \end{bmatrix}$$

(f)

$$A = \begin{bmatrix} -2 & -8 & -8 \\ -5 & 6 & 2 \\ -3 & 5 & 3 \end{bmatrix}$$

(g)

$$A = \begin{bmatrix} 3 & -1 & 0 & 2 \\ 0 & -2 & 0 & 1 \\ 4 & 3 & 2 & -1 \\ 1 & -1 & 0 & 3 \end{bmatrix}$$

(h)

$$A = \begin{bmatrix} -2 & 5 & 0 & -1 \\ 1 & 3 & -3 & -2 \\ 0 & 0 & 1 & 1 \\ 3 & -2 & 4 & 0 \end{bmatrix}$$

Problem 8.31. *Use Cramer's Rule to find the solution of the given system of linear equations.*

(a)

$$3x + 5y = 11$$
$$4x - 2y = -4$$

(b)

$$2x + 3y = -1$$
$$4x + 7y = 5$$

(c)

$$2x + 2y - 5z = -3$$
$$x - 4y + 2z = 4$$
$$3x - 6y - 4z = 0$$

(d)

$$-x + 2y + z = -1$$
$$x + y + 2z = 2$$
$$3x + 2y + z = 1$$

Problem 8.32. *Find the areas or volumes of the given objects.*

a. *The parallelogram with corners* $(0,0)$, $(2,-1)$, $(6,2)$, *and* $(8,1)$.

b. *The parallelogram with corners* $(2,-1)$, $(-2,3)$, $(5,0)$, *and* $(1,4)$.

c. *The parallelepiped with one corner at* $(0,0,0)$, *and with* $(1,-2,0)$, $(0,3,0)$, *and* $(3,0,-1)$ *as the corners adjacent to* $(0,0,0)$.

d. *The parallelepiped with one corner at* $(1,-1,1)$, *and with* $(3,-4,2)$, $(4,1,-1)$, *and* $(0,-2,1)$ *as the corners adjacent to* $(1,-1,1)$.

Chapter 9.

Sequences, Counting Principles, and Probability

9.1. Introduction to Sequences

Consider the following two offers from a prospective employer. The question is: Which one do you choose, and why?

Offer 1: $25,000 for the first year, and a $3,100 raise per year for the next four years.?
Offer 2: $25,000 for the first year, and a 12% increase per year for the next four years?

It is typical to choose an offer on the basis of which pays the largest amount for the life of the contract. The following table allows you to compare the yearly salary under these two contracts, over the course of a five year period. It also allows you to compare the total five year salary under the two given offers.

Year	Offer 1	Offer 2
1	$25,0000	$25,000
2	$28,300	$28,000
3	$31,600	$31,600
4	$34,900	$35,392
5	$38,200	$42,470
Five year total	**$158,200**	**$162,462**

The above table represents two functions, S_1 and S_2. These two function can be represented by the following formulas:

$$S_1(y) = 25000 + (y - 1)3200$$
$$S_2(y) = 25000(1 \cdot 12)^{(y-1)}$$

Where y represents the year of a given salary.

These functions are special because their **domain** is a **subset** of the **set of positive integers**. Such functions are called a **sequence**. The first sequence, S_1, is an example of an **Arithmetic Sequence** and the second sequence, S_2, is an example of **Geometric Sequence**.

Definition

Infinite and finite sequence

A **sequence** is a **function** in which the domain is a set of consecutive positive integers starting with 1.

A sequence is called a **finite sequence**, if its domain is finite.

An **infinite sequence** has an infinite domain.

Example **9.1.** *Define each of the following sequences as either a finite or infinite sequence.*

(a) *The sequence of positive odd integers that are less than* 20 .

(b) *The sequence of all even positive integers.*

(c) *The sequence of all odd numbers.*

- **Solution**: Listing the elements of the given sequences shows that the first sequence is a finite sequence and the other two sequences are infinite sequences.

  (a) $1, 3, 5, 7, 9, 11, 13, 15, 17, 19$.

  (b) $2, 4, 6, 8, 10, 12, 14, 16, 18, 20, 22...$

  (c) $1, 3, 5, 7, 9, 11, 13, 15, 17, 19...$

Sequences are usually given by describing their general, or n-th, terms.
For example, the general term of the sequence defined by $f(x) = 2x$, for the domain $D = \{1,2,3,4,5,6\}$, is $a_n = 2n$.

Example 9.2. Identify the range values of the sequence given by $b_n = n + 3$, for the domain $D = \{1,2,3,4,5\}$.

- **Solution**: The range values are as follows.

$b_1 = 1 + 3 = 4$

$b_2 = 2 + 3 = 4$

$b_3 = 3 + 3 = 6$

$b_4 = 4 + 3 = 7$

$b_5 = 5 + 3 = 8$

Example 9.3. Find the first five terms of the sequence $a_n = (-2)^n$.

- **Solution**: The first five terms are as follows.

$a_1 = (-2)^1 = -2$

$a_2 = (-2)^2 = 4$

$a_3 = (-2)^3 = -8$

$a_4 = (-2)^4 = 16$

$a_5 = (-2)^5 = -32$

Example 9.4. Write an expression for the n-th term of each of the following sequences.

(a) $\frac{1}{2}, 2/3, 3/4...$

(b) $4, 8, 12, 16...$

(c) $5, 25, 125, 625...$

- **Solution**: The n-th terms can be written as follows.

(a)

$n =$	1	2	3	n
a_n	$a_1 = 1/2$	$a_2 = 2/3$	$a_3 = 3/4$	$a_n = ?$

It is clear that each term has a numerator that is 1 less than the denominator. This implies that $a_n = n/(n + 1)$.

(b)

$n =$	1	2	3	n
a_n	$a_1 = 4$	$a_2 = 8$	$a_3 = 12$	$a_n = ?$

It is apparent that each term is 4 times its position in the sequence. This implies that $a_n = 4n$.

(c)

$n =$	1	2	3	n
a_n	$a_1 = 5 = 5^1$	$a_2 = 25 = 5^2$	$a_3 = 125 = 5^3$	$a_n = ?$

It is clear that each term is 5 to the power of its position in the sequence. Therefore, $a_n = 5^n$.

Factorial Notation

Some sequences involve the product of the first n positive integers. The symbol $n!$ (n **factorial**) is used to name this product.

Example 9.5.

Let $a_1 = x^0$ and $a_n = na_{n-1}$. Find a_2, a_3, a_4 and a_n in term of x_0.

- **Solution**:

$$a_2 = 2a_1 = 2x_0$$
$$a_3 = 3a_2 = 3 \cdot 2x_0$$
$$a_4 = 4a_3 = 4 \cdot 3 \cdot 2 \cdot x_0$$
and
$$a_n = n \cdot a_{n-1}$$
$$= n \cdot (n-1) \cdot a_{n-2}$$
$$= n \cdot (n-1) \cdot (n-2) \cdot a_{n-3}$$
$$= n \cdot (n-1) \cdot (n-2) \cdot (n-3) \cdot a_{n-4}$$
$$\ldots$$
$$= n \cdot (n-1) \cdot (n-2) \cdot (n-3) \cdot (n-4)\ldots3 \cdot 2 \cdot x_0 = n!x_0$$

-
Definition
Factorial: If n is a positive integer, then n **factorial** ($n!$)   is defined to be $n! = 1 \cdot 2 \cdot 3 \cdot 4\ldots(n-1) \cdot n$ Zero factorial is defined as $0! = 1$

Example 9.6.

Compute the first four terms of the sequence $a_n = \dfrac{(n+1)!}{n!}$.

- **Solution**:

$$a_1 = \frac{(1+1)!}{1!} = \frac{2!}{1} = \frac{1 \cdot 2}{1} = 2$$
$$a_2 = \frac{(2+1)!}{2!} = \frac{3!}{2!} = \frac{1 \cdot 2 \cdot 3}{1 \cdot 2} = 3$$
$$a_3 = \frac{(3+1)!}{3!} = \frac{4!}{3!} = \frac{1 \cdot 2 \cdot 3 \cdot 4}{1 \cdot 2 \cdot 3} = 4$$
$$a_3 = \frac{(4+1)!}{4!} = \frac{5!}{4!} = \frac{4! \cdot 5}{4!} = 5$$

Example **9.7**. Find the first four terms of the sequence defined by $a_n = \dfrac{n!}{(2n)!}$

- Solution:

$$a_1 = \frac{1!}{2!} = \frac{1}{2}$$

$$a_2 = \frac{2!}{4!} = \frac{1 \cdot 2}{1 \cdot 2 \cdot 3 \cdot 4} = \frac{1}{12}$$

$$a_3 = \frac{3!}{6!} = \frac{1 \cdot 2 \cdot 3}{1 \cdot 2 \cdot 3 \cdot 4 \cdot 5 \cdot 6} = \frac{1}{120}$$

$$a_4 = \frac{4!}{8!} = \frac{4!}{4! \cdot 5 \cdot 6 \cdot 7 \cdot 8} = \frac{1}{1680}$$

Summation Notation:

Many applications in real life involve computing the sum of the first n terms of a sequence.

Example **9.8**. Consider a newly married couple saving for a house down payment of at least $10,000$. They deposit 300 per month in an account that earns 7% interest compounded monthly for 36 months. The balance for the n-th deposit in the account is

$$a_n = 300\left(1 + \frac{0 \cdot 07}{12}\right)^{(36-n)}$$

(a) Compute the first four terms of this sequence.

(b) Compute the balance in this account after 12 months, 24 months, and 36 months.

(c) How long would it take for this couple to accumulate $10,000$ in their account?

- Solution:

(a)

$$a_1 = 300\left(1 + \frac{0 \cdot 07}{12}\right)^{(35)} = 367 \cdot 73$$

$$a_2 = 300\left(1 + \frac{0 \cdot 07}{12}\right)^{(34)} = 365 \cdot 60$$

$$a_3 = 300\left(1 + \frac{0 \cdot 07}{12}\right)^{(33)} = 363 \cdot 48$$

$$a_4 = 300\left(1 + \frac{0 \cdot 07}{12}\right)^{(32)} = 361 \cdot 37$$

(b) Lets use S_n for the sum of the first n balances in their account, then

$$S_{12} = a_1 + a_2 + a_3 \ldots + a_{12}$$
$$= 367 \cdot 73 + 365 \cdot 60 + 363 \cdot 48 + \ldots + 344 \cdot 91$$
$$= \$3719 \cdot 90$$

(c) Computing S_{31} and S_{30} shows that it would take this couple 31 months to accumulate $\$10,161$.
You can write the above sums using the summation notation, where the Greek letter $\sum$ (capital sigma) is used to indicate that the sum of the first n terms of a sequence are added together:

$$S_{12} = \sum_{k=1}^{12} a_k \quad S_{24} = \sum_{k=1}^{24} a_k \quad S_{30} = \sum_{k=1}^{30} a_k \quad S_{36} = \sum_{k=1}^{36} a_k$$

Example **9.9.** Compute $\sum_{k=1}^{6} a_k$ where $a_k = \frac{n-1}{n}$.

- **Solution:**

$$\sum_{k=1}^{6} a_k = a_1 + a_2 + a_3 + a_4 + a_5 + a_6$$

$$= \frac{0}{1} + \frac{1}{2} + \frac{3}{4} + \frac{4}{5} + \frac{5}{6}$$

$$= \frac{0 + 30 + 40 + 45 + 48 + 50}{60}$$

$$= \frac{213}{60} = \frac{71}{20}$$

Exercise **9.1.** Evaluate the sum $\sum_{i=1}^{3} i^2$.

Solution:

Example **9.10.** Compute the sum of the first six terms of the sequence $a_k = 2k - 1$.

- **Solution:**

$$\sum_{k=1}^{6} a_k = (2 \cdot 1 - 1) + (2 \cdot 2 - 1) + (2 \cdot 3 - 1)$$

$$+ (2 \cdot 4 - 1) + (2 \cdot 5 - 1) + (2 \cdot 6 - 1)$$

$$= 1 + 3 + 5 + 7 + 9 + 11$$

$$= 36$$

End of Section 9.1. Exercises

Exercise 9.2. First write the first five terms for each of the following sequences and then graph the terms.

(a) $a_n = \frac{n-1}{n-2}$

Solution:

(b) $a_n = \frac{1}{n}$

Solution:

(c) $a_n = (-1)^n \cdot n$

Solution:

Exercise 9.3. First write the first five terms for each of the following sequences and then graph the terms.

(a) $a_1 = 3$; for $n \geq 2$, $a_n = 3a_{n-1}$

Solution:

(b) $a_1 = 100$; for $n \geq 2$, $a_n = \frac{a_{n-1}}{n+1}$

Solution:

Exercise 9.4. Write a short paragraph about some similarities and some differences of functions and sequences.

Solution:

Exercise 9.5. A friend of yours has difficulty with the concept of the general terms or the n-th term of a sequence. How would you explain this idea to your friend?

Solution:

Exercise 9.6. *Evaluate each of the following sums.*

(a) $\displaystyle\sum_{k=1}^{3} 3k - 1$

Solution:

(b) $\displaystyle\sum_{k=1}^{4} (-1)^k \cdot 2k$

Solution:

(c) $\displaystyle\sum_{k=1}^{9} \frac{1}{k}$

Solution:

(d) $\displaystyle\sum_{k=1}^{7} k^2$

Solution:

(e) $\displaystyle\sum_{k=1}^{5} (k-1)^3$

Solution:

Exercise 9.7. *Evaluate each factorial.*

(a) $7!$

Solution:

(b) $1!$

Solution:

(c) $5!$

Solution:

(d) $9!$

Solution:

(e) $12!$

Solution:

Exercise 9.8. *Evaluate each of the following.*

(a) $\frac{12!}{10}$

Solution:

(b) $\frac{8!7!}{5!14}$

Solution:

(c) $\frac{7!}{2!4!}$

Solution:

(d) $\frac{7!}{5!2!}$

Solution:

(e) $\displaystyle\sum_{k=1}^{5} (k-1)^3$

Solution:

9.2. Arithmetic Sequences

We have defined that a sequence is a function where the **domain** is a **set of consecutive positive integers**.

Consider the function $f(n) = 3 + dn$ defined over the set of positive integers. The **range values** of this function for the **domain** $\{1, 2, 3, 4, ...\}$ are as follows.

Domain:	1	2	3	n
Range:	$a_1 = f(1) = 3 + d,$	$a_2 = f(2) = 3 + 2d$	$a_3 = f(3) = 3 + 3d$	$a_n = f(n) = 3 + nd?$

The **range values** of this function create a **special sequence**. The **difference between each two consecutive terms** is d. Meaning that the n-th term, a_n, can be computed by adding d to the $(n-1)$th term, $a_n - 1$. This relationship can be described by the recursive formula:

$$a_n = a_n - 1 + d$$

or the explicit formula

$$a_n = a_1 + (n-1)d$$

where d is a fixed value called the common difference.

A sequence with this property is called an **Arithmetic Sequence**.

Definition

Arithmetic Sequence:

A sequence a_n is an Arithmetic Sequence if the difference between its consecutive terms is the same.

It means that

$a_{n+1} - a_n = d$

for all $n = 1, 2, ...$

$\mathcal{Example}$ **9.11.** *Show that the sequence where the n-th term* **is** $3n - 1$ *is an arithmetic sequence.*

- **Solution**: For this sequence the difference between two consecutive terms is
$$a_{n+1} - a_n = [3(n + 1) - 1] - [3(n) - 1]$$
$$= 3n + 3 - 1 - 3n + 1 = 3 \bullet$$

Since the **difference between two consecutive terms** for this sequence is a **fixed number**, the sequence **is** an arithmetic sequence.

$\mathcal{Example}$ **9.12.** *Show that the sequence where the n-th term is $3n^2 - 1$ is* **not** *an arithmetic sequence.*

- **Solution**: For this sequence the difference between two consecutive terms is
$$a_{n+1} - a_n = [3(n + 1)^2 - 1] - [3n^2 - 1]$$
$$= [3(n^2 + 2n + 1) - 1] - 3n^2 + 1$$
$$= 3n^2 + 6n + 3 - 1 - 3n^2 + 1$$
$$= 6n + 3$$

This shows that the differences between two consecutive terms for this sequence depends on the position of the consecutive terms. Consequently the differences between the consecutive terms are not the same.

Thus this sequence is not an arithmetic sequence.

Example 9.13. (a) Find a formula for the n-th term of the arithmetic sequence where the first two terms are 4 and 11.

(b) Find the 19-th term of this sequence.

- **Solution:** (a) The difference between the first two terms is

$$d = a_2 - a_1 = 11 - 7 = 4$$

We know that the explicit formula for the n-th term is $a_n = a_1 + (n-1)d$. Replacing 4 and 7 for a and d respectively in this formula gives

$$a_n = 4 + (n-1)7$$
$$a_n = 4 + n - 7$$
$$a_n = n - 3$$

(b) The 19-th term of this sequence is

$$a19 = 7(19) - 3$$
$$a19 = 133 - 3$$
$$a19 = 130$$

Example 9.14. Find the next three terms of the arithmetic sequence where the first two terms are 12 and 18.

- **Solution:** The common difference for this sequence is

$$d = a_2 - a_1 = 18 - 12 = 6$$

The use of the recursive formula, $a_n = a_n - 1 + d$, for the n-th term of this sequence produces the following results.

$$a_3 = a_2 + d = 18 + 6 = 24$$
$$a_4 = a_3 + d = 24 + 6 = 30$$
$$a_5 = a_3 + d = 30 + 6 = 36$$

It is noteworthy here that the n-th term of each of the above arithmetic sequences fits the form $a_n = (n-1)d + c$ where d is the common difference between consecutive terms and $c = a_1$.

Definition

- The n-th **term of an Arithmetic Sequence,**

  having the common difference d and the first term a_1, is

  $a_n = (n-1)d + a_1$

Example 9.15. The third term of an arithmetic sequence is 11, and the common difference is 4. Find a formula for the n-th term of this sequence.

- **Solution:** The n-th term of an arithmetic sequence can be computed from the formula

$$a_n = (n-1)d + a_1$$

Replacing 3 for n, and 4 for d in this formula gives us:

$$a_3 = (3-1)4 + a_1$$
$$= 2 \cdot 4 + a_1$$
$$= 8 + a_1$$
$$11 = 8 + a_1$$
$$\Rightarrow a_1 = 3$$

Now, replacing 4 for d and 3 for a_1 in the formula for the n-th term of an arithmetic sequence gives
$$a_n = (n-1)4 + 3$$
$$= 4n - 4 + 3$$
$$= 4n - 1$$

The Sum of the first n terms of an Arithmetic Sequence

Adding the first n terms of an arithmetic sequence , when n is large, can be very time consuming and tedious. Fortunately, we are able to develop a shortcut. Let us consider a general situation, where S_n represent the sum of the first n terms of the arithmetic sequence given by $ak = a_1 + (k-1)d$. Thus, S_n can be written as follows:

$$S_n = \sum_{k=1}^{n} a_1 + (k-1)d$$
$$= a_1 + (a_1 + d) + (a_1 + 2d) + ... + [a_1 + (n-1)d$$

Writing S_n in reverse order, and adding the two equations together gives us:
$$S_n = [a_1 + (n-1)d] + [a_1 + (n-2)d] + [a_1 + (n-3)d] ... + a_1$$
$$\Rightarrow 2S_n = [2a_1 + (n-1)d] + [2a_1 + (n-2)d] +$$
$$+ [2a_1 + (n-3)d] ... + 2a_1$$

The right hand side of this equation is the sum of n terms, each having the form
$$2a_1 + (n-1)d$$

Therefore,

$$2S_n = n[2a_1 + (n-1)d]$$

Solving this equation for S_n gives:
$$S_n = [2a_1 + (n-1)d]$$

Now, rewriting the right hand side of this equation as $[a_1 + a_1 + (n-1)d]$, and replacing a_n for $a_1 + (n-1)d$, we can get another shortcut formula for S_n:
$$S_n = \frac{n}{2}[a_1 + a_n]$$

Now, using the sigma notation, we can summarize our results as follows.

| Definition |

The sum of the first n terms of an arithmetic sequence,

a_n, with common difference d is

$$S_n = \sum_{k=1}[a_1 + (k-1)d]$$
$$S_n = \frac{n}{2}[2a_1 + (n-1)d]$$
$$S_n = \frac{n}{2}[a_1 + a_n].$$

Example **9.16**. *Find the sum of the first* 1000 *positive even integers.*

- **Solution**: Since $a_1 = 2, d = 2$, and $n = 1000$, replacing these values in the formula
$$S_n = \frac{n}{2}[2a_1 + (n-1)d]$$

results in

$$S_{1000} = \frac{1000}{2}[2 \cdot 2 + (1000 - 1)d]$$
$$= 1001000$$

Example **9.17**. *Find the sum of the first* 100 *terms of the arithmetic sequence where the* n-*th term is* $4n - 1$.

- **Solution**: Since $a_1 = 4(1) - 1 = 3$, and $a_{100} = 4(100) - 1 = 399$, replacing 100 for n, 3 for a_1 and 399 for a_n in the formula $S_n = \frac{n}{2}[a_1 + a_n]$ returns :

$$S_n = \frac{100}{2[3 + 399]}$$
$$= 50(402) = 20100$$

End of Section 9.2. Exercises

Exercise **9.9**. *For each of the following arithmetic sequences, compute the indicated term.*

(a) $a_3 = 9, d = 3, a_5$

Solution:

(b) $a_2 = -4, d = 5, a_5$

Solution:

(c) $a_5 = 12, d = -5, u_8$

Solution:

Exercise **9.10**. *Find* a_{10} *and* a_n *for each of the following arithmetic sequences.*

(a) $a_2 = 6, a_9 = 34$

Solution:

(b) $a_4 = 2k, a_5 = 4k$

Solution:

(c) $a_2 = -5, d = -5$

Solution:

Exercise 9.11. *Which of the following sequences is not an arithmetic sequence.*

(a) $2, 4, 8, 10...$

Solution:

(b) $6, 12, 18, 24...$

Solution:

(c) $2, 4, 8, 16...$

Solution:

Exercise 9.12. *A stack of canned soup in a grocery store has 50 in the bottom row, 49 in the next, and so on, with one can soup on the top row. The number of cans in the rows form an arithmetic sequence. How many are there?*

Solution:

Exercise 9.13. *The population of wild horses in a country was 2000, ten years ago. Each year the wild animal commission permits an increase of 500 in the population. What will be the population of wild horses in this country 3 years from now?*

Solution:

9.3. Geometric Sequences

Example **9.18.** *Suppose that someone offered you a job that pays $3 for the first day, $6 for the second day, $12 for the third day...each day paying you double what you were paid the preceding day. How much money would you be paid for seven days work?*

- **Solution**: On the first day you would be paid $3, on the second day you would be paid $6, on the third day you would be paid $12, and so on. These values form the following sequence of seven terms.

$$3, 6, 12, 24, 48, 96, 192$$

The differences between the consecutive terms of this sequence are not the same. But the ratios between consecutive terms of this sequence are the same.

Each successive term can be obtained by multiplying its preceding term by 2; it means that 6 is twice 3, and 12 is twice 6, and so on. Sequences with this property are called **geometric sequences**.

The total amount of money you would be paid for seven days work is the sum of the first seven terms of this sequence, as follows:

$$3 + 6 + 12 + 24 + 48 + 96 + 192 = 381$$

- | Definition |

  Geometric Sequence:

  A sequence a_n is a **geometric sequence** if the **ratios between** its **consecutive terms** are the **same**. That is,

  $\frac{a_{n+1}}{a_n} = r$ for all $n = 1, 2...$

Example **9.19.** *Show that the sequence where the n-th term is $4(2)^{n-1}$ is a geometric sequence.*

- **Solution**: The first few terms of this sequence are:

$$4, 8, 16, 32, 64, ...$$

The ratios between the consecutive terms of this sequence are: .

$$\frac{8}{4} = 2, \; \frac{16}{8} = 2, \; \frac{32}{16} = 2, \; \frac{64}{32} = 2.$$

Since these ratios are the same this sequence is a **geometric sequence**.

Example **9.20.** *Show that the sequence where the n-th term is $4(2)^{n-1} + 5$ is **not** a geometric sequence.*

- **Solution**: The first few terms of this sequence are:

$$9, 13, 21, 37, 69, ...$$

The ratios between the consecutive terms of this sequence are:

$$\frac{13}{9} = 1 \cdot 44, \; \frac{21}{13} = 1 \cdot 61, \; \frac{37}{21} = 1 \cdot 76, \; \frac{69}{37} = 1 \cdot 86.$$

Since these ratios are not the same this sequence is **not** a geometric sequence.

Example 9.21. *Find a formula for the n-th term of the geometric sequence where the first term is $a_1 = 3$ and where the common ratio is $r = 2$.*

- **Solution:**

$$a_1 = a_1 = 3$$
$$a_2 = a_1 r = 3 \cdot 2 = 6$$
$$a_3 = a_2 \cdot r = a_1 r \cdot r = a_1(r)^2 = 3(2)^2 = 12$$
$$a_4 = a_3 \cdot r = a_1(r)^2 \cdot r = a_1(r)^3 = 3(2)^3 = 24$$
$$a_5 = a_4 \cdot r = a_1(r)^3 \cdot r = a_1(r)^4 = 3(2)^4 = 48$$
$$\dots$$
$$a_n = a_n - 1 \cdot r = a_1(r)^{n-2} \cdot r = a_1(r)^{n-1} = 3(2)^{n-1}$$

Example 9.22. *Express the n-th term of the geometric sequence, where the first term is $a_1 = 8$ and where the common ratio is $r = \frac{1}{2}$, in terms of a_1 and r.*

- **Solution:**

$$a_1 = a_1 = 8$$
$$a_2 = a_1 r = 8 \cdot \left(\frac{1}{2}\right) = 4$$
$$a_3 = a_2 r = a_1(r)^2 = 8\left(\frac{1}{2}\right)^2 = 2$$
$$a_4 = a_3 r = a_1(r)^3 = 8\left(\frac{1}{2}\right)^3 = 1$$
$$a_5 = a_4 r = a_1(r)^4 = 8\left(\frac{1}{2}\right)^3 =$$
$$\dots$$
$$a_n = a_n - 1r = a_1(r)^{n-1} = 8\left(\frac{1}{2}\right)^{n-1}$$

In the above two examples, it is noteworthy that the n-th term of each of the geometric sequences, fits the form $a_n = a_1(r)^{n-1}$, where r is the common ratio between consecutive terms and a_1 is the first term of each sequence.

| Definition |

The n-th **term of a Geometric Sequence**, having the **common ratio** r, and the **first term** a_1, is

$$a_n = a_1(r)^{n-1}$$

Example 9.23. *The fifth term of a geometric sequence is 162, and the common ratio is 3. Find a formula for the n-th term of this sequence.*

- **Solution:** The n-th term of a geometric sequence can be computed from the formula

$$a_n = a_1(r)^{n-1}$$

Replacing 5 for n, and 3 for r in this formula gives us:

$$a_5 = a_1(3)^{5-1}$$
$$= a_1(3)^4$$
$$= 81a_1 = 162$$

Solving this equation for a_1 gives

$$a_1 = \frac{162}{81} = 2$$

Now, replacing 2 for a_1 and 3 for r in the formula for the n-th term of geometric sequence gives

$$a_n = 2(3)^{n-1}$$

The Sum of the first n terms of a Geometric Sequence:

Adding the first n terms of a **geometric sequence**, when n is large, is tedious. As with **arithmetic sequences**, we are able to develop a shortcut.

Let us consider a general situation, where S_n represents the sum of the first n terms of the geometric sequence given by $a_n = a_1(r)^{n-1}$.

Thus S_n can be written as follows:

$$S_n = \sum_{k=1}^{n} a_1(r)^{k-1}$$

$$= a_1 + a_1 r + a_1(r)^2 + \dots + a_1(r)^{n-1}$$

Writing $r \bullet S_n$, and subtracting S_n from it produces the following equality.

$$r \bullet Sn = a_1 r + a_1(r)^2 + \dots + a_1(r)^{n-1}$$

$$- Sn = -a_1 - a_1 r - a_1(r)^2 - \dots - a_1(r)^{n-1}$$

$$add$$

$$\Rightarrow r \bullet Sn - Sn = -a_1 + a_1(r)^n$$

$$\Rightarrow Sn(r-1) = a_1(r^n - 1)$$

Solving the above equation for S_n gives the following formula for the sum of the first n-terms of a geometric sequence where the common ratio is r and where the first term is a_1.

$$S_n = \frac{a_1(r^n - 1)}{r - 1}$$

Now, using the sigma notation, we can summarize our results as follows.

| Definition |

The **sum of the first** n terms of a **geometric sequence** a_n, with common ratio of r and the first term a_1, is

$$S_n = \sum_{k=1}^{n} a_1(r)^{k-1} = \frac{a_1(r^n - 1)}{r - 1}$$

Example **9.24.** *Find the sum of the first* 1000 *terms of the geometric sequence where the common ratio is* 2 *and where the first term is* 1.

- **Solution**: Since $a_1 = 1$, $r = 2$, and $n = 1000$, replacing these values in the formula

$$S_n = \frac{a_1(r^n - 1)}{r - 1}$$

results in

$$S_{1000} = \frac{1(2^{1000} - 1)}{2 - 1}$$

$$= 2^{1000} - 1$$

Example **9.25.** *Find the sum of the first* 100 *terms of the geometric sequence where the* n-*th term is* $a_n = 128(\frac{1}{2})^{n-1}$

- **Solution**: Since $a_1 = 128$, and $r = \frac{1}{2}$, replacing 100 for n, 128 for a_1 and $\frac{1}{2}$ for r in the formula

$$S_n = \frac{a_1(r^n - 1)}{r - 1}$$

returns

$$S_{100} = \frac{1024[(\frac{1}{2})^{10} - 1]}{(\frac{1}{2} - 1)}$$

$$= \frac{1024[(\frac{1}{10242}) - 1]}{(-\frac{1}{2})}$$

$$= \frac{1 - 1024}{(-\frac{1}{2})}$$

$$= (-1023)\frac{(-2)}{1}$$

$$= 2046$$

End of <u>Section</u> 9.3. Exercises

Exercise 9.14. *Use the appropriate sum formula to find the sum of the first six terms of each of the following sequences.*

(a)$a_3 = 4$,$d = 3$,

Solution:

(b)$a_3 = 8$,$r = 2$

Solution:

(c)$3, 9, 15...$

Solution:

Exercise 9.15. *Your father wants to start a college account for your newborn sister. If at the beginning of each month he puts $100 into an account that pays 6% interest compounded monthly, how much will your sister have in her account at the time she is 1 year old; 2 years old; 3 years old; 16 years old? Graph a sequence of the successive sums to display each years growth.*

Solution:

Exercise 9.16. *Place one penny on the first square of a chessboard, two pennies on the second square, four pennies on the third square. If you continued in this manner until you filled the chessboard, compute the total number of pennies needed to do this.*

Compute the total number of dollars needed to do this. Compare this amount with the annual budget of U.S.A. Which one is bigger? How many times?

Solution:

Exercise 9.17. *Assume that you have accepted a job at an annual salary of $25000 and expect to receive an annual increase of 6%. What will your salary be after four years of service in this job?*

Solution:

Exercise 9.18. *Find the n-th term of each of the following sequences.*

(a) $3, 9, 27...$ (b) $3, 6, 9...$ (c) $1, \frac{1}{2}, \frac{1}{4}, \frac{1}{8}...$ (d) $12, 18, 24...$

Solution: Solution: Solution: Solution:

Exercise 9.19. *Find the value of the first n-th terms of each of the following sequences.*

(a) $1, 3, 5...(2n-1)$ (b) $2, 4, 8...2^{(n-1)}$ (c) $1, 2, 3, 4...n$

Solution: Solution: Solution:

9.4. Counting Principles

Counting plays a major role in many areas of our daily life. In this section we shall consider special types of problems and develop general formulas for solving these problems.

When you count the number of cookies in a jar, you are really matching each, on a one-to-one basis, with the counting numbers $1, 2, 3, 4, ..., n$, for some counting number n. If the set of cookies in the jar, J, match up with the $set\{1, 2, 3, 4, ..., 50\}$, you can conclude that there are 50 cookies in the jar, J. The notation $n(J) = 50$ is often used to indicate that there are 50 elements in the jar or the $set J$.

Now, lets look at an example that will demonstrate a general counting principle.

Example **9.26**. *A fixed-price menu has the following choices*:

Appetizers	Entrees	Deserts	Drinks
Salad	Roast beef,	Pecan pie	Water
Soup	Fried Chicken	Cheese cake	
	T-bone steak		

■ How many possible meals can you order from this menu? List all of them.

Solution: Ordering a meal requires four different decisions:

Choose an Appetizer	2 Options
Choose an Entree	3 Options
Choose a Desert	2 Options
Choose a Drink	1 Option

The following tree diagram indicates that for each option of appetizer there are 3 choices of entree. And for each option of entree, there are two choices of desert. And for each of these $12 (= 2 \cdot 3 \cdot 2)$ choices there is one choice of drink. Therefore, there are a total of $2 \cdot 3 \cdot 2 \cdot 1 = 12$ different ways of ordering a meal.

This diagram gives the set of all possible orders:

1	2	3	4	5	6	7	8	9	10	11	12
Soup	Soup	Soup	Salad	Salad	Salad	Soup	Soup	Soup	Salad	Salad	Salad
Chicken	Beef	Steak	Chicken	Beef	Steak	Chicken	Beef	Steak	Chicken	Beef	Steak
Pie	Pie	Pie	Pie	Pie	Pie	Cake	Cake	Cake	Cake	Cake	Cake
Water	Water	Water	Water	Water	Water	Water	Water	Water	Water	Water	Water

Now, let us take a look at another example.

Example **9.27**. *A computer club consists of four girls and three boys. The members wish to elect a president from the girls and a Vice-President from the boys. Compute the number of different ways of electing these two officers and then list the set of all possible ways of electing these officers.*

- **Solution**: Let $G = \{g_1, g_2, g_3, g_4\}$ represent the set of girls, and $B = \{b_1, b_2, b_3\}$ represent the set of boys in this club. Now, it is easy to see that there are 4 choices for electing a girl president and there are 3 ways of electing a boy vice-president. The tree diagram below shows that there are 12 $(= 4 \bullet 3)$ different ways of electing a girl president and a boy vice-president.

↓		↓		← **START** →	↓			↓			
g_1		g_2			g_3			g_4			President
↓	↓	↓	↓	↓	↓	↓	↓	↓	↓	↓	
b_1	b_2	b_3	b_1	b_2	b_3	b_1	b_2	b_3	b_1	b_2	b_3	Vice-President

- Now, the above tree diagram gives the following list (the set of all possible ways of choosing a girl president and a boy vice-president for this club).

1	2	3	4	5	6	7	8	9	10	11	12
g_1b_1	g_2b_1	g_3b_1	g_4b_1	g_1b_2	g_2b_2	g_3b_2	g_4b_2	g_1b_3	g_2b_3	g_3b_3	g_4b_3

Multiplication Principle of Counting

If a task consists of a finite sequence of options in which there are n_1 selections for the first option, n_2 selections for the second option, n_3 choices for the third option, and so on, then the total number of distinct ways of making selections will be $n_1 \bullet n_2 \bullet n_3....$

Example **9.28**. *The International Communication Association (ICA) assigns three digit codes to regional districts. For example* 607 *code represents Cortland, Ithaca, and Homer. How many different regional codes are possible ?*

- **Solution**: The problem calls for three selections. Each selection requires choosing a digit out of $\{0, 1, 2, 3, 4, 5, 6, 7, 8, 9\}$ (10 options). Therefore, using the multiplication principle, there are $10 \bullet 10 \bullet 10 = 1000$ different regional codes.

Example **9.29**. *Suppose you decide to create a three-digit code without duplications of digit to represent regional districts. How many different regional codes can you create?*

- **Solution**: The task consists of three selections. The first selection can be done in ten different ways. Since repetition of digits are not allowed, the second selection requires choosing one digit out of the nine remaining digits. The third selection calls for choosing one digit from the eight remaining digits. Thus, by multiplication principle, there are $10 \cdot 9 \cdot 8 = 720$ different three digit codes with no digit repeated.

The above examples demonstrate a type of counting problem known as a **permutation**.

> | Definition |
>
> **Permutation**:
>
> An ordered arrangement of n distinct objects without repetitions, taken r objects at a time, is called a **permutation**.
>
> The symbol $P(n, r)$ represents the number of permutations of n distinct objects, taken r at a time, where $r \leq n$.

Example **9.30.** *How many three-letter codes, composed from the set of alphabet letters with no duplication of letters, are possible?*

Solution: Since duplication is not allowed, once a letter is selected, it may not be selected again. Therefore, the first letter can be chosen from the 26 letters of the alphabet, the second letter may oe selected from the remaining set of 25 letters. And the third letter can be chosen from the remaining set of 24 letters. Thus, the total number of different codes is $26 \cdot 25 \cdot 24 = 15,600$. Since we have taken 3 letters out of the set of 26 letters, without any duplications of a letter in a code, we conclude that there are $15,600$ permutations.

That is $P(26, 3) = 26 \cdot 25 \cdot 24 = 15,600$.

It is noteworthy that $p(26, 3) = 26 \cdot 25 \cdot 24 = 15,600$, has **three** factors beginning with 26 and each successive factor is decreased by 1. In general, for $P(n, r)$, there will be r factors beginning with n, as follows:

$$
\begin{aligned}
P(n, r) &= n \cdot (n - 1) \cdot (n - 2)...[n - (r - 1)] \\
&= n \cdot (n - 1) \cdot (n - 2)...(n - r + 1) \\
&= n \cdot (n - 1) \cdot (n - 2)...(n - r - 1) \cdot \frac{(n - r) \cdot (n - r + 1)...3 \cdot 2 \cdot 1}{(n - r) \cdot (n - r + 1)...3 \cdot 2 \cdot 1} \\
&= \frac{n!}{(n - r)!}
\end{aligned}
$$

Example **9.31.** *In how many ways can three numbers, from the set of digits $\{0, 1, 2, 3, 4, 5, 6, 7, 8, 9\}$, be arranged if the zero digit cannot be used in the middle position, and repetition of digits are allowed?*

Solution: First of all, we cannot use $P(10, 3)$ here, because of the restriction on the middle digit. In the middle position, we can use any digit but zero. This means that, we have just nine options for the middle digit. For the first and the third positions, we can choose any of the ten digits. Thus, using the multiplication counting principle, there are $10 \cdot 9 \cdot 10 = 900$ ways to arrange the digits according to the given specifications.

Combinations

You have learned how to find the number of ways of arranging r objects taken from a set of n objects. However, there are counting problems in which the order of arrangement is not significant.

Example **9.32**. *Suppose five applicants apply for three identical jobs. How many ways can the human resource's office fill these positions.*

- **Solution**: Selecting Applicant 1, Applicant 2, and Applicant 3; is the **same** as selecting any of the following combinations:

$$\text{Applicant1, Applicant3, and Applicant2} \quad = \quad \text{Applicant2, Applicant1, and Applicant3}$$
$$= \quad \text{Applicant2, Applicant3, and Applicant1}$$
$$= \quad \text{Applicant3, Applicant1, and Applicant2}$$
$$= \quad \text{Applicant3, Applicant2, and Applicant1}$$

- Therefore, there are only $\frac{5 \cdot 4 \cdot 3}{6} = 10$ ways to fill these positions. These ten selections are called the **Combinations** of five objects, taken three at a time. And is shown by $C(5,3)$. In this example, each combination of three applicants forms $3!$ **Permutations**.

  Thus, the number of combinations $C(5,3)$ can be found by dividing $P(5,3)$ by $3!$. That is,

$$C(5,3) = \frac{P(5,3)}{3!} = \frac{\frac{5!}{(5-3)!}}{3!} = \frac{5!}{(5-3)! \cdot 3!}$$

  Similarly, the number of combinations of n objects taken r at a time can be computed by dividing $P(n,r)$ by $r!$. That is,

$$C(n,r) = \frac{\frac{n!}{(n-r)!}}{r!} = \frac{n!}{(n-r)!r!}$$

- > **Definition**
  >
  > **Combination**:
  >
  > An arrangement of r objects without regarding the order, without repetition, of n distinct objects is called the **Combination** of n distinct objects, taken r objects at a time. Shown by $C(n,r)$ and computed by the following formula.
  >
  > $C(n,r) = \dfrac{n!}{(n-r)!r!}$

Example **9.33**. *How many different groups of five officers can be selected from a pool of* 12 *members?*

- **Solution**: Since the order of being selected in the group is not significant the problem can be solved by using the combination formula as follows:

$$C(12,5) = \frac{12!}{(12-5)!5!} = \frac{12 \cdot 11 ... 3 \cdot 2 \cdot 1}{7 \cdot 6 ... 3 \cdot 2 \cdot 1 \cdot 5 \cdot 4 \cdot 3 \cdot 2 \cdot 1} = 264$$

Example **9.34.** *From a class of* 20 *students,* 3 *are to be selected to work on a presentation. In how many different ways can such a group be selected.*

- **Solution:** Since the selection order is not significant the problem can be solved by using the combination formula as follows:

$$C(20,3) = \frac{20!}{(20-3)!3!} = \frac{20!}{17!3!} = 1140$$

End of <u>Section</u> 9.4. Exercises

Exercise **9.20.** *Find the value of each permutation.*

(a) $P(5,5)$

Solution:

(b) $P(6,2)$

Solution:

(c) $P(8,5)$

Solution:

(d) $P(9,7)$

Solution:

Exercise **9.21.** *Find the value of each combination.*

(a) $C(5,5)$

Solution:

(b) $C(6,2)$

Solution:

(c) $C(8,5)$

Solution:

(d) $C(9,7)$

Solution:

Exercise **9.22.** *How many three-character codes can be formed from the set*
$S = \{A,B,4,C,D\}$?

Solution:

Exercise 9.23. *A teenage girl has 7 blouses and 5 skirts. How many different outfits can she wear?*

Solution:

Exercise 9.24. *There are five people attending a movie and all want to sit in the same row.*

(a) *Find the number of ways the group can sit together if three people in the group want to sit together?*

Solution:

(b) *Find the number of ways the group can sit together if two people in the group refuse to sit together?*

Solution:

Exercise 9.25. *A warehouse receives a shipment of 20 televisions, four of which are defective. Six televisions randomly were chosen from the shipment and delivered to a showroom.*

(a) *In how many ways can the showroom receive no defective television?*

Solution:

(b) *In how many ways can the showroom receive all four defective televisions?*

Solution:

9.5. Probability

Probability is a major part of our everyday lives. For example, when we read a report on the stock market telling of an 80 percent increase for the last two years, we might change our investment plan from buying real estate to buying stocks. When we hear a weather forecast of 75 percent chance of rain for the next two days, we may change our plans from having a picnic to playing indoor soccer.

Managers who deal with inventories of teenager's clothing are counting on the chance that sales will exceed a certain level. When you begin to study for a test you may ask yourself, what are the chances the professor will ask a certain type of question.

Forecasting the future with perfect certainty is almost impossible. Consequently, we need to learn to cope with uncertainty. This is where the study of **probability** comes in. In dealing with social, economic and many other types of problems, we should have at least a basic understanding about the probable outcome of a decision. The following discussion will help develop a better understanding of the laws that govern probability, and hopefully reveal the separation between the possible and the probable.

■ Lets consider an experiment that has one or more possible outcomes, each of which is equally likely to occur.

▶ Rolling a die once, has six possible outcomes: landing with one dot face up (1), two face up (2), three face up (3), four face up (4), five face up (5), or six face up (6).

▶ Tossing a coin has two equally likely outcomes; these outcomes are: Head (H) and Tail (T).

Definition
Sample space:
The set S of **all possible outcomes** of a given
experiment is called the **sample space**
for the experiment.

Example **9.35.** *Consider the experiment of tossing a coin once. List the set of* **sample space** S *for this experiment.*

• **Solution**: $S = \{Head, Tail\} = \{H, T\}$.

Example **9.36.** *Consider the experiment of rolling a die once. List the set of sample space* S *for this experiment.*

• **Solution**: $S = \{1, 2, 3, 4, 5, 6\}$.

Example **9.37**. *Consider the experiment of tossing a coin first and then rolling a die. List the set of sample space S for this experiment.*

- **Solution**:

		$\leftarrow\leftarrow$ **START** $\rightarrow\rightarrow$		
	$\downarrow$			$\downarrow$
	H			T
$\downarrow$ $\downarrow$ $\downarrow$ $\downarrow$ $\downarrow$ $\downarrow$			$\downarrow$ $\downarrow$ $\downarrow$ $\downarrow$ $\downarrow$ $\downarrow$	
1 2 3 4 5 6			1 2 3 4 5 6	

- The tree diagram above shows that the **sample space** S consists of the outcomes:

  $S = \{H1, H2, H3, H4, H5, H6, T1, T2, T3, T4, T5, T6\}$.

Definition
Event:
Any **subset** of the sample space S is called an **event**.

Example **9.38**. *A single die is rolled. Write the following events in set notation and give the of elements for each event.*

    (a) $E6$: the number showing is less than 6.
    (b) $E1$: the number showing is greater than 1.
    (c) $E3$: the number showing is a multiple of 3.

- **Solution**: (a) $E6 = \{1, 2, 3, 4, 5\}$, $n(E6) = 5$.

  (b) $E1 = \{2, 3, 4, 5, 6\}$, $n(E1) = 5$.

  (c) $E3 = \{3, 6\}$, $n(E3) = 2$.

Definition
Probability of an event:
If an experiment has n equally likely outcomes, and if the number of ways that an event E can occur is m, then the **probability of the event** $E = P(E)$, is defined as follows.
$P(E) = \dfrac{Number\ of\ ways\ that\ E\ can\ happen}{Number\ of\ all\ possible\ outcomes\ of\ the\ experiment}$
$P(E) = \dfrac{n(E)}{n(S)}$
$P(E) = \dfrac{m}{n}$

Example **9.39.** *Assume you have drawn a card from a well-shuffled deck. Find the probability of the following events.*

    (a) E : the card is an ace.
    (b) E: the card is a diamond.
    (c) E : the card is an ace or a two.
    (d) E: the card is 5 diamond.

- **Solution**: (a) Since there are four aces in the deck of 52 cards, $n(E) = 4$ and $n(S) = 52$. Thus,

$$P(E) = \frac{n(E)}{n(S)} = \frac{4}{52} = \frac{1}{13}$$

(b) Since there are 13 diamonds in the deck of 52 cards, $n(E) = 13$ and $n(S) = 52$. Thus,

$$P(E) = \frac{n(E)}{n(S)} = \frac{13}{52} = \frac{1}{4}$$

(c) Since there are four aces and four twos in the deck of 52 cards, $n(E) = 8$ and $n(S) = 52$. Thus,

$$P(E) = \frac{n(E)}{n(S)} = \frac{8}{52} = \frac{2}{13}$$

(d) Since there are four fives in the deck of 52 cards and one of them is diamond $n(E) = 1$ and $n(S) = 52$ Thus,

$$P(E) = \frac{n(E)}{n(S)} = \frac{14}{52}$$

Example **9.40.** *A lottery game requires you to pick 6 different numbers from 1 to 99.*

    (a) How many ways are there to choose 6 numbers, if order is not important?
    (b) How many ways are there if order is important?
    (c) Assume order is important. What is the probability of picking all 6 numbers correctly to win the big prize?
    (d) Assume order is not significant. What is the probability of picking all 6 numbers correctly, regardless of their order, to win the big prize?

- **Solution**: (a) Since order is not important, using the combination formula will solve our problem as follows.

$$C(99,6) = \frac{99!}{(99-6)!6!} = \frac{99!}{93!6!} = \frac{99 \cdot 98 \cdot 97 \cdot 96 \cdot 95 \cdot 94}{6 \cdot 5 \cdot 4 \cdot 3 \cdot 2 \cdot 1} = 1120529256$$

(b) Since order is important using the permutation formula will solve this problem as follows.

$$P(99,6) = \frac{99!}{(99-6)!} = \frac{99!}{93!} = \frac{99 \cdot 98 \cdot 97 \cdot 96 \cdot 95 \cdot 94}{1} = 806781064320$$

(c) Let E be the event of picking all 6 numbers correctly in order, then $n(E) = 1$. Since order is significant, therefore the total number of outcomes in this experiment is $P(99,6) = 806781064320$. Thus, the probability of wining the big prize is as follows.

$$P(E) = \cdot \frac{n(E)}{n(S)} = \frac{1}{806781064320} = 0.0000000000124$$

(d) Let E be the event of picking all 6 numbers correctly regardless of their order, then

$$n(E) = 6! = 6 \cdot 5 \cdot 4 \cdot 3 \cdot 2 \cdot 1 = 6! = 720$$

Thus the probability of picking all 6 numbers correctly is as follows

$$P(E) = \frac{n(E)}{n(S)} = \frac{720}{806781064320} = 0.00000000089$$

> **Definition**
>
> The **complement** of E:
>
> The set of all outcomes in the **sample space** S
> that are not in the event E is called the **complement** of E.
> This set is written as E' and has the following two properties.
> $E \cup E' = S$. ($\cup$ is the union of sets)
> $E \cap E' = \emptyset = Empty\ set$. ($\cap$ is the intersection of sets)

Example **9.41**. *A single die is rolled. Write the following events in set notation and give the probability for each event.*

(a) $E6$: the number showing is less than 6.
(b) $E1$: the number showing is less than 1.
(c) $E3$: the number showing is a multiple of 3.
(d) E: the number showing is not a multiple of 3.

Solution: (a) $E_6 = \{1,2,3,4,5\}$, and $P(E_6) = \dfrac{n(E_6)}{n(S)} = \dfrac{5}{6}$

(b) $E_1 = \emptyset$, and $n(E_1) = 0$. Thus, $P(E_1) = 0$

(c) $E_3 = \{3,6\}$, and $P(E_3) = 2$. Therefore, $P(E_3) = \dfrac{2}{6} = \dfrac{1}{3}$

(d) Since the event E is the complement of the event E_3, therefore, $n(E) = 4$ and $P(E) = \dfrac{4}{6} = \dfrac{2}{3}$

> **Definition**
>
> **Mutually exclusive events**:
>
> If E and F are two events, then events E and F
> are called **mutually exclusive events**
> if $E \cap F = \emptyset$.

Example **9.42**. *Suppose you draw one card from a well-shuffled deck of 52 cards. What is the probability of the following outcomes?*

(a) *The card is a four or a diamond?*

(b) *The card is a 5 or a 7 ?*

Solution: (a) The events "drawing a four" and "drawing a diamond" are not mutually exclusive because it is possible to draw the four of diamond, an event satisfying both events. The probability of this event is:

$$P(four\ or\ diamond) = P(four) + P(diamond)$$
$$- P(four\ and\ diamond)$$
$$P(four\ or\ diamond) = \frac{4}{52} + \frac{13}{52} - \frac{1}{52}$$
$$= \frac{16}{52}$$
$$= \frac{4}{13}$$

(b) Since drawing a 5 and a 7 at the same time is not possible, therefore, these two events are mutually exclusive events and the probability of this event is

$$P(5 \text{ or } 7) = \frac{4}{52} - \frac{4}{52}$$
$$= \frac{8}{52}$$
$$= \frac{2}{13}$$

The above example suggests that for any events E and F, the probability of event E or event F can be computed by using the following formula

$$P(E \text{ or } F) = P(E \cup F) = P(E) + P(F) - P(E \cap F)$$

End of Section 9.5. Exercises

Exercise **9.26.** *For each of the following experiments, construct the sample space S in set notation:*

(a) A coin is tossed three times and sequences of H (heads) and T (tails) is recorded.

Solution:

(b) A die is rolled two times and sequences of numbers shown is recorded.

Solution:

Exercise **9.27.** *A die is tossed and the number on the top face is recorded. Let E, F, and G be the events:*

E: the number tossed is odd. F: the number tossed is divisible by 2. G: the number tossed is not greater than 4. Describe the following events in words:

(a) $F \cup G$

Solution:

(b) $E \cap F$

Solution:

(c) $E \cap E$

Solution:

(d) $F' \cap G$

Solution:

(e) $E \cup F \cup G$

Solution:

Exercise 9.28. *An organization that conducts political polls classifies people according to sex, income and political registration, thus:* **Sex**: *male(m), female(f).* **Income**: *high(h), average(a), low(l).* **Political registration**: *democrat (d), republican (r), independent (i). Thus a male with low income who is a democrat would be classified (m,l,d).*

(a) *A person is randomly chosen and classified according to the above classification. Describe the sample space for this experiment in set notation.*

Solution:

(b) *Let E be the event that the person chosen is either a republican or female. Express G in set notation.*

Solution:

(c) *Let F be the event that the person selected is a male registered as a democrat. Write F in set notation.*

Solution:

Exercise 9.29. *A pair of dice is tossed and the sum of the numbers on faces is recorded.*

(a) *What is the probability that the sum of numbers is 8?*

Solution:

(b) *What is the probability that the sum of numbers is 2?*

Solution:

(c) *What is the probability that the sum of numbers is even?*

Solution:

(d) *What is the probability that the sum of numbers is odd?*

Solution:

(e) *What is the probability that the sum of numbers is 0?*

Solution:

(f) *What is the probability that the sum of numbers is even or 5?*

Solution:

Exercise **9.30**. *A card is picked at random from a well-shuffled deck of* 52 *cards.*

(a) *What is the probability that the card is either a diamond or a club?*

> **Solution:**

(b) *What is the probability that the card is either a two or a ten?*

> **Solution:**

(c) *What is the probability that the card is either a five or a club?*

> **Solution:**

(d) *What is the odds in favor of drawing an ace?*

> **Solution:**

(e) *What is the odds against of drawing an ace?*

> **Solution:**

Exercise **9.31**. *A golf ball is chosen at random from a container that contains* 9 *white balls,* 8 *green balls, and* 5 *orange balls. Find the probability of the following events.*

(a) *The golf ball is green or white or black.*

> **Solution:**

(b) *The golf ball is not black.*

> **Solution:**

(c) *The golf ball is green or white and green.*

> **Solution:**

(d) *The golf ball is green or white or orange.*

> **Solution:**

Exercise **9.32**. *A baseball player with a batting average of* 0.425 *comes to bat. What are the odds in favor of his getting a hit?*

> **Solution:**

Exercise **9.33**. *A given family has* 6 *children. Compute the probabilities that the family has the following children. Assume that the probability of having a boy is* 0.5.

(a) *Exactly* 4 *girls.* (b) *At least* 3 *boys.* (c) *No more than* 5 *boys.*

| Solution: | Solution: | Solution:

(d) *No girls.* (e) *No boys.* (f) *Exactly* 4 *boys and* 2 *girls.*

| Solution: | Solution: | Solution:

Chapter 9. Summary

Definition

Infinite and finite sequence

A **sequence** is a **function** in which the domain is a set of consecutive positive integers starting with 1.

A sequence is called a **finite sequence**, if its domain is finite. An **infinite sequence** has an infinite domain.

The n-th **term of an Arithmetic Sequence**, having the common difference d and the first term a_1, is

$$a_n = (n-1)d + a_1$$

Arithmetic Sequence:

A sequence a_n is an Arithmetic Sequence if the difference between its consecutive terms is the same. It means that

$$a_{n+1} - a_n = d$$

for all $n = 1, 2, \ldots$

The **sum of the first** n **terms of an arithmetic sequence**, a_n, with common difference d is

$$S_n = \sum_{k=1}^{n} [a_1 + (k-1)d]$$

$$S_n = \frac{n}{2}[2a_1 + (n-1)d]$$

$$S_n = \frac{n}{2}[a_1 + a_n].$$

Sigma Notation:

The **sum of the first** n **terms of a sequence** with the general term a_k, is

$$\sum_{k=1}^{n} a_k = a_1 + a_2 + a_3 \ldots + a_n$$

Where k is called the **index of summation**, 1 is the lower limit of summation, and n is the upper limit of summation.

Geometric Sequence:

A sequence a_n is a **geometric sequence** if the **ratios between** its **consecutive terms** are the **same**. That is,

$$\frac{a_{n+1}}{a_n} = r \text{ for all } n = 1, 2 \ldots$$

The n-th **term of a Geometric Sequence**, having the **common ratio** r, and the **first term** a_1, is

$$a_n = a_1(r)^{n-1}$$

The **sum of the first** n terms of a **geometric sequence** a_n, with common ratio of r and the first term a_1, is

$$S_n = \sum_{k=1}^{n} a_1(r)^{k-1} = \frac{a_1(r^n - 1)}{r - 1}$$

Factorial:

If n is a positive integer, then n **factorial** ($n!$) is defined to be $n! = 1 \cdot 2 \cdot 3 \cdot 4 ... (n-1) \cdot n$

Zero factorial is defined as $0! = 1$

Permutation:

An ordered arrangement of n distinct objects without repetitions, taken r objects at a time, is called a **permutation**. The symbol $P(n, r)$ represents the number of permutations of n distinct objects, taken r at a time, where $r \le n$.

Combination:

An arrangement of r objects without regarding the order, without repetition, of n distinct objects is called the **Combination** of n distinct objects, taken r objects at a time. Shown by $C(n, r)$ and computed by the following formula.

$$C(n, r) = \frac{n!}{(n-r)! r!}$$

Sample space:

The set S of **all possible outcomes** of a given experiment is called the **sample space** for the experiment.

Probability of an event:

If an experiment has n equally likely outcomes, and if the number of ways that an event E can occur is m, then the **probability of the event $E = P(E)$**, is defined as follows.

$$P(E) = \frac{Number\ of\ ways\ that\ E\ can\ happen}{Number\ of\ all\ possible\ outcomes\ of\ the\ experiment}$$

$$P(E) = \frac{n(E)}{n(S)}$$

$$P(E) = \frac{m}{n}$$

The complement of E:

The set of all outcomes in the **sample space** S that are not in the event E is called the **complement** of E. This set is written as $E`$ and has the following two properties.

$E \cup E` = S$.   ($\cup$ is the union of sets)

$E \cap E` = \emptyset = Empty\ set$.   ($\cap$ is the intersection of sets)

Mutually exclusive events:

If E and F are two events, then events E and F are called **mutually exclusive events** if $E \cap F = \emptyset$.

End of Chapter 9. Problems

Problem 9.1. *Determine which of the following functions are sequences. For those that are, state the domain. For those that are not, explain why?*

(a) $f(x) = 2 - \frac{1}{x}$, where x is a positive integer.

(b) $f(x) = \frac{1}{x}$, where x is a natural number.

(c) $f(x) = x + 2$, where x is a real number between 1 and 10 .

(d) $f(x) = \frac{2}{x}$, where x is a non-zero number.

(e) $f(x) = \left(\frac{1}{2}\right)^x$, where x is a natural number.

Problem 9.2. *Find a formula for the general term of each sequence.*

(a) $a_1 = 2$, $a_2 = 5$, $a_3 = 8$,...a_n
(b) $a_1 = 0.3$, $a_2 = 0.03$, $a_3 = 0.003$,

(c) $a_1 = 5$, $a_2 = 11$, $a_3 = 17$,...
(d) $a_1 = 7$, $a_2 = \frac{7}{3}$, $a_3 = \frac{7}{9}$,...

(e) $a_1 = 2$, $a_2 = 3$...a_n
(f) $a_1 = -1$, $a_2 = 1$, $a_3 = -1$,

(g) $a_1 = 1$, $a_2 = 1 + 2$, $a_3 = 1 + 2 + 3$...a_n
(h) $a_1 = 1$, $a_2 = -2$, $a_3 = 4$,...a_n

(i) $a_1 = \frac{1}{3}$, $a_2 = \frac{4}{5}$, $a_3 = \frac{9}{7}$,...a_n

Problem 9.3. *Find the next two terms for the following sequences:*

(a) $1.5, 1.51, 1.515, 1.5151...$
(b) $\frac{2}{3}, \frac{4}{3}, \frac{8}{3}...$

(c) $\frac{1}{2}, \frac{2}{3}, \frac{3}{4}...$
(d) $\frac{1}{1}, \frac{-1}{4}, \frac{1}{9}, \frac{-1}{16}...$

(e) $1, 2, 3, 5, 8...$
(f) $1, -3, 9, -27...$

(g) $-1, 3, -9, 27...$
(h) $2, 5, -8, 11...$

(i) $6, 12, 20, 30, 42...$
(j) $\frac{1}{2}, \frac{1}{8}, \frac{1}{32}, \frac{1}{128}, \frac{1}{512}...$

Problem 9.4. *Compute the following:*

(a) $\sum_{n=1}^{4} 2^n$ (b) $\sum_{n=1}^{4} 2^{-n}$ (c) $\sum_{n=1}^{6} \left(\frac{1}{2}\right)^n$ (d) $\sum_{n=1}^{5} (-2)^n$ (e) $\sum_{n=1}^{10} (-1)^n$

Problem 9.5. *Express each of the following series using the* $\sum_{n=1}^{l}$ *notation:*

(a) $1 + 2 + 3 + 4 + 5 + ... + 10$
(b) $2 + 4 + 8 + 16 + 32...$
(c) $0.2 + 0.02 + 0.002 + 0.0002...$

(d) $\frac{1}{2} + \frac{2}{3} + \frac{3}{4} + \frac{4}{5} + \frac{5}{6}...$
(e) $1 + \frac{4}{5} + \frac{9}{7} + \frac{16}{9} + \frac{25}{11}...$

Problem 9.6. *Graph each of the following sequences:*

(a) $1, \frac{4}{3}, \frac{6}{4}, \frac{8}{5}, \frac{10}{6}, ... \frac{2n}{n+1}...$
(b) $1, \frac{1}{2}, \frac{1}{4}, \frac{1}{8}, ... \frac{1}{2^{n-1}}...$

(c) $2, \frac{8}{3}, 3, \frac{16}{5}, \frac{22}{6}, ...2 + \frac{2(n-1)}{n+1}...$ (d) $1, 2, 4, ...2^{n-1}, ...$

(e) $1, \frac{1}{2}, \frac{1}{3}, \frac{1}{4}, ... \frac{1}{n}...$

Problem 9.7. *Determine which of the following are arithmetic sequences and which are geometric sequences. Give the common difference of each arithmetic sequence; and the common ratio of each geometric sequence.*

(a) $5, 8, 11, 14...$ (b) $1, 4, 16, 64...$ (c) $3, 2, 1, 0...$

(d) $2, 1, \frac{1}{2}, \frac{1}{4}...$ (e) $2, 4, 6, 8...$ (f) $a, ab, ab^2, ab^3...$

Problem 9.8. *Find the sum of the following sequences:*

(a) $2 + 4 + 6 + \dots + 217$ (b) $3 + 5 + 7 + \dots + (2n-1)$ (c) $1 + 3 + 5 + \dots + (2n-1)$

(d) $3 + \frac{3}{2} + \frac{3}{4} + \dots + \frac{3}{2^n}$ (e) $\frac{1}{2} + 1 + 2 + \dots \left(\frac{1}{2}\right) \cdot (2)^{n-1}$ (f) $1 + \frac{1}{3} + \frac{1}{9} + \dots + \frac{1}{3^{n-1}}$

Problem 9.9. *Find the sum of each of the following infinite series:*

(a) $1 + \frac{1}{3} + \frac{1}{9} + \dots$ (b) $6 - \frac{1}{3} + \frac{1}{12} - \frac{1}{48} + \dots$ (c) $1 + \frac{1}{2} + \frac{1}{4} + \dots$

(d) $12 + 6 + 3 + \dots$ (e) $\sum_{n=1}^{\infty} 3\left(\frac{1}{2}\right)^{n-1}$ (f) $\sum_{k=1}^{\infty} \left(-\frac{1}{2}\right)^{k-1}$

Problem 9.10. *Find the number of all possible permutations of the given letters.*

(a) A (b) A, B (c) A, B, C

(d) A, B, C, D (e) A, B, C, D, E, F

Problem 9.11. *A bag contains* 8 *chips numbered* 1 *through* 8. *Two chips are selected and their numbers are written down.*

(a) *These numbers are added. How many ways can the sum of* 10 *be obtained.*

(b) *These numbers are subtracted. How many ways can a difference of* 2 *be obtained.*

Problem 9.12. *A combination lock will open when the right choice of four numbers from* 1 *to* 40 *inclusive, is selected. What is the chance of opening the combination lock the first try?*

Problem 9.13. *The officers of a students club will be filled from a group of* 12 *students. In how many ways can a president, a vice-president, a secretary and a treasurer be selected for the club?*

Answers to Exercises, Chapter 1.

1(a) $x^3, -2$ (b) $\frac{1}{2}y, -3x$ (c) $\sqrt{x}, 2x, -3x^5, 19$. **3**(a) $10^{-2} = \frac{1}{10^2}$. (b) $\frac{1}{x^{-2}} = x^2$. (c) $\frac{x^{-4}}{x^{-3}} = \frac{x^3}{x^4}$. **5**(a) $\frac{1}{x^4y^6}$. (b) $\frac{16}{x^4}$. **7**(a) $16 = 2^4$. (b) $9 = 3^2$. (c)

$90 = 2 \times 3^2 5$. **9**(a) $\frac{3 \times 5^2 7 \times 11}{7 \times 11} = 3 \times 5^2$, yes. (b) $\frac{3 \times 107}{5}$, no. **11**(a) $3 \times 5^2 7 \times 11$. (b) $\frac{x}{y^3}z^3$. (c) $\frac{5}{2^3}4^3 = 40$. (d) $\frac{3 \times 5^2 7 \times 11}{7 \cdot 5} = 3 \times 5 \times 11$.

(e) $\frac{3 \times 5^2 7 \times 11}{2^5}$. (f) $\frac{2^4 3 \times 5 \times 7}{3^2} = \frac{2^4 \times 5 \times 7}{3}$, no. **13**(a) No. For $x = -1$, $x^3 + 1 = 0$. Therefore x is restricted. (b) Yes. x is not restricted in

$\frac{1}{x^2 + 1}$. **15**(a) 2. (b) 1. (c) 4. **17**(a) $x^2 - 9x - 4$. (b) $-5x^3 - 2x^2 - x$. **19**(a) $-24x^2 - 66x - 27$. (b) $2x^2 - 2$. **21**(a) $x^2(yx + z)$. (b) $2x - 2 = 2 \cdot x^{-1}$(a

prime polynomial). **23**(a) $3(x + y)$. (b) $2x^3(z - 3)$. (c) $(3x + 5)(x + 1)^2$. (d) $\frac{2x + 3}{\sqrt{x}}$. (e) $2x - 4 = 2 \cdot x^{-2}$. (f) $-\frac{-3x + 4 + 3x^2}{\sqrt{x}}$. (g) $\frac{x}{\sqrt{(x^2 + 1)}}$. (h)

$x^2(x + 1)^2$. (i) $3\frac{x}{\sqrt{(x + 3)}}$. (j) $4x^2(x + 1)$. (k) $\frac{1}{4}(x - 7)$. (l) $(-x + 2)(3x - 2)$. **25**(a) 16. Not prime, factorable $(x + 1)(3x - 1)$. (b) 81. Not prime,

factorable $(x + 5)(x - 4)$. **27**. $b^2 - 4ac = 13 = -\frac{5}{6} - \frac{1}{6}\sqrt{13} = r$ and $-\frac{5}{6} + \frac{1}{6}\sqrt{13} = s \Rightarrow 3\left(x - \left(-\frac{5}{6} - \frac{1}{6}\sqrt{13}\right)\right)\left(x - \left(-\frac{5}{6} + \frac{1}{6}\sqrt{13}\right)\right)$.

29(a) -8. Prime and factored. (b) $7.0 = 3 + \sqrt{7} = r$ and $3 - \sqrt{7} = s \Rightarrow \frac{1}{2}\left(x - \left(3 - \sqrt{7}\right)\right)\left(x - \left(3 + \sqrt{7}\right)\right)$. (c) -20. Prime and factored. (d)

$25 \Rightarrow -(x + 1)(x - 4)$.

Answers to Problems, Chapter 1.

1.1(a) Terms: $3x$ and -1. (Coefficients 3 and -1.). (b) $1x^2y$, $-2xy^{-1}$, and $-x$. (1, -2, and -1.). (c) $2y^{-1}\sqrt{x + 4}$ and $3x^2y$. (2 and 3.). (d) $-3x^2$, x, and 4. $(-3, 1, \text{ and } 4.)$. (e) $\sqrt{x}\sqrt{y}, -3$, and x^2. $(1, -3, \text{ and } 1.)$. **1.3**(a) No. (b) Yes 2^4x. (c) No. (d) Yes 3^2x^4. (e) No. (f) No. (g) Yes x^3. **1.5**(a)

$3 \cdot 3 \cdot 2 \cdot 2 \cdot 2 \cdot 2$. (b) $3 \cdot 43$. (c) $-1 \cdot 2 \cdot 151$. (d) $13 \cdot 3 \cdot 2$. (e) $11 \cdot 3 \cdot 7 \cdot 2 \cdot 2$. (f) $13 \cdot 11 \cdot 3 \cdot 3$. (g) $3 \cdot 29 \cdot 41$. **1.7**. $\frac{17 \cdot 3 \cdot 3 \cdot 2}{17} = 2 \cdot 3 \cdot 3$.

1.9. Yes, since $105 = 7 \cdot 15 = 7 \cdot 3 \cdot 5$. **1.11**(a) $\frac{2s^2}{13xy}$. (b) $4x^8r^3$. (c) $\frac{x^2y^2}{12rs^3}$. **1.13**(a) Yes. (b) Yes. (c) Yes. (d) Yes. (e) Yes. (f) Yes. (g) No. (h)

No. (i) No. (j) No. **1.15**(a) $-x^2 - 3x$. (b) $\frac{1}{2}x^3 + 1$. (c) $x^2 + x$. (d) $\frac{1}{2}x^3 + x - 2$. (e) $-2x^2 + 1$. **1.17**(a) $4x^2 - 3x + 1$. (b) $2x^2 - 2x$. (c) $4x^3 - x$. (d) -4.

(e) $5x^2 - 2x + 4$. **1.19**(a) 1. (b) 3. (c) 0. (d) 0. (e) 5. **1.21**. Failed to apply the distributive property correctly. $-(2x + 1) = -2x - 1$.

Therefore, $3x - (2x + 1) = 3x - 2x - 1 = x - 1$. **1.23**(a) $x^2 - 2x - 3$. (b) $a^2 - b^2$. (c) $9x^2 + 6x + 1$. (d) $-x^2 - 3x + 10$. (e) $-2x^2 + 9x + 5$. (f) $2x^2 - 1$.

(g) $(1 - u)(u + 1) = 1 - u^2$. (h) $3x^2 - 7x + 4$. (i) $x^2 - 2x + 1$. (j) $(u + w)(r + s) = ur + us + wr + ws$. (k) $(z - 1)(3 - 2z) = -2z^2 + 5z - 3$. (l)

$-2x^2 - 7x + 4$. **1.25**. $2(3x^2 + 1)$. **1.27**. $-1(x - 3)$. **1.29**. $x(x + 2)(x^2 + 1)$. **1.31**. $x^2\left(-1 + \frac{1}{x}\right)$. **1.33**. $x^3\left(-1 - \frac{3}{x} + \frac{1}{x^3}\right)$. **1.35**(a) Discriminant:

$256 = 16^2$. Factorization $[(2x - 4)(2x + 4)]$. (b) $0 = 0^2$. $[(x - 1)(x - 1)]$. (c) $64 = 8^2$. $[(-x + 1)(x + 7)]$. (d) $1 = 1^2$. $[(2x - 7)(x - 3)]$. (e)

$0 = 0^2$. $[-1(3x - 1)(3x - 1)]$. (f) $361 = 19^2$. $[(5x + 4)(x - 3)]$. (g) $49 = 7^2$. $[(3x - 2)(2x + 1)]$. (h) $289 = 17^2$. $[(6x - 1)(x - 3)]$. (i) $169 = 13^2$.

$[(-2x + 1)(x - 7)]$. (j) $0 = 0^2$. $[(2x - 3)(2x - 3)]$. **1.37**(a) $:r = -2$; $s = -3$: $(x + 2)(x + 3)$. (b) $r = \frac{1}{2}; s = 3$: $-2\left(x - \frac{1}{2}\right)(x - 3)$. (c)

$r = 3; s = -3$; $(x - 3)(x + 3)$. (d) $r = \frac{1}{5}; s = -\frac{1}{2}$: $10\left(x - \frac{1}{5}\right)\left(x + \frac{1}{2}\right)$. (e) $r = \sqrt{5}; s = -\sqrt{5}$: $10\left(x - \sqrt{5}\right)\left(x + \sqrt{5}\right)$. **1.39**(a) $\frac{x^2}{3^4}$. (b) $5^2x^8y^6$.

(c) $5^{\frac{1}{4}}x^{\frac{5}{4}}$. (d) $\frac{5}{x^{\frac{1}{2}}}$. (e) $\frac{3^5 \cdot 5}{x^4}$. (f) $\frac{w^2}{2xz^3}$. (g) $\frac{2p^3}{w^3}$. **1.41**. $2340 = 13 \cdot 5 \cdot 3 \cdot 3 \cdot 2 \cdot 2$. (a) Yes. (b) Yes. (c) Yes. (d) Yes. (e) No. (f) Yes. **1.43**.

$\frac{x^3y^{-1}z^2}{xy^2z} = \frac{x^2z}{y^3}$. **1.45**. $x^2 + 2 - x$. **1.47**. $-x + 6 - x^2$. **1.49**(a) $9x^2 + 6x + 1$. (b) $-28x^2 + x + 2$. (c) $\frac{1}{4}x^2 - \frac{1}{9}$. (d) $8x^2 - 2x - 21$. (e) $-2x^2 + 4x + 6$.

(f) $x^2 + 2x + 1$. (g) $4x^2 - 4x + 1$. **1.51**. $(x + 4)(2x - 1)$. **1.53**(a) $x^4 + 2x^3 - 2x^2 - 2x + 1$. (b) $x^3 + 3x^2 + 3x + 1$. **1.55**. $-1(2 - x)$. **1.57** $\frac{1}{2}x(1 + 2x)$.

1.59 $2x^3\left(1 - \frac{1}{2x^2} + \frac{1}{2x^3}\right)$. **1.61**. $(2x - 1)^{-1}[3 + (2x - 1)^3]$. **1.63**(a) Common base: y. Least degree: $[-1.]$ (b) $(x + 1)$ y. [1]. (c) x. $[\frac{1}{2}]$. (d) x.

$[-\frac{3}{2}]$. (e) xyz. $[x = -1, y = 3, z = -1]$. **1.65**(a) $x = 1$, a double root. (b) $x = \sqrt{3}$ and $x = -\sqrt{3}$. (c) $x = \frac{1}{2}$ and $x = \frac{1}{3}$. (d) $x = \frac{3}{2}$ and $x = 1$. (e)

$x = 3$ and $x = \frac{1}{2}$. **1.67**(a) $(-5x + 1)(x - 2)$. (b) $(3x + 1)(3x - 1)$. (c) $(6x - 1)(x + 2)$. (d) Prime. (e) $(2x - 1)(4x - 2)$.

Answers to Exercises, Chapter 2.

2.1(a) 5. (b) $-\frac{43}{5}$. **2.3**(a) $S = \left\{-\frac{1}{7}\right\}$. (b) $S = \phi$. **2.5**(a) $x = \frac{2a + 2b}{a + b - c}$. (b) $x = \frac{2s - 2a + d}{d}$. **2.7**(a) $-6 + 5i$. (b) $-2 + 53i$. **2.9**(a) $-\frac{43}{41} + \frac{23}{41}$.

(b) $\frac{13}{58} - \frac{47}{58}$. **2.11**(a) $\left\{0, -\frac{7}{9}\right\}$. (b) $\left\{\pm\frac{\sqrt{15}}{3}i\right\}$. **2.13**(a) $1 \pm \sqrt{5}$. (b) $-1 \pm \frac{\sqrt{6}}{2}$. **2.15**(a) $1 \pm \frac{\sqrt{6}}{2}$. (b) $1 \pm \sqrt{5}i$. (c) $\left\{\frac{2}{3}, -1\right\}$. (d) $\left\{\frac{1}{2}, -\frac{5}{2}\right\}$.

2.17(a) $(-\infty, 2)$. (b) $[10, \infty)$. (c) $\left(-\infty, \frac{1}{4}\right)$. (d) $(-3, \infty)$. **2.19**(a) $(-\infty, -4] \cup [3, \infty)$. (b) $(-\infty, -2) \cup \left(\frac{1}{3}, \infty\right)$. **2.21**(a) $\left[-\frac{5}{3}, 3\right]$. (b) $(-4, -1)$. **2.23**.

8, 12, 16. **2.25**. 56mph; 76mph. **2.27**. 3hrs. **2.29**. $20' \times 20'$.

Answers to Problems, Chapter 2.

2.1(a) T. (b) F. (c) T. (d) F. (e) T. (f) T. **2.3**(a) $\frac{10}{7}$. (c) all reals. (e) ϕ. **2.5**(a) 1. (c) 6. (e) ϕ. **2.7**(a) $1 + 3i$. (c) $4 + i\sqrt{2}$. (e) $3 + 4i$. **2.9**(a)

$-\frac{17}{10} + \frac{11}{10}i$. (c) -34. (e) $\frac{84}{205} - \frac{187}{205}i$. **2.11**(a) $\{5, -2\}$. (c) $\left\{\frac{3}{4}, \frac{1}{5}\right\}$. (e) $\left\{3, -\frac{1}{2}\right\}$. **2.13**(a) $\left\{0, -\frac{2}{3}\right\}$. (c) $\left\{-\frac{1}{2}, -2\right\}$. (e) $\left\{-1 \pm \frac{\sqrt{10}}{2}\right\}$.

2.15(a) $\left\{-\frac{1}{2}, -2\right\}$. (c) $\left\{\frac{1}{6}, -3\right\}$. (e) $\left\{\frac{4}{3}, \frac{2}{3}\right\}$. **2.17**(a) $(-\infty, -2]$. (c) $\left(\frac{22}{3}, \infty\right)$. (e) $\left[\frac{13}{8}, \infty\right)$. **2.19**(a) $(-\infty, -2]$. (c) $\left(-\infty, \frac{3}{2}\right)$. (e) $\left(-\infty, -\frac{5}{2}\right)$.

2.21(a) $(-2, 5) \cup (5, \infty)$. (c) $(-\infty, -7] \cup [-3, \frac{1}{3}]$. (e) $(-3, 1)$. **2.23**(a) $\{8, -2\}$. (c) $\{-2, 5\}$. (e) $\left\{-2, 4, 1 \pm i\sqrt{7}\right\}$. **2.25**(a) $(-\infty, -5) \cup (-1, \infty)$. (c)

$(-\infty, \frac{2}{3}] \cup [2, \infty)$. (e) $(-\infty, -\frac{4}{3}] \cup [4, \infty)$. **2.27**. 6, 7 and 8. **2.29**. 8 and 9. **2.31**. 75 mph and 85 mph. **2.33**. 20ml @ 30%; 30ml @ 70%. **2.35**. 12 lbs @

$1.20; 24 lbs. @ $1.80. **2.37**. $8,000 @ 8%; $4,000 @ 10%. **2.39**. Aaron 104 hrs; Ben 65 hrs. **2.41**. 80 lbs. @ $1.20; 40 lbs. @ $1.8. **2.43**. The

square is 11×11; rectangle is 7×14. **2.45**(a) $\left\{-2, \frac{2}{3}\right\}$. (c) $-\frac{1}{2} \pm \frac{\sqrt{11}}{2}$. (e) $\pm\frac{1}{2}$. **2.47**(a) $(-\infty, 4)$. (c) All real numbers. (e) $[1, 6]$. **2.49**(a)

$\{2,-5\}$. (c) $[-1,4]$. (e) $(-\infty,-1) \cup (5,\infty)$. **2.51**. $\$7,000$. **2.52**. $\frac{9}{2}$ days; 9 days.

Answers to Exercises, Chapter 3.

3.1 a).independent variable: $f(x)$. dependent variable $[x]$. (b) $f(x)$. $[x]$. (c)$H(u)$. $[u]$. **3.3**(a) S: Given a number, produce a number which is the given number times itself added to one. (b). H: Given a number, produce twice the given number plus three. (c) F: Given a number, produce the negative of the given number added to the given number times itself.

3.5.

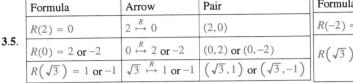

Formula	Arrow	Pair
$R(2) = 0$	$2 \overset{R}{\mapsto} 0$	$(2,0)$
$R(0) = 2$ or -2	$0 \overset{R}{\mapsto} 2$ or -2	$(0,2)$ or $(0,-2)$
$R(\sqrt{3}) = 1$ or -1	$\sqrt{3} \overset{R}{\mapsto} 1$ or -1	$(\sqrt{3},1)$ or $(\sqrt{3},-1)$

Formula	Arrow	Pair
$R(-2) = 0$	$-2 \overset{R}{\mapsto} 0$	$(-2,0)$
$R(\sqrt{3}) = 1$ or -1	$\sqrt{3} \overset{R}{\mapsto} 1$ or -1	$(\sqrt{3},1)$ or $(\sqrt{3},-1)$

3.7. $S\left(-\frac{1}{2}\right) = -1\frac{1}{2}$, $S(0) = -1$, $S(4) = 3$, $S(6) = 5$. **3.9**(a) $f(0) = 4$. (b) $f(1) \approx 2$. (c) $f(3) \approx -4$. (d) $z = -3$ or $z = 2$. (e)$-3 < z < 2$ or $(-3,2)$. (f) $z < -3$ or $2 < z :(-\infty,-3)$ or $(2,\infty)$. **3.11**(a) -1 and $\frac{1}{2}$. (b) $(-\infty,-1)$ and $\left(\frac{1}{2},\infty\right)$ (c) $\left(-1,\frac{1}{2}\right)$. (d) $(-\infty,-1]$ and $\left[\frac{1}{2},\infty\right)$. (e) $\left[-1,\frac{1}{2}\right]$. (f) 1. (g) 0 (h) 2. (i) -4.

3.13.

3.15(a)

3.15(b)

There are many possible correct answers to the first two questions.

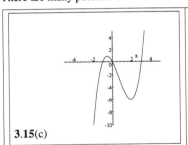

3.15(c)

3.15(d)

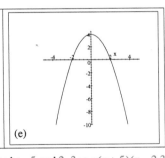

(e)

3.15.(f)

3.17. Prime quadratic polynomial. **3.19.** The zeros of the function appear to be -5 and $2.3. \Rightarrow a(x+5)(x-2.3)$ looking at a third point on the graph. $(0,35)$. Then $35 = -a(0-5)(0-2.3) \Rightarrow a$ is -3.0435 If we expand$-3.0435(x+5)(x-2.3) = -3.0435x^2 - 8.2175x + 35.0$. This verifies that the approximation is close. To get an accurate factorization in this case it is sufficient to use "foil" to get $(-3x+7)(x+5)$. **3.21.** $(x-4)(x+4)$

3.23.

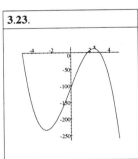

3.25. Graph $x^2 - 12x$. Therefore, $0 < x < 12$ or interval $(0,12)$.

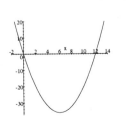

3.27(a) The graph of $(x-1)(x-3)(x-5)$.is:

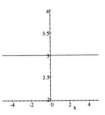

The graph of $.|(x-1)(x-3)(x-5)|$ is:

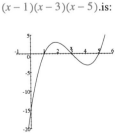

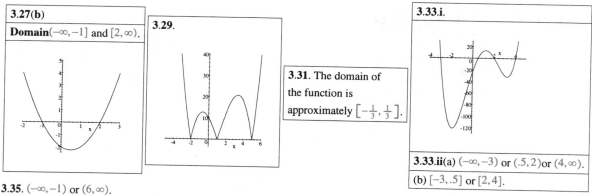

3.27(b)

Domain$(-\infty, -1]$ and $[2, \infty)$.

3.29.

3.31. The domain of the function is approximately $\left[-\frac{1}{3}, \frac{1}{3}\right]$.

3.33.i.

3.33.ii(a) $(-\infty, -3)$ or $(.5, 2)$ or $(4, \infty)$.

(b) $[-3, .5]$ or $[2, 4]$.

3.35. $(-\infty, -1)$ or $(6, \infty)$.

Answers to <u>Problems</u>, Chapter 3.

3.1.

$0 \xrightarrow{T} \frac{0}{-2}$	$(0,0)$	$T(0) = 0$
$-1 \xrightarrow{T} \frac{-1}{-3}$	$(-1, \frac{1}{3})$	$T(-1) = \frac{1}{3}$
$-2 \xrightarrow{T} \frac{-2}{-4}$	$(-2, \frac{1}{2})$	$T(-2) = \frac{1}{2}$
$\frac{1}{4} \xrightarrow{T} \frac{\frac{1}{4}}{\frac{1}{4}-2}$	$(\frac{1}{4}, -\frac{1}{7})$	$T(\frac{1}{4}) = -\frac{1}{7}$
$x \xrightarrow{T} \frac{x}{x-2}$	$(x, \frac{x}{x-2})$	$T(x) = \frac{x}{x-2}$

3.3.

$0 \xrightarrow{Pro} 0$	$(0,0)$	$Pro(0) = 0$
$-1 \xrightarrow{Pro} -1$	$(-1,-1)$	$Pro(-1) = -1$
$-2 \xrightarrow{Pro} -8$	$(-2,-8)$	$Pro(-2) = -8$
$\frac{1}{4} \xrightarrow{Pro} -\frac{11}{16}$	$(\frac{1}{4}, -\frac{11}{16})$	$Pro(\frac{1}{4}) = -\frac{11}{16}$
$x \xrightarrow{Pro} -3x^2 - 2x$	$(x, -3x^2 - 2x)$	$Pro(x) = -3x^2 - 2x$

3.5. The entries in the chart may be determined in any manner. **3.7. Given a number, produce**....the given number plus 2. **3.9. Given a number, produce**... the absolute value of the number. **3.11. Given a number, produce**... the number that is half the given number plus three. **3.13 Given a number, produce** (a) ...twice the given number plus one. (b) ...three minus the given number. (c) ...zero times the given number and then add ten. (d) ...twice the given number plus one. (e) ...twice the product of the given number times itself. (f)... four times the given number. (g) ...the number ten. (h) ...negative two times the given number squared and then add two. (i) ...negative one times the given number. (j) ...three divided by the given number. (k) ...the given number times itself minus one.

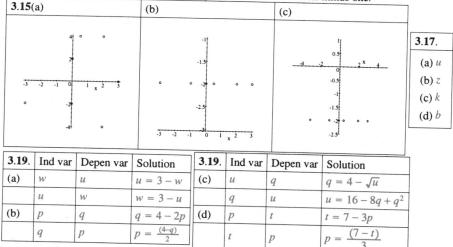

3.15(a) **(b)** **(c)**

3.17.

(a) u

(b) z

(c) k

(d) b

3.19.	Ind var	Depen var	Solution
(a)	w	u	$u = 3 - w$
	u	w	$w = 3 - u$
(b)	p	q	$q = 4 - 2p$
	q	p	$p = \frac{(4-q)}{2}$

3.19.	Ind var	Depen var	Solution
(c)	u	q	$q = 4 - \sqrt{u}$
	q	u	$u = 16 - 8q + q^2$
(d)	p	t	$t = 7 - 3p$
	t	p	$p = \frac{(7-t)}{3}$

3.21(a) i. $x = -\frac{1}{2}$, 1, 4. ii. $x = 0$. iii. $x = -1$, 2. iv. $x = -1, \frac{-1}{2}, 1, 2, 4$. v. $x = \frac{-1}{2}, 0, 1$, 4. vi. $x = -1, \frac{-1}{2}, 1, 2, 4$. vii. No solutions. viii. No solutions. ix. No solutions. x. $-.5, 0, 1, 4$. **3.21(b)** i. $x = -\frac{9}{2}, -3, -1, 1, 4$. ii. No solutions. iii. No solutions. iv. $x = -\frac{9}{2}, -3, -1, 1, 4$. v. : $x = -\frac{9}{2}, -3, -1, 1, 4$. vi. $x = -\frac{9}{2}, -3, -1, 1, 4$ vii. No solutions. viii. No solutions. ix. No solutions. x. $x = -\frac{9}{2}, -3, -1, 1, 4$.. **3.21(c)** i. -3, 1, 4. ii. $-2, -1, 0$. iii. 2, 3. iv. $-3, 1, 2, 3, 4$. v. $-3, -2, -1, 0, 4$. vi. $-3, -2, 0, 1, 2, 3, 4$. vii. No solution. viii. No solution. ix. No solution. x. $-3, -2, -1, 0, 1, 2, 4$. **3.23(a)** Independent variable:x. Dependent variable: $[j]$. (b) u. $[w]$. (c) x. $[P]$. (d) no zeros. **3.25(a)** $x = -2, -1, 4$. (b) $x = 0, 2$. (c) no zeros. **3.27.**(a) -5. (b) -2. (c) No value. (d) No value. (e) No values. (f) $(-\infty, \infty)$. (g) $(-\infty, \infty)$. (h) Approximately -17. (i) No. (j) Yes.

3.29.	3.31(a)	
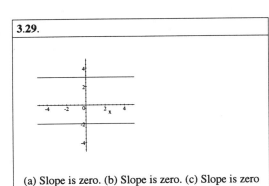 (a) Slope is zero. (b) Slope is zero. (c) Slope is zero	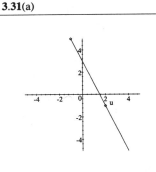	(b) Slope is -2. (c) $u = \frac{3}{2}$ produces the height 0, so we add the point $\left(\frac{3}{2}, 0\right)$ to the plot. (d) $\left(-\infty, \frac{3}{2}\right)$ or $u < \frac{3}{2}$. (e) $\left(\frac{3}{2}, \infty\right)$ or $u > \frac{3}{2}$

3.33.(a) Degree is zero. (b) Slope is zero. (c) $(4, -3)$ and $\left(\frac{7}{2}, -3\right)$ are two possible answers. (d) $m = 0$. (e) $c = -3$. (f) $mx + c = 0$ has no solutions.

3.35.	(d)
(a) The degree is 2. (b) The graph is not a line. (c) $x = 3$, $x = 5$. (e) $x = 3$, $x = 5$ (f) $(3, 5)$ or equivalently $3 < x < 5$	(graph)

3.37.	i.	
	(a)	$g(u) = -\frac{1}{2}u - 3 = -\frac{1}{2}(u - (-6))$
	(b)	$h(z) = 2z - 6 = 2(z - 3)$
	ii.	
	(a)	$u = -6$
	(b)	$z = 3$

3.41.(a)	3.45(b) $(-\infty, 1]$ or $[2, \infty)$
(graph)	(graph)

3.39.	i.	ii.
(a)	$1, -1, 3, -3$	$f(x) = 1(x-1)(x+1)(x-3)(x+3)$
(b)	$-1, 3, \frac{1}{2}$	$g(x) = -2\left(x - \frac{1}{2}\right)(x+1)(x-3)$
(c)	$-2, -\frac{1}{2}, \frac{1}{2}, 2, 3$	$f(x) = 4(x+2)\left(x + \frac{1}{2}\right)\left(x - \frac{1}{2}\right)(x-2)(x-3)$
(d)	$\frac{1}{3}, \frac{4}{3}, 3$	$h(u) = 9\left(x + \frac{1}{3}\right)\left(x - \frac{4}{3}\right)(x-3)$
(e)	$-1, -1, -1;$	$f(x) = 1(x+1)(x+1)(x+1)$

(b) $x = 3$.	(c) $(-\infty, 1]$ or $[2, \infty)$

3.43(a) The graph is **V-shaped**, points below the x-axis moved to positions of the same magnitude above the axis. **3.47**(a) $(-\infty, \infty)$. (b) No values of x. (c) No values of x.

3.49.				3.51.
$1 \xrightarrow{E} -2$	$(1, -2)$	$E(1) = -2$		(a) Given a number, produce the reciprocal of the given number.
$\frac{-1}{64} \xrightarrow{E} -\frac{129}{32}$	$\left(\frac{-1}{64}, -\frac{129}{32}\right)$	$E\left(\frac{-1}{64}\right) = -\frac{129}{32}$		(b) Solution: Given a number, produce the number which when
$3 \xrightarrow{E} 2$	$(3, 2)$	$E(3) = 2$		multiplied by itself yields the given number.
$x \xrightarrow{E} 2x - 4$	$(x, 2x - 4)$	$E(x) = 2x - 4$		

3.53(a) $h(0) = 0$. (b) $1 \xrightarrow{h} 84$. (c) $h(2) = 136$. (d) $\left(\frac{1}{2}sec, 46ft\right)$. (e) $h(t) = -16t^2 + 100t$.

3.55.
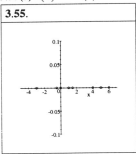

3.57(a) v. (b) x. (c) w. **3.59**(a) $(-\infty, -5]$ or $(5, \infty]$. (b) $(-5, 5]$ (c) No solution. (d) $(-\infty, -5]$ or $(5, \infty]$..(e) $(\infty, -\infty)$ (f) $(\infty, -\infty)$. (g) No solution. (h) No solution.(i) No solution. (j) $(\infty, -\infty)$. **3.61**(a) Independent variable: u. Dependent variable: $[w]$. (b) x. $[y]$. (c) I. $[T]$. (d) A. $[I]$. (e) K.$[C]$.

3.63(a)

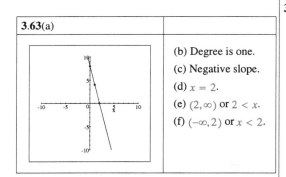

(b) Degree is one.

(c) Negative slope.

(d) $x = 2$.

(e) $(2, \infty)$ or $2 < x$.

(f) $(-\infty, 2)$ or $x < 2$.

3.65(a) $y = 5 = 0x + 5$. (b) y is a polynomial. (c) Degree of y is zero.

(d)

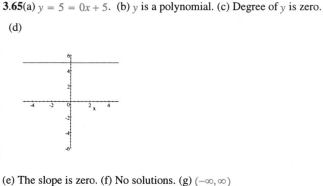

(e) The slope is zero. (f) No solutions. (g) $(-\infty, \infty)$

3.67(a) True. (b) True. (c) True. (d) True. (e) True. (f) True. (g) False. (h) False. (i) False. (j) True. **3.69**.(a) $F(u)$ has two zeros. (b) From the graph, it is apparent that the zeros are both $u = 2$. (c) For the factorization, the constant factor is a and the linear factors are identical: $u - 2$. Consequently, $F(u) = au^2 + bu + c = a(u - 2)(u - 2)$. **3.71**.(a) $\frac{9}{5}C + 32 = 0, C = -\frac{160}{9} = -17.778$. (b) $f(C) = \frac{9}{5}\left(C + \frac{160}{9}\right)$. **3.73**(a) $x = -\frac{1}{4}, x = 1, x = 2$. (b) $4\left(x + \frac{1}{4}\right)(x - 1)(x - 2)$. **3.75**(a) $x = -1$. (b) $x = \sqrt{2}$, $x = -4$, $x = \sqrt{3}$.

3.77.(a) The graph of $g(x) = (2x - 3)(x + 2)$:

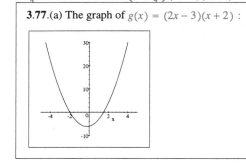

(b) The graph of $h(x) = |(2x - 3)(x + 2)|$:

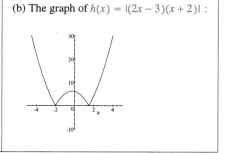

Answers to <u>Exercises</u>, Chapter 4.

4.1(a) Yes. (b) No. **4.3**(a) Yes. (b) Yes.

4.5(a)

Years owned	2	3	4	5
Value of House	$25,000	$22,500	$20,000	$17,500

(d) $C(t) = 30000 - 2500t, t = 0, 1, 2, 3, 4, 5$.

(e) $\{(0, 30000), (1, 27500), (2, 25000)(3, 22500), (4, 20000), (5, 17500)\}$

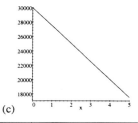

(c)

4.7(a)

Time(in seconds)	8
Height (in feet)	8

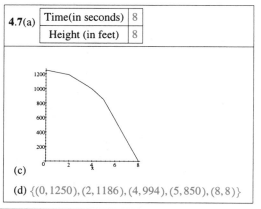

(c)

(d) $\{(0, 1250), (2, 1186), (4, 994), (5, 850), (8, 8)\}$

4.9.	Strictly increasing	Strictly decreasing	**4.9.**	Strictly increasing	Strictly decreasing
(a)	$[-4, -1]$ and $[1, 4]$	$[-1, 1]$	(d)	$[3, 4]$	$[-4, 4]$
(b)	$[-1, 1]$		(e)	$[-2, 2]$	$[-4, -2]$ and $[2, 4]$
(c)	$[4, 10]$	$[-4, -1]$ and $[1, 4]$	(f)	$[-3, 0]$ and $[3, 4]$	$[-5, -3]$ and $[0, 3]$

4.11. a, b, d, e, f. **4.13**(a) Domain: $(-\infty, 7]$. (b) Domain: $[-3, 3]$. (c) Domain: $(4, \infty)$. (d) Domain: R. (e) Domain: $(-\infty, 7]$. (f) Domain: R(g) Domain: $[0, \infty)$. (h) Domain: R. (i) Domain: $(-3, 3)$. **4.15**(a) $f(g(x)) = x$ and $g(f(x)) = x$. (b) $f(g(x)) = x$ and $g(f(x)) = x$. (c) $f(g(x)) = x$ and $g(f(x)) = x$. (d) $f(g(x)) = x$ and $g(f(x)) = x$. (e) $f(g(x)) = x$ and $g(f(x)) = x$. (f) $f(g(x)) = x$ and $g(f(x)) = x$.

4.17(b).Use the horizontal line test

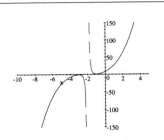

4.19.	Degree	Polynomial
(a)	3	Trinomial
(b)	5	Trinomial
(c)	n/a	not
(d)	3	Binomial
(e)	n/a	not
(f)	9	Monomial

4.21(a) ... (b) ... (c) ... (d)

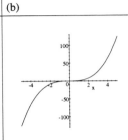

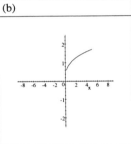

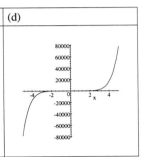

4.23.(a) ... (b)

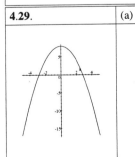

4.25(a) $4x^4 - 10x^2 + 13x + 16$. (b) $2x^4 + 13x - 24$. (c) $-8x^2 + 3$. (d) $15x^4 + 35x^3 - 35x^2 + 240x - 80$. (e) $x^5 - 6x^4 + 22x^3 - 69x^2 + 52x$. (f) $2y^4 - 10y^2 - 3y + 3y^3 + 18$.

4.27(a)	(b)	(c)	(d)
$D = R\backslash\{-3\}$. HA: $x = -3$ VA: $y = 0$	$D = R\backslash\{-1,1\}$ HA: $x = -1, 1$ VA: *none*. SA: $y = x$	$D = R\backslash\{-3\}$ HA: $x = -3$ VA: $y = 1$	$D = R\backslash\{-2,2\}$ HA: $x = -2, 2$ VA: $y = 0$

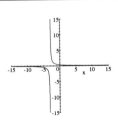

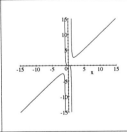

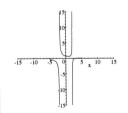

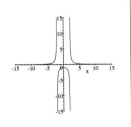

4.29. (a) (b) (c) (d)

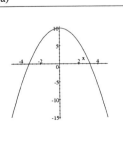

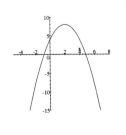

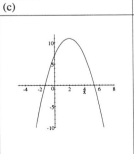

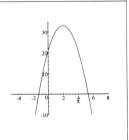

Answers to Problems, Chapter 4.

4.1(a) Domain = R. (b) Domain = $R\backslash(-2,2)$. (c) Domain = $R\backslash\{-\sqrt{2},\sqrt{2}\}$. (d) Domain = $[-5,R)$. (e) Domain = $[2,\infty)\backslash\{0\}$. (f) Domain = $R\backslash\{0\}$. (g) Domain = R. (h) Domain = $[0,\infty)$.

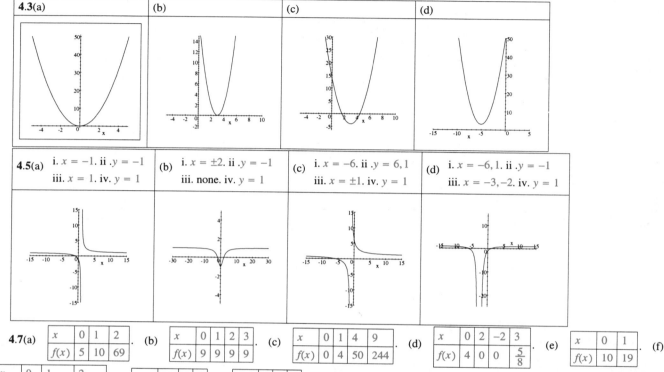

| 4.3(a) | (b) | (c) | (d) |

| 4.5(a) | i. $x = -1$. ii .$y = -1$ iii. $x = 1$. iv. $y = 1$ | (b) | i. $x = \pm 2$. ii .$y = -1$ iii. none. iv. $y = 1$ | (c) | i. $x = -6$. ii .$y = 6, 1$ iii. $x = \pm 1$. iv. $y = 1$ | (d) | i. $x = -6, 1$. ii .$y = -1$ iii. $x = -3, -2$. iv. $y = 1$ |

4.7(a)

x	0	1	2
$f(x)$	5	10	69

(b)

x	0	1	2	3
$f(x)$	9	9	9	9

(c)

x	0	1	4	9
$f(x)$	0	4	50	244

(d)

x	0	2	-2	3
$f(x)$	4	0	0	$\frac{5}{8}$

(e)

x	0	1
$f(x)$	10	19

(f)

x	0	1	2
$f(x)$	10	$\frac{219}{20}$	$\frac{243}{10}$

(g)

x	-1	0	1
$f(x)$	-9	0	9

(h)

x	0	4	9
$f(x)$	0	2	3

4.9(a)$x^3 - 8$. (b) $y^3 + 8$. (c) $x^4 - 4x^2 + 10x - 25$. (d) $z^6 - 9z^2 + 30z - 25$.

Answers to Exercises, Chapter 5.

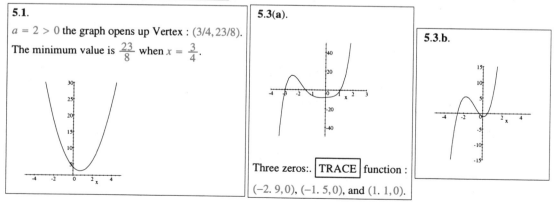

5.1.

$a = 2 > 0$ the graph opens up Vertex : $(3/4, 23/8)$.

The minimum value is $\frac{23}{8}$ when $x = \frac{3}{4}$.

5.3(a).

Three zeros:. TRACE function :

$(-2.9, 0)$, $(-1.5, 0)$, and $(1.1, 0)$.

5.3.b.

5.5(a)	(b)
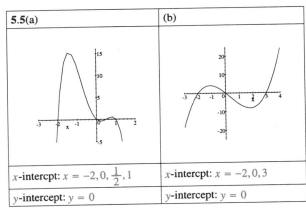	
x-intercpt: $x = -2, 0, \frac{1}{2}, 1$	x-intercpt: $x = -2, 0, 3$
y-intercept: $y = 0$	y-intercept: $y = 0$

5.7. Vertical asymptote: $x = -1$. No missing point. Horizontal asymptote: $y = \frac{3x}{x} = 3$. No slant asymptote. x-intercept: $x = \frac{5}{3}$ y-intercept: $y = -5$. The line $x = -1$ is an asymptote and helps us sketch the graph but is not actually a part of the graph.

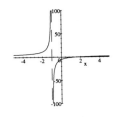

Answers to Problems, Chapter 5.

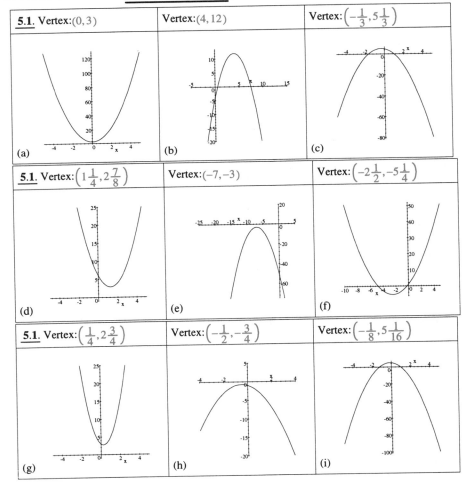

5.1. Vertex: $(0, 3)$

(a)

Vertex: $(4, 12)$

(b)

Vertex: $\left(-\frac{1}{3}, 5\frac{1}{3}\right)$

(c)

5.1. Vertex: $\left(1\frac{1}{4}, 2\frac{7}{8}\right)$

(d)

Vertex: $(-7, -3)$

(e)

Vertex: $\left(-2\frac{1}{2}, -5\frac{1}{4}\right)$

(f)

5.1. Vertex: $\left(\frac{1}{4}, 2\frac{3}{4}\right)$

(g)

Vertex: $\left(-\frac{1}{2}, -\frac{3}{4}\right)$

(h)

Vertex: $\left(-\frac{1}{8}, 5\frac{1}{16}\right)$

(i)

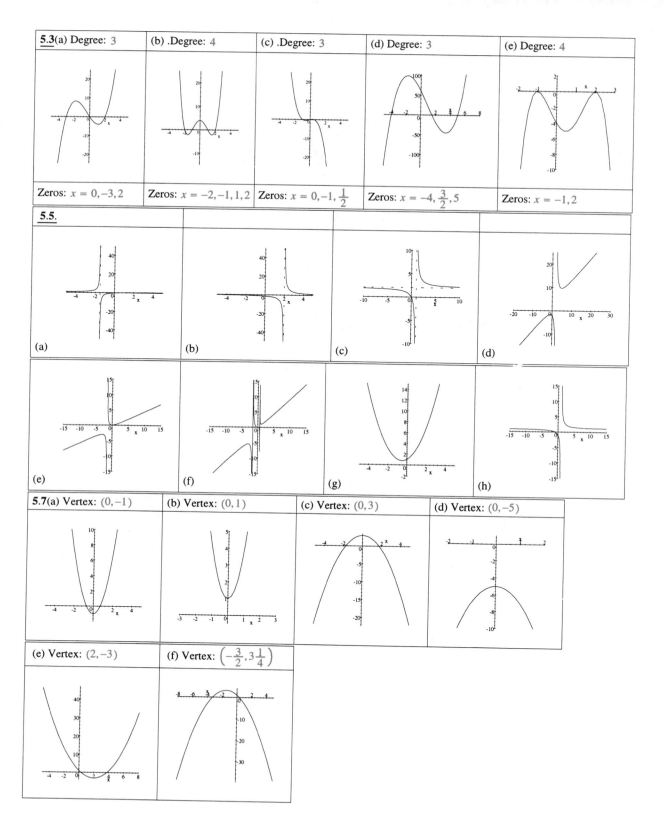

5.3(a) Degree: 3   (b) .Degree: 4   (c) .Degree: 3   (d) Degree: 3   (e) Degree: 4

Zeros: $x = 0, -3, 2$ Zeros: $x = -2, -1, 1, 2$ Zeros: $x = 0, -1, \frac{1}{2}$ Zeros: $x = -4, \frac{3}{2}, 5$ Zeros: $x = -1, 2$

5.5.

(a)   (b)   (c)   (d)

(e)   (f)   (g)   (h)

5.7(a) Vertex: $(0, -1)$ (b) Vertex: $(0, 1)$ (c) Vertex: $(0, 3)$ (d) Vertex: $(0, -5)$

(e) Vertex: $(2, -3)$ (f) Vertex: $\left(-\frac{3}{2}, 3\frac{1}{4}\right)$

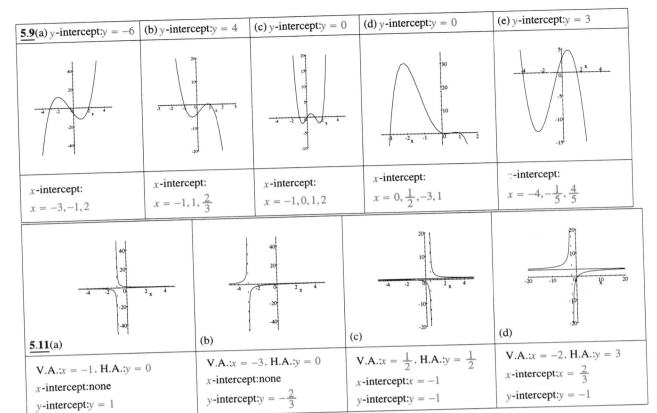

5.9(a) y-intercept:$y = -6$ (b) y-intercept:$y = 4$ (c) y-intercept:$y = 0$ (d) y-intercept:$y = 0$ (e) y-intercept:$y = 3$

x-intercept:
$x = -3, -1, 2$

x-intercept:
$x = -1, 1, \frac{2}{3}$

x-intercept:
$x = -1, 0, 1, 2$

x-intercept:
$x = 0, \frac{1}{2}, -3, 1$

x-intercept:
$x = -4, -\frac{1}{5}, \frac{4}{5}$

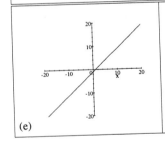

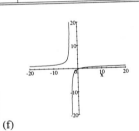

5.11(a)

V.A.:$x = -1$. H.A.:$y = 0$
x-intercept:none
y-intercept:$y = 1$

(b)

V.A.:$x = -3$. H.A.:$y = 0$
x-intercept:none
y-intercept:$y = -\frac{2}{3}$

(c)

V.A.:$x = \frac{1}{2}$. H.A.:$y = \frac{1}{2}$
x-intercept:$x = -1$
y-intercept:$y = -1$

(d)

V.A.:$x = -2$. H.A.:$y = 3$
x-intercept:$x = \frac{2}{3}$
y-intercept:$y = -1$

(e)

(f)

5.13. Max: $(1, -2)$. **5.15**. No.

Answers to <u>Exercises</u>, Chapter 6.

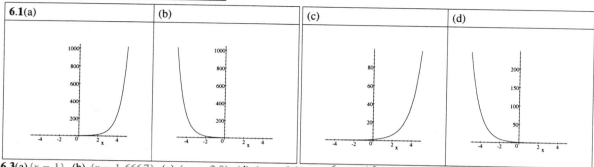

6.3(a) $\{x = 1\}$. (b) $\{x = 1.6667\}$. (c) $\{x = -2.0\}$. (d) $\{x = -2.0\}$. (e) $\{x = -\frac{1}{2}\}$. (f) $\{x = 0\}$. **6.5**(a) $3^4 = 81$. (b): $z^m = 5$. (c) : $10^{1.386} = 4$.
6.7(a) $\{x = 3.729\}$. (b) $\{x = 3\}$. (c) $\{x = -3\}, \{x = 2\}$. (d) $\{x = 1\}$. (e) $\{x = -1\}, \{x = 2\}$. (f) $\{x = 19\}$. **6.9**(a) monthly.
$10000\left(1 + \frac{.0825}{12}\right)^{120} = \$22754$. (b) continuously. $10000e^{0.0825*10} = \$22819$. **6.11**. $20000e^{0.07*21} = \$86985$. **6.13**(a) 1 year after purchase.
$12500(\frac{3}{4})^1 = \$9375$. (b) 3 years after purchase. $12500(\frac{3}{4})^3 = \$5273$. (c) 10 years after purchase. $12500(\frac{3}{4})^{10} = \$703$.

Answers to <u>Problems</u>, Chapter 6.

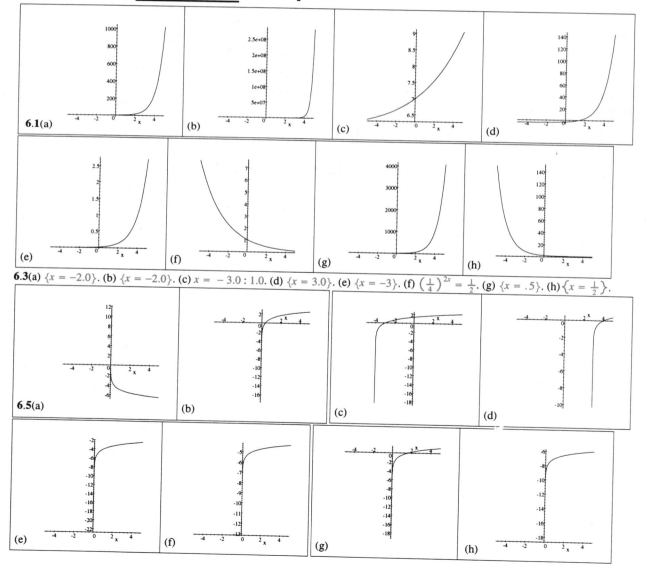

6.3(a) $\{x = -2.0\}$. (b) $\{x = -2.0\}$. (c) $x = -3.0 : 1.0$. (d) $\{x = 3.0\}$. (e) $\{x = -3\}$. (f) $\left(\frac{1}{4}\right)^{2x} = \frac{1}{2}$. (g) $\{x = .5\}$. (h) $\{x = \frac{1}{2}\}$.

6.7(a) $\{x = 3\}$. **(b)** $\{x = 2\}$. **(c)** $\{x = 2\}$. **(d)** $\{x = -\frac{4}{3}\}$. **(e)** $\{x = 6\}$. **(f)** $\{x = 0\}$. **(g)** $\{x = 137\}$. **(h)** $\{x = -4.6261\}$. **6.9(a)** $\{x = -.68454\}$. **(b)** $\{x = .64041\}$. **(c)** $\{x = .94591\}$. **(d)** $\{x = 18.876\}$. **(e)** $x = -1.1774 : x = 1. 1774$. **(f)** $\{x = -2.2276 \times 10^{-2}\}$. **(g)** $\{x = 5.2304\}$. **(h)** $\{x = 2.9459\}$. **6.11** $2500e^{-0.04*6} = 1966.6$. **6.13.** $500 = 100e^{k4}$, $\{k = .40236\}$, $1500 = 100e^{.40236t}$, $\{t = 6.7304\}$. **6.15.** $t = \frac{10\ln 3}{\ln 67 - \ln 51} = 40.262$. **6.17(a)** $t = 8.4018$. **(b)** $t = 27.91$.

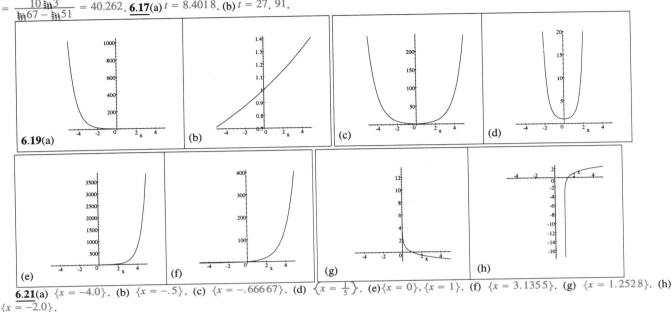

6.19(a) (b) (c) (d)

(e) (f) (g) (h)

6.21(a) $\{x = -4.0\}$. **(b)** $\{x = -.5\}$. **(c)** $\{x = -.66667\}$. **(d)** $\{x = \frac{1}{5}\}$. **(e)** $\{x = 0\}, \{x = 1\}$. **(f)** $\{x = 3.1355\}$. **(g)** $\{x = 1.2528\}$. **(h)** $\{x = -2.0\}$.

Answers to Exercises, Chapter 7.

7.1(a) $(3, -1)$. **(b)** $(10, 12)$. **7.3** $\{(t, 2t - 3)/t \in \mathbf{R}\}$. **7.5(a)** $(-1, 2)$. **(b)** $\{(\frac{4}{3}t + 2, t)/t \in \mathbf{R}\}$. **7.7(a)** ϕ (i.e. no solution). **(b)** $\{(t, \frac{2t+1}{2})/t \in \mathbf{R}\}$. **7.9.** The first number is 13, the second number is 9. **7.11.** The speeds were 56 mph and 76 mph. **7.13.** $(2, 0)$ and $(-\frac{2}{3}, \frac{8}{3})$. **7.15.** Max of z is 129 @ $(3, \frac{3}{2})$; minimum of z is 96 @ $(0, 3)$.

Answers to Problems, Chapter 7.

7.1(a) $(-1, 2)$. **(b)** $(3, 1)$. **(c)** $(6, -4)$. **(d)** ϕ (i.e. no solution) **(e)** $(\frac{1}{2}, 1)$. **(f)** $(1, 0)$. **7.3(a)** $(\frac{1}{11}, \frac{17}{11})$. **(b)** $(\frac{11}{7}, \frac{6}{7})$. **(c)** $(3, 3)$. **(d)** $(\frac{23}{26}, \frac{12}{13})$. **(e)** ϕ (i.e. no solution). **(f)** $(0, 2)$. **7.5.** The numbers are 17 and 29, respectively. **7.7.** The number is 27. **7.9.** The speeds of the plane and wind are 320 mph and 65 mph, respectively. **7.11.** There are 29 \$10 bills and 12 \$20 bills. **7.13.** John is 32 years old; his father is 46 years old. **7.15.** 20 ml @ 30% and 30 ml @ 70%. **7.17.** 25 ml. **7.19.** The length is 95 m and the width is 75 m. **7.21(a).** $(2, 0); (-3, 5)$ **(b).** $(3, 0)$ **(c).** $(\pm 1, \pm 2)$ i.e. four solutions. **(d).** $(3, \pm 2); (-6, \pm 2\sqrt{2}\, i)$ i.e. four solutions. **(e).** $(\frac{1+\sqrt{5}}{2}, \frac{-1+\sqrt{5}}{2}); (\frac{1-\sqrt{5}}{2}, \frac{-1-\sqrt{5}}{2})$ **(f).** $(\frac{1\pm\sqrt{13}}{2}, \frac{7\pm\sqrt{13}}{2})$ i.e. four solutions. **7.23(a).** $(0, 0); (1, 1); (-t - 1, t)$ where $t^2 + t + 1 = 0$ **(b).** $(\pm 2, 0)$ i.e. two solutions. **(c).** $(-2, 0)$ two identical solutions. **(d).** $(\pm\frac{3\sqrt{10}}{5}, \pm\frac{\sqrt{10}}{5}i)$ i.e. four solutions. **(e).** $(-1, -3); (3, 1)$ **(f).** $(-2, 4); (4, -2)$.

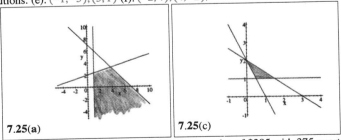

7.25(a)   **7.25(c)**

7.27. $\max z = 336$ @ $(4, 6)$. **7.29.** Maximum sales of \$285 with 375 quarts of ice cream and 150 quarts of yogurt. **7.31.** Minimum cost of \$390 with 5 loads of mid-size truck, 3 loads of large truck. **7.33(a)** ϕ. **(b)** $(14, -7)$. **(c)** $\{(5t - 1, t)/t \in \mathbf{R}\}$. **(d)** ϕ. **(e)** $(2, 1)$. **(f)** $\{(\frac{7}{2}t + 3, t)/t \in \mathbf{R}\}$. **7.35(a)** $(-4, \frac{3}{2})$. **(b)** $(5, -\frac{2}{3})$. **(c)** $(\frac{84}{17}, -\frac{12}{17})$. **(d)** $(4, -1)$. **(e)** $(38, 123)$. **(f)** $(-\frac{3}{2}, 3)$. **7.37(a)** $(-2, -3); (2, 3)$. **(b)** $(\pm 3, \pm 2)$ i.e. four solutions. **(c)** $(\pm 2, \pm 1)$ i.e. four solutions. **(d)** $(\pm 2, \pm i)$ i.e. four solutions. **(e)** $(4 - t, t)$ where $t^2 - 7t + 24 = 0$. **(f)** $(\pm 3i, \pm 2)$ i.e. four solutions. **7.39.** $\max z = 840$ @ $(40, 180)$. **7.41.** $\min z = 16$ @ $(0, 4)$. **7.43(a)** $\max z = 28$ @$(12, 8)$; $\min z = 14$ @ $(6, 4)$. **(b)** $\max z = 4$ @ $(12, 8)$; $\min z = -\frac{40}{3}$ @ $(\frac{20}{3}, \frac{20}{3})$. **7.45.** The number is 37. **7.47.** The speed of plane is 600 mph; the speed of the wind is 100 mph. **7.49.** The maximum weekly profit of \$3,760 is obtained with 320 premium locks and 40 dead bolts.

Answers to Exercises, Chapter 8.

8.1. $x = -16$, $y = -9$, $z = 14$. **8.3.** $x = 1$, $y = -1$, $z = 4$, $w = -2$. **8.5.** $AB = \begin{bmatrix} -9 & 1 \\ -12 & 8 \end{bmatrix}$, $BA = \begin{bmatrix} 4 & 8 \\ 5 & -5 \end{bmatrix}$.

8.7. $A = \begin{bmatrix} 3 & -1 & 7 \\ 2 & 3 & -5 \\ 1 & 1 & -1 \end{bmatrix}$, $B = \begin{bmatrix} 4 \\ -3 \\ 11 \end{bmatrix}$, $X = \begin{bmatrix} x \\ y \\ z \end{bmatrix}$. **8.9(a)** A does not have a inverse. **(b)** $A^{-1} = \begin{bmatrix} 28 & -19 & 3 \\ 2 & 1 & 0 \\ -7 & -5 & 1 \end{bmatrix}$

8.11(a) $A^{-1} = \begin{bmatrix} -8 & -3 \\ -3 & -1 \end{bmatrix}$, **(b)** $A^{-1} = \begin{bmatrix} \frac{8}{3} & -\frac{7}{3} \\ 1 & -1 \end{bmatrix}$, **(c)** $A^{-1} = \begin{bmatrix} 6 & 5 & -5 \\ -6 & -\frac{9}{2} & 5 \\ -1 & -1 & 1 \end{bmatrix}$

8.13(a)–5. **(b)** 13. **(c)**. –12. **(d)** 15. **8.15.** $\frac{1}{2}|\det A| = \frac{1}{2}|(-1)(-1) - (4)(-2)| = \frac{1}{2}|9| = \frac{9}{2}$. **8.17(a)** $x = \dfrac{\det A_x(B)}{\det A} = 2$, $y = \dfrac{\det A_y(B)}{\det A} = -1$.
(b) $x = \dfrac{\det A_x(B)}{\det A} = 5$, $y = \dfrac{\det A_y(B)}{\det A} = 1$, $z = \dfrac{\det A_z(B)}{\det A} = -3$. **8.19.** $\frac{1}{2}|\det A| = \frac{1}{2}|-66| = 33$.

Answers to Problems, Chapter 8.

8.1(a) $x = 7$, $y = 2$. **(b)** $x = 1$, $y = 2$, $z = -1$ **(c)** $x = 2$, $y = 1$, $z = -1$. **8.3(a)** $x = 3$, $y = 1$. **(b)** $x = 1$, $y = -2$. **(c)** $x = -2$, $y = 4$. **(d)** $x = 4$, $y = -1$, $z = 2$. **(e)** $x = 3$, $y = -2$, $z = -2$. **(f)** $x = 1$, $y = 0$, $z = -1$. **(g)** $x = 3$, $y = -2$, $z = -1$, $w = 4$. **(h)** $x = -1$, $y = 0$, $z = 3$, $w = 5$. **(i)** $x = -3$, $y = 4$, $z = -\frac{1}{2}$, $w = \frac{5}{2}$.

8.5. $A + B = \begin{bmatrix} 11 & -5 \\ -7 & -2 \end{bmatrix}$; $A - B = \begin{bmatrix} -7 & 7 \\ -3 & -14 \end{bmatrix}$; $-3A + 2B = \begin{bmatrix} 12 & -15 \\ 11 & 36 \end{bmatrix}$.

8.7(a) $AB = \begin{bmatrix} -50 & 54 \\ -29 & -33 \end{bmatrix}$; $BA = \begin{bmatrix} -23 & 68 \\ -27 & -60 \end{bmatrix}$. **(b)** $AB = \begin{bmatrix} 60 & 108 & -18 \\ -10 & 54 & -44 \end{bmatrix}$; BA is not defined. **(c)** $AB = \begin{bmatrix} -30 & 59 & -79 \\ 18 & 81 & -37 \end{bmatrix}$;

BA is not defined. **(d)** $AB = \begin{bmatrix} 83 & 5 & 21 & 35 \\ -81 & -77 & -57 & -46 \\ 104 & 28 & -12 & 114 \end{bmatrix}$; BA is not defined. **8.9(a)** $\begin{bmatrix} -5 & -3 \\ -2 & -1 \end{bmatrix}$. **(b)** $\begin{bmatrix} \frac{7}{3} & -\frac{2}{3} \\ -\frac{2}{3} & \frac{1}{3} \end{bmatrix}$. **(c)** $\begin{bmatrix} \frac{2}{3} & \frac{5}{3} & -1 \\ -\frac{4}{3} & -\frac{7}{3} & 2 \\ -\frac{1}{3} & -\frac{4}{3} & 1 \end{bmatrix}$

(d) $x = -13$, $y = 29$, $z = -9$. **8.11 (a)** $\begin{bmatrix} \frac{2}{3} & \frac{5}{3} & -1 \\ -\frac{4}{3} & -\frac{7}{3} & 2 \\ -\frac{1}{3} & -\frac{4}{3} & 1 \end{bmatrix}$, **(b)** $\begin{bmatrix} -1 & -2 & 11 \\ 1 & 3 & -15 \\ 0 & -1 & 5 \end{bmatrix}$, **(c)** $\begin{bmatrix} 0 & 1 & 2 \\ -1 & 3 & 3 \\ 2 & -6 & -7 \end{bmatrix}$, **(d)** This matrix is not invertible.

8.13(a) $\det A = 3$. **(b)** $\det A = -6$. **(c)** $\det A = 16$. **(d)** $\det A = -32$. **8.15(a)** $x = -\frac{13}{5}$, $y = \frac{6}{5}$. **(b)** $x = 1$, $y = -1$, $z = 4$. **8.17.** 26. **8.19(a)** 5. **(b).** 15 **(c).** 9. **8.21.** 42. **8.23.** $\frac{13}{2}$.

8.25(a) $x = 4$, $y = 1$. **(b)** $x = -1$, $y = 7$. **(c)** $x = 3$, $y = 1$. **(d)** $x = 1$, $y = -3$, $z = 2$. **(e)** $x = 0$, $y = -4$, $z = 7$. **(f)** This system is inconsistent. **(g)** $x = 10$, $y = -3$, $z = -1$. **(h)** $x = 4$, $y = -5$, $z = 3$, $w = 1$. **(i)** This system is inconsistent.

8.27(a) $AB = \begin{bmatrix} -12 & 40 \\ -1 & -15 \end{bmatrix}$; $BA = \begin{bmatrix} -38 & 58 \\ -11 & 11 \end{bmatrix}$. **(b)** AB is not defined; $BA = \begin{bmatrix} 5 & -24 & 5 \\ -40 & 48 & -40 \end{bmatrix}$. **(c)** $AB = \begin{bmatrix} 8 & -36 \\ 1 & -17 \\ 5 & 40 \end{bmatrix}$; BA is not

defined. **(d)** $AB = \begin{bmatrix} 14 & -26 & 54 \\ -14 & 41 & -27 \\ 2 & 7 & 27 \end{bmatrix}$; $BA = \begin{bmatrix} 34 & -36 \\ 4 & 48 \end{bmatrix}$. **8.29(a)** $x = -11$, $y = 19$, $z = 10$. **(b)** $x = -16$, $y = -10$, $z = 8$. **(c)** $x = 11$, $y = -8$, $z = 13$. **(d)** $x = 34$, $y = 19$, $z = -9$. **8.31(a)** $x = \frac{1}{13}$, $y = \frac{28}{13}$. **(b)** $x = -11$, $y = 7$. **(c)** $x = \frac{34}{23}$, $y = -\frac{1}{23}$, $z = \frac{27}{23}$. **(d)** $x = \frac{1}{2}$, $y = -\frac{5}{6}$, $z = \frac{7}{6}$.

Answers to Exercises, Chapter 9.

9.1. 14. **9.3**(a) $a_1 = 3, a_2 = 9, a_3 = 27, a_4 = 81, a_5 = 243$. (b) $a_1 = 100, a_2 = \frac{100}{3}, a_3 = \frac{100}{12}, a_4 = \frac{100}{60}, a_5 = \frac{100}{360}$. **9.7**(a) 5040. (b) 1. (c) 120. (d) 362,880. (e) 479,001,600. **9.9**(a) $a_5 = 15$. (b) $a_5 = 11$. (c) $a_8 = -3$. **9.11**. a and c are not. **9.13**. $S_{13} = 65,000$.

9.15.

S_{12}	S_{24}	S_{36}	S_{192}
$1,240.	$2,558. 59	$3,963. 24	$34,660. 38

9.17. $a_4 = \$29,775.4$. **9.19**(a) $a_1 = 1, a_2 = 3, a_3 = 5, a_4 = 7, a_5 = 9$. (b) $a_1 = 2, a_2 = 4, a_3 = 8, a_4 = 16, a_5 = 32$. (c) $a_1 = 1, a_2 = 2, a_3 = 3, a_4 = 4, a_5 = 5$. **9.21**(a) 1. (b) 15. (c) 56. (d) 36. **9.23**. 35. **9.25**(a) 8,008. (b) 120. **9.27**(a) ...is odd or it is not greater than four. (b) ...is odd and divisible by two. (c) ...is odd. (d) ...is divisible by two and it is not greater than four. (e) ...is odd or is divisible by two or is not greater than four. **9.29**(a) $\frac{5}{36}$. (b) $\frac{1}{36}$. (c) $\frac{1}{2}$. (d) $\frac{1}{2}$. (e) 0. (f) $\frac{11}{18}$. **9.31**(a) $\frac{17}{24}$. (b) 1. (c) $\frac{17}{22}$. (d) 1. **9.33**(a) $\frac{15}{64}$. (b) $\frac{37}{64}$. (c) $\frac{63}{64}$. (d) $\frac{1}{64}$. (e) $\frac{15}{64}$.

Answers to Problems, Chapter 9.

9.1(a) *Domain = N*. (b) *Domain = N*. (c) Not a sequence. (d) Not a sequence. (e) *Domain = N*. **9.3**(a) 1. 51515, 1. 515151. (b) $\frac{16}{3}, \frac{32}{3}$. (c) $\frac{4}{5}, \frac{5}{6}$. (d) $\frac{1}{25}, \frac{-1}{36}$. (e) 12, 17. (f) 81, -243. (g) -81, 243. (h) 14, 17. (i) 56, 72. (j) $\frac{1}{2048}, \frac{1}{8192}$. **9.5**(a) $\sum_{i=1}^{10} i$. (b) $\sum_{n=1}^{5} 2^n$. (c) $\sum_{n=1}^{4} (0.1)^n 2$. (d) $\sum_{n=1}^{5} \frac{n}{n+1}$. (e) $\sum_{n=1}^{5} \frac{n^2}{2n+1}$. **9.7**(a) $d = 3$. (b) $r = 4$. (c) $d = -1$. (d) $r = \frac{1}{2}$. (e) $d = 2$. (f) $r = 6$. **9.9**(a) $S_{n\to\infty} = \frac{3}{2}$. (b) $S_{n\to\infty} = 9$. (c) $S_{n\to\infty} = 2$. (d) $S_{n\to\infty} = 24$. (e) $S_{n\to\infty} = 6$. (f) $S_{n\to\infty} \frac{2}{3}$. **9.11**(a) 3. (b) 6. **9.13**. 495.

INDEX

INDEX

INDEX